Lecture Notes Electrical Engineering

Volume 37

Juan Andrade-Cetto · Jean-Louis Ferrier
Joaquim Filipe (Eds.)

Informatics in Control, Automation and Robotics

Selcted Papers from the International Conference on Informatics in Control, Automation and Robotics 2008

Springer

Juan Andrade-Cetto
Institut de Robòtica i
Informàtica Industrial
CSIC-UPC
Llorens i Artigas, 4-6
08028 Barcelona
Spain
E-mail: cetto@iri.upc.edu

Joaquim Filipe
INSTICC
Av. D. Manuel I, 27A 2 Esq.
2910-595 Setubal
Portugal
E-mail: jfilipe@insticc.org

Jean-Louis Ferrier
Institut des Sciences et Techniques de
l'Ingénieur d'Angers (ISTIA)
Labo. d'Ingénierie des Systémes
Automatisés (LISA)
62 avenue Notre Dame du Lac
49000 Angers
France
E-mail: ferrier@istia.univ-angers.fr

ISBN 978-3-642-10131-1 e-ISBN 978-3-642-00271-7

DOI 10.1007/978-3-642-00271-7

Typeset & Coverdesign: Scientific Publishing Services Pvt. Ltd., Chennai, India.

Printed in acid-free paper

9 8 7 6 5 4 3 2 1

springer.com

Preface

The present book includes a set of selected papers from the Fifth International Conference on Informatics in Control Automation and Robotics (ICINCO 2008), held in Funchal, Madeira - Portugal, from 11 to 15 May 2008. The conference was organized in three simultaneous tracks: *Intelligent Control Systems and Optimization*, *Robotics and Automation*, and *Systems Modeling, Signal Processing and Control*. The book is based on the same structure.

ICINCO 2008 received 392 paper submissions, from more than 50 different countries in all continents. From these, after a blind review process, only 33 where accepted as full papers, of which 18 were selected for inclusion in this book, based on the classifications provided by the Program Committee. The selected papers reflect the interdisciplinary nature of the conference. The diversity of topics is an important feature of this conference, enabling an overall perception of several important scientific and technological trends. These high quality standards will be maintained and reinforced at ICINCO 2009, to be held in Milan, Italy, and in future editions of this conference.

Furthermore, ICINCO 2008 included 4 plenary keynote lectures given by Ronald C. Arkin (Georgia Institute of Technology), Miguel Ayala Botto (Instituto Superior Técnico, Lisbon), Marco Dorigo (IRIDIA, Université Libre de Bruxelles) and Peter Simon Sapaty (Institute of Mathematical Machines and Systems, National Academy of Sciences of Ukraine). We would like to express our appreciation to all of them and in particular to those who took the time to contribute with a paper to this book.

On behalf of the conference organizing committee, we would like to thank all participants. First of all to the authors, whose quality work is the essence of the conference and to the members of the Program Committee, who helped us with their expertise and diligence in reviewing the papers. As we all know, producing a conference requires the effort of many individuals. We wish to thank also all the members of our organizing committee, whose work and commitment were invaluable.

November 2008

Juan Andrade-Cetto
Jean-Louis Ferrier
Joaquim Filipe

Organization

Conference Co-chairs

Jorge Cardoso	University of Madeira (UMa), Madeira, Portugal
Joaquim Filipe	Polytechnic Institute of Setúbal / INSTICC, Portugal

Program Co-chairs

Juan Andrade Cetto	Institut de Robòtica i Informàtica industrial, Spain
Jean-Louis Ferrier	University of Angers, France

Organizing Committee

Paulo Brito	INSTICC, Portugal
Marina Carvalho	INSTICC, Portugal
Helder Coelhas	INSTICC, Portugal
Vera Coelho	INSTICC, Portugal
Andreia Costa	INSTICC, Portugal
Bruno Encarnação	INSTICC, Portugal
Vitor Pedrosa	INSTICC, Portugal
Vera Rosário	INSTICC, Portugal

Program Committee

Arturo Hernandez Aguirre, Mexico
Eugenio Aguirre, Spain
Hyo-Sung Ahn, Korea
Frank Allgower, Germany
Fouad Al-Sunni, Saudi Arabia
Bala Amavasai, UK
Francesco Amigoni, Italy
Yacine Amirat, France
Nicolas Andreff, France
Stefan Andrei, USA
Plamen Angelov, UK
Luis Antunes, Portugal

Peter Arato, Hungary
Helder Araújo, Portugal
Gustavo Arroyo-Figueroa, Mexico
Marco Antonio Arteaga, Mexico
Vijanth Sagayan Asirvadam, Malaysia
Wudhichai Assawinchaichote, Thailand
Robert Babuska, The Netherlands
Ruth Bars, Hungary
Adil Baykasoglu, Turkey
Laxmidhar Behera, India
Maren Bennewitz, Germany
Karsten Berns, Germany
Arijit Bhattacharya, India
Robert Bicker, UK
Sergio Bittanti, Italy
Stjepan Bogdan, Croatia
Jean-Louis Boimond, France
Djamel Bouchaffra, USA
Patrick Boucher, France
Guy Boy, France
Bernard Brogliato, France
Edmund Burke, UK
Kevin Burn, UK
Clifford Burrows, UK
Dídac Busquets, Spain
Luis M. Camarinha-Matos, Portugal
Marc Carreras, Spain
Jorge Martins de Carvalho, Portugal
Alessandro Casavola, Italy
Riccardo Cassinis, Italy
Chien Chern Cheah, Singapore
Tongwen Chen, Canada
YangQuan Chen, USA
Albert M. K. Cheng, USA
Graziano Chesi, China
Yiu-ming Cheung, Hong Kong
Sung-Bae Cho, Korea
Ryszard S. Choras, Poland
Carlos Coello Coello, Mexico
Patrizio Colaneri, Italy
António Dourado Correia, Portugal
Yechiel Crispin, USA
Danilo De Rossi, Italy
Elena De Santis, Italy
Matthias Dehmer, Austria
Angel P. del Pobil, Spain
Mingcong Deng, Japan
Guilherme DeSouza, USA
Jorge Dias, Portugal
Rüdiger Dillmann, Germany
Denis Dochain, Belgium
Tony Dodd, UK
Alexandre Dolgui, France
Marco Dorigo, Belgium
Petr Ekel, Brazil
Sebastian Engell, Germany
Simon Fabri, Malta
Sergej Fatikow, Germany
Jean-Marc Faure, France
Jean-Louis Ferrier, France
Limor Fix, USA
Juan F. Flores, Mexico
Georg Frey, Germany
Manel Frigola, Spain
Colin Fyfe, UK
Dragan Gamberger, Croatia
Leonardo Garrido, Mexico
Nicholas Gans, USA
Ryszard Gessing, Poland
Lazea Gheorghe, Romania
Maria Gini, USA
Alessandro Giua, Italy
Luis Gomes, Portugal
John Gray, UK
Dongbing Gu, UK
Guoxiang Gu, USA
Jason Gu, Canada
José J. Guerrero, Spain
Jatinder (Jeet) Gupta, USA
Thomas Gustafsson, Sweden
Maki K. Habib, Japan
Hani Hagras, UK
Wolfgang Halang, Germany
Riad Hammoud, USA
Uwe D. Hanebeck, Germany
John Harris, USA
Dominik Henrich, Germany
Francisco Herrera, Spain
Victor HInostroza, Mexico
Wladyslaw Homenda, Poland

Alamgir Hossain, UK
Dimitrios Hristu-Varsakelis, Greece
Guoqiang Hu, USA
Nor Ashidi Mat Isa, Malaysia
Ray Jarvis, Australia
Odest Jenkins, USA
Ping Jiang, UK
Agustin Jimenez, Spain
Ivan Kalaykov, Sweden
Michail Kalogiannakis, France
Dimitrios Karras, Greece
Fakhri Karray, Canada
Dusko Katic, Serbia
Graham Kendall, UK
Bart Kosko, USA
George L. Kovács, Hungary
Krzysztof Kozlowski, Poland
Gerhard Kraetzschmar, Germany
H. K. Lam, UK
Cecilia Laschi, Italy
Jean-Claude Latombe, USA
M. Kemal Leblebicioglu, Turkey
Loo Hay Lee, Singapore
Soo-Young Lee, Korea
Graham Leedham, Singapore
Kauko Leiviskä, Finland
Kang Li, UK
Yangmin Li, China
Zongli Lin, USA
Vincenzo Lippiello, Italy
Honghai Liu, UK
Luís Seabra Lopes, Portugal
Brian Lovell, Australia
Peter Luh, USA
Jose Tenreiro Machado, Portugal
Anthony Maciejewski, USA
N. P. Mahalik, USA
Bruno Maione, Italy
Frederic Maire, Australia
Om Malik, Canada
Jacek Mandziuk, Poland
Hervé Marchand, France
Philippe Martinet, France
Aníbal Matos, Portugal
Rene V. Mayorga, Canada
Barry McCollum, UK
Ken McGarry, UK
Gerard McKee, UK
Seán McLoone, Ireland
Patrick Millot, France
José Mireles Jr., Mexico
Masoud Mohammadian, Australia
Pieter Mosterman, USA
Vladimir Mostyn, Czech Republic
Rafael Muñoz-Salinas, Spain
Kenneth Muske, USA
Fazel Naghdy, Australia
Tomoharu Nakashima, Japan
Andreas Nearchou, Greece
Luciana Porcher Nedel, Brazil
Sergiu Nedevschi, Romania
Maria Neves, Portugal
Anton Nijholt, The Netherlands
Hendrik Nijmeijer, The Netherlands
Juan A. Nolazco-Flores, Mexico
Urbano Nunes, Portugal
Tsukasa Ogasawara, Japan
José Valente de Oliveira, Portugal
Manuel Ortigueira, Portugal
Djamila Ouelhadj, UK
Christos Panayiotou, Cyprus
Stefano Panzieri, Italy
Evangelos Papadopoulos, Greece
Michel Parent, France
Igor Paromtchik, Japan
Mario Pavone, Italy
Witold Pedrycz, Canada
Carlos Eduardo Pereira, Brazil
Duc Pham, UK
J. Norberto Pires, Portugal
Marios Polycarpou, Cyprus
Marie-Noëlle Pons, France
Raul Marin Prades, Spain
Libor Preucil, Czech Republic
José Ragot, France
A. Fernando Ribeiro, Portugal
Robert Richardson, UK
Rodney Roberts, USA
Kurt Rohloff, USA
Juha Röning, Finland
Agostinho Rosa, Portugal
António Ruano, Portugal

Fariba Sadri, UK
Carlos Sagüés, Spain
Mehmet Sahinkaya, UK
Priti Srinivas Sajja, India
Antonio Sala, Spain
Abdel-Badeeh Salem, Egypt
Medha Sarkar, USA
Nilanjan Sarkar, USA
Jurek Sasiadek, Canada
Daniel Sbarbaro, Chile
Carsten Scherer, The Netherlands
Matthias Scheutz, USA
Klaus Schilling, Germany
Carla Seatzu, Italy
Rodolphe Sepulchre, Belgium
Michael Short, UK
Bruno Siciliano, Italy
João Silva Sequeira, Portugal
Silvio Simani, Italy
Dan Simon, USA
Michael Small, China
Cyrill Stachniss, Germany
Burkhard Stadlmann, Austria
Tarasiewicz Stanislaw, Canada
Olaf Stursberg, Germany
Chun-Yi Su, Canada
Raúl Suárez, Spain
Ryszard Tadeusiewicz, Poland
Tianhao Tang, China
Adriana Tapus, USA
József K. Tar, Hungary
Daniel Thalmann, Switzerland
Gui Yun Tian, UK
Avgoustos Tsinakos, Greece
Antonios Tsourdos, UK
Nikos Tsourveloudis, Greece
Ivan Tyukin, UK
Anthony Tzes, Greece
Masaru Uchiyama, Japan
Dariusz Ucinski, Poland
Nicolas Kemper Valverde, Mexico
Marc Van Hulle, Belgium
Gerrit van Straten, Netherlands
Eloisa Vargiu, Italy
Annamaria R. Varkonyi-Koczy, Hungary
Laurent Vercouter, France
Luigi Villani, Italy
Bernardo Wagner, Germany
Axel Walthelm, Germany
Dianhui Wang, Australia
Lipo Wang, Singapore
Zidong Wang, UK
Vincent Wertz, Belgium
Dirk Wollherr, Germany
Sangchul Won, Korea
Peter Xu, New Zealand
Bin Yao, USA
Xinghuo Yu, Australia
Marek Zaremba, Canada
Janan Zaytoon, France
Du Zhang, USA
Changjiu Zhou, Singapore
Dayong Zhou, USA
Primo Zingaretti, Italy
Argyrios Zolotas, UK

Auxiliary Reviewers

Hyo-Sung Ahn, Korea
Prasanna Balaprakash, Belgium
Majid Chauhdry, USA
Ying Chen, USA
Pedro Fernandes, Portugal
Matteo De Felice, Italy
Michele Folgheraiter, Germany
Jun Fu, Canada
Andrea Gasparri, Italy
Emmanuel Godoy, France
Che Guan, USA
Istvan Harmati, Hungary
Abhinaya Joshi, USA
Balint Kiss, Hungary
Gabor Kovacs, Hungary
Roland Lenain, France
Nikolay Manyakov, Germany
Philippe Martinet, France
Sandro Meloni, Italy
Eduardo Montijano Muñoz, Spain

A. C. Murillo, Spain
Gonzalo Lopez Nicolas, Spain
Sorin Olaru, France
Federica Pascucci, Italy
Karl Pauwels, Germany
Paulo Peixoto, Portugal
Jun Peng, China
Luis Puig, Spain
Maurizio di Rocco, Italy
Marco Montes De Oca Roldan, Belgium
Joerg Stueckler, Germany
Jin Sun, China
Emese Szadeczky-Kardos, Hungary
Sihem Tebbani, France
Benoit Thuilot, France
Guoyu Tu, China
Peng Wang, USA
Weihua Wang, USA
Bingjie Zhang, USA
Yige Zhao, USA
Ying Zhao, USA

Invited Speakers

Miguel Ayala Botto	Instituto Superior Técnico, Portugal
Peter Simon Sapaty	Institute of Mathematical Machines and Systems, National Academy of Sciences, Ukraine
Ronald C. Arkin	Georgia Institute of Technology, USA
Marco Dorigo	IRIDIA, Université Libre de Bruxelles, Belgium

Contents

Invited Papers

Distributed Technology for Global Control

Peter Simon Sapaty

Institute of Mathematical Machines and Systems, National Academy of Sciences
Glushkova Ave 42, 03187 Kiev, Ukraine
sapaty@immsp.kiev.ua

Abstract. A flexible, ubiquitous, and universal solution for management of distributed dynamic systems will be presented. It allows us to grasp complex systems on a higher than usual, semantic level, penetrating their infrastructures, also creating and modifying them, while establishing local and global dominance over the system organizations and coordinating their behavior in the way needed. The approach may allow the systems to maintain high runtime integrity and automatically recover from indiscriminate damages, preserving global goal orientation and situation awareness in unpredictable and hostile environments.

Keywords: Global dominance, global control, spatial scenarios, world processing language, distributed interpretation, emergency management, sensor networks, directed energy systems, avionics, electronic warfare, distributed objects tracking, collective behavior.

1 Introduction

We are witnessing rapid growth of world dynamics caused by global warming, globalization of economy, numerous ethnic, religious and military conflicts, and international terrorism. To match this dynamics and withstand threats and adversaries, effective integration of available human and technical resources is crucial. These resources may be scattered and emergent, lacking infrastructures and authorities for the solutions needed, in real time and ahead of it.

Just communication between predetermined parts and systems with sharing common vision, often called as *interoperability*, may not be sufficient. The whole distributed system (or system of systems) should rather represent a dynamic and integral organism, in which parts are defined and interlinked dynamically in subordination to system goals, which can vary at runtime, with the term *overoperability* [1] becoming more appropriate in such situations.

A related ideology and accompanying information and control technology, allowing us to provide a higher than usual level of distributed system understanding and control, will be outlined in this paper.

2 The World Processing Paradigm

Within the developed approach, a network of intelligent modules (marked as U in Fig.1) embedded into important system points, collectively interprets mission scenarios in a special high-level language, which can start from any node and cover the networked system at runtime.

J.A. Cetto et al. (eds.): Informatics in Control, Automation and Robotics, LNEE 37, pp. 3–24.
springerlink.com

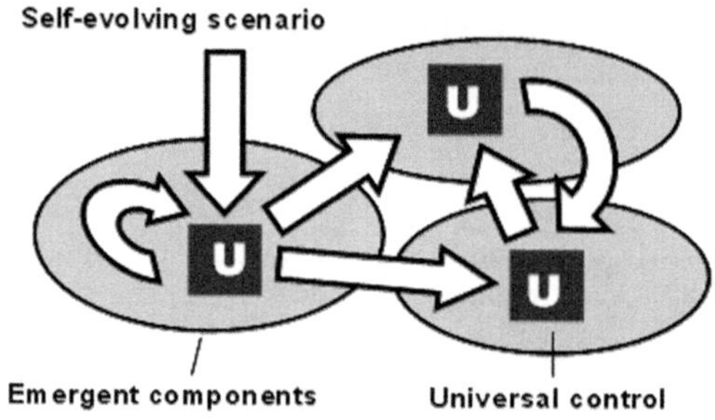

Fig. 1. Runtime coverage of a distributed system

The scenarios may be integral and compact, capable of self-recovery after damages; they may also be created on the fly, as traditional synchronization, data, code, and agents handling and exchanges routines are effectively shifted to the automatic implementation. This parallel and distributed, without any central resources spatial process can constantly take into account peculiarities of the environment in which mission scenarios evolve.

Originally represented in a unified and compact form, the scenario and resources needed for its evolution can start from any system point (as shown in Fig. 2).

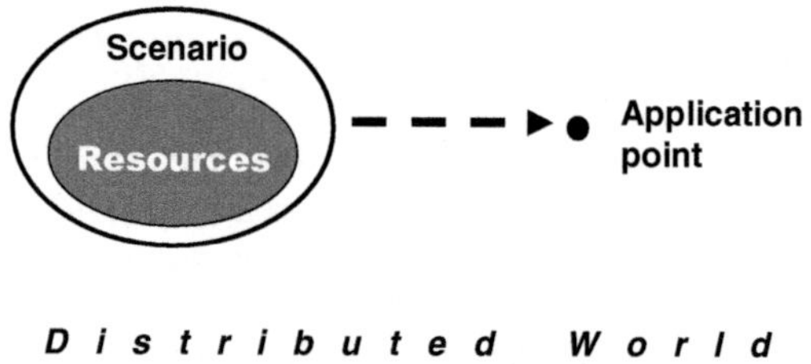

Fig. 2. The initial state

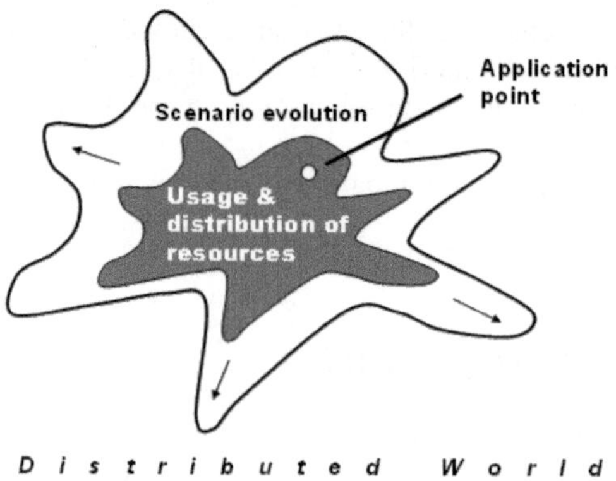

Fig. 3. Spreading operations and resources

The scenarios can self-split, replicate and modify while covering the distributed world or its part(s) at runtime, bringing operations and (virtual as well as physical) resources into different points. They can also lift, activate, and spread further other scenarios and resources already accumulated in the navigated world, as in Fig. 3.

This is causing movement of both information and physical matter, and initiating interactions between manned and unmanned components, command and control (C2) including, as in Fig. 4 (where S is for spatial scenarios or their parts, and R – for resources supporting the scenarios).

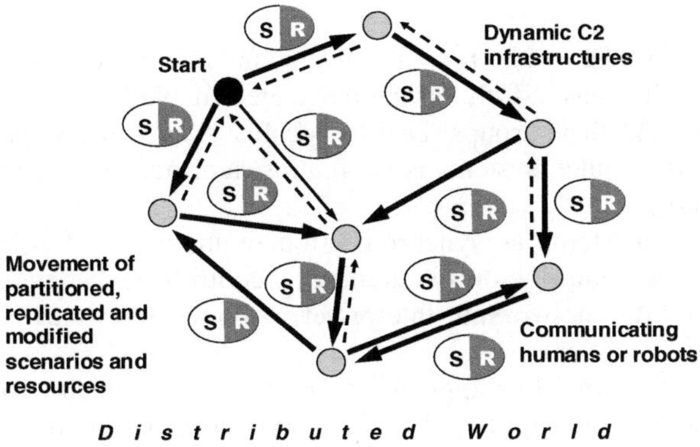

Fig. 4. Resultant interactions between system parts

The main difference of this approach with other works in this area is that it describes (on a higher level, in a concise way) of what the system should do or how should behave as a whole, while delegating numerous routines of partitioning into components (agents), with their interaction and synchronization to the effective automatic level. Other approaches are handling the latter manually, and from the start. The approach can, however, describe and implement peculiarities of system organization and its behavior at any other levels needed, which may include:

- Most general, semantic, task formulation.
- Explicit projecting intelligence, information, matter, and power into particular physical or virtual locations, with doing jobs directly in the places reached, and if needed, cooperatively.
- Creating new active physical, virtual, or combined worlds, and organizing and coordinating their activity.
- Setting up implementation details, at any levels, say, for optimization of usage of restricted resources.

3 The World Processing Language (WPL)

This ideology and technology are based on the World Processing Language, WPL [2], describing what to do in distributed spaces rather than how to do, and by which resources (or even system organization), leaving these to the automatic interpretation in networked environments. The WPL main features include:

- Association of any action with a position in physical, virtual, or combined space.
- Working with both information and physical matter.
- Runtime creation of distributed knowledge networks.
- Unlimited parallelism.
- Free movement or navigation in physical, virtual, or combined worlds.
- Fully distributed decision making with high integrity as a whole.
- Automatic command and control.

It is a higher-level language to efficiently command and control emergent human teams and armies. It is also a fully formal language suitable for automatic interpretation by mobile robots and their groups. Due to peculiar syntax and semantics, its parallel interpretation in distributed systems is straightforward, transparent, and does not need any central resources.

Such complex problems as synchronization of multiple activities and collective (swarm as well as centrally or hierarchically controlled) behavior can be solved automatically by the networked interpreter, without traditional load on human managers and programmers.

This dramatically simplifies application programming, which is often hundreds of times more concise (and simpler) than in traditional programming languages. WPL allows for a direct access to the distributed world, performing any operations in any its points over local or remote data, which may represent both information and matter. Navigating in the world, WPL can modify it or even create from scratch, if required. Different movements and operations can be performed simultaneously and in parallel, and these may be free or may depend on each other.

WPL has a recursive syntax which can be expressed on the top level as follows (square brackets are for an optional construct, braces mean construct repetition with a delimiter at the right, and vertical bar separates alternatives).

wave → *constant* | *variable* | [*rule*] ({*wave* , })
constant → *information* | *matter*
variable → *nodal* | *frontal* | *environmental*
rule → *evolution* | *fusion* | *verification* | *essence*
evolution → *expansion* | *branching* | *advancing* | *repetition* | *granting*
fusion → *echoing* | *processing* | *constructing* | *assignment*
verification → *comparison* | *membership* | *linkage*
essence → *type* | *usage*

A rule is a very general construct, which, for example, can be:

- Elementary arithmetic, string or logic operation
- Hop in physical, virtual, or combined space
- Hierarchical fusion and return of (remote) data
- Parallel and distributed control
- Special context for navigation in space
- Sense of a value for its proper interpretation

Different types of variables, especially when used together, allow us to create efficient spatial algorithms which work *in between* components of distributed systems rather than *in* them. The variables called *nodal* can store and access local results in the system points visited, while others ones can move data in space together with the evolving control (*frontal* variables) or can access and impact the world navigated (*environmental* variables).

4 Elementary Examples

4.1 Setting Global Dominance

Let us assume that a node in a distributed system (see Fig. 5) wants to establish a field of dominance over the nodes that have lower rank that itself (considering here the content, or name, of each node as its rank).

The following parallel and distributed program, applied in this node, spreads its own rank throughout the whole system in frontal variable Rank. This puts the rank into nodal variable Dominance in each visited node, if Rank exceeds the already existing value in Dominance. (By the first access, Dominance is assigned the value of the personal rank of each node.)

```
frontal Rank = CONTENT;
nodal Dominance;
repeat (
   if (Dominance == nil, Dominance = CONTENT);
   if (Dominance < Rank,
         (Dominance = Rank; hop to all neighbors),
            otherwise stop))
```

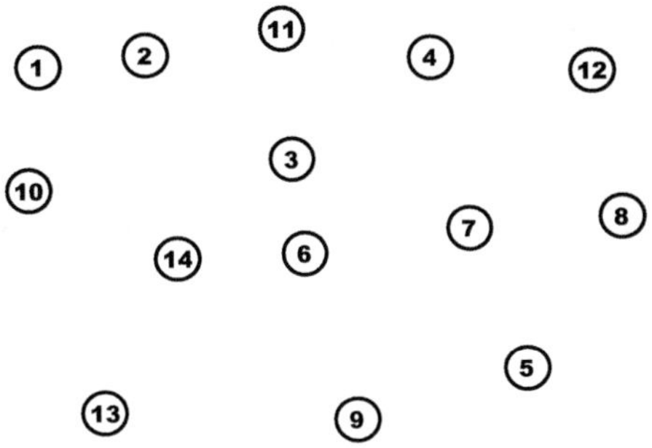

Fig. 5. Distributed system nodes

If applied, say, in node 11, this distributed program establishes only a partial dominance in the system, as shown in Fig. 6.

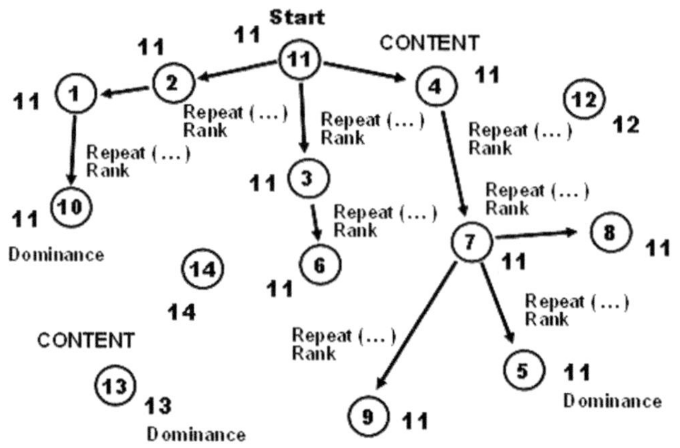

Fig. 6. Resulting in partial dominance

That will not be the case for node 14, which will set up its absolute dominance over the whole system by the program above, as in Fig. 7.

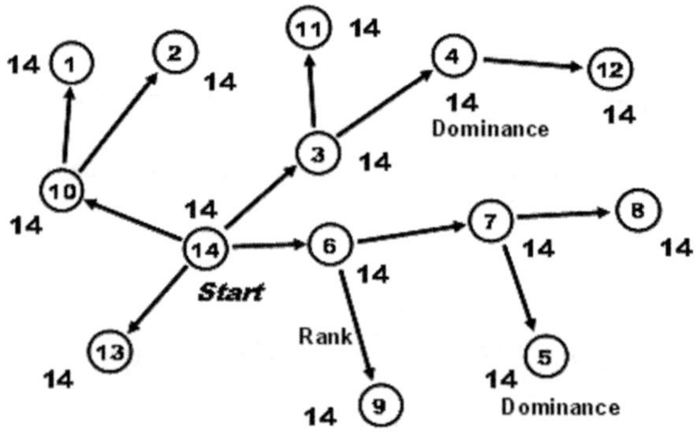

Fig. 7. Setting absolute dominance of node 14

4.2 Creating Infrastructures in the Distributed Space

It is easy to set up any infrastructures in the distributed space by the approach presented, with any topology. The following program, starting from node 3, will create (in parallel and distributed way) the networked structure shown in Fig. 8 (over the set of already existing nodes).

```
hop node 3; create links ((L6# 2; L8#14),
           (L7#12; L5#7; L4#3), (L1#14; L2#9; L3#7))
```

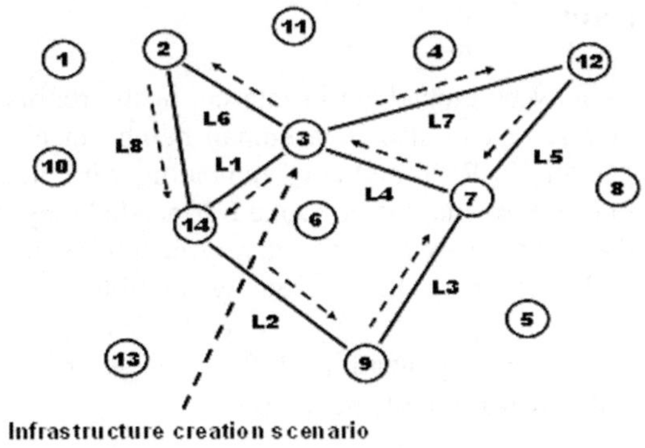

Fig. 8. Creating a distributed infrastructure

Any functionality can be associated with both nodes and links of the obtained infrastructure at runtime, which will be operating together as an integral system for the purpose needed.

4.3 Finding Patterns in the Infrastructure

It is easy to find any patterns in the distributed infrastructures in WPL. Let such a pattern be a triangle, and we would like to find all of them in the infrastructure created. The following spatial program starting in any node does this, with listing resultant nodes of the triangles in their descending ranks.

```
hop to all nodes; frontal (Triangle) = CONTENT;
do twice (hop by all links; CONTENT < BACK;
           Triangle &= CONTENT);
hop by all links; element (Triangle, first) == CONTENT;
output Triangle
```

The result, issued in the node where the program was injected, will be as: (14, 3, 2), (12, 7, 3), see Fig.9.

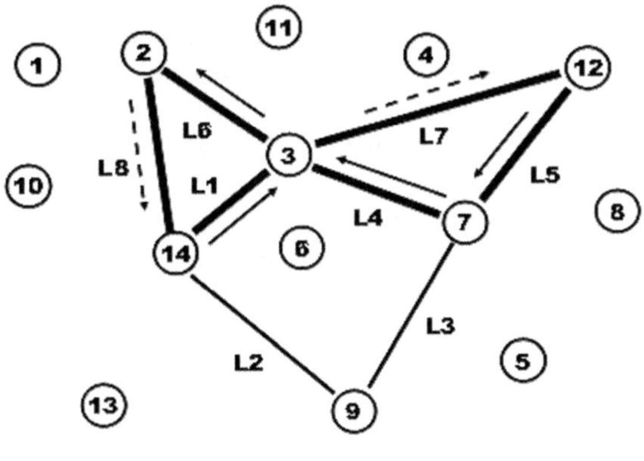

Fig. 9. Finding all triangles in the infrastructure

5 WPL Interpreter

The WPL interpreters may be embedded in internet hosts, robots, mobile phones, or smart sensors (an interpreter can also be a human being herself, understanding and executing high-level orders in WPL and communicating with other humans or robots via WPL too). The interpreters may be concealed, if needed (say, to work in a hostile system); they can also migrate freely, collectively executing (also mobile) mission scenarios, resulting altogether in the extremely flexible and ubiquitous system organization.

The basic WPL interpreter organization [2, 3, 4] is shown in Fig. 10, which may have both software and hardware implementation.

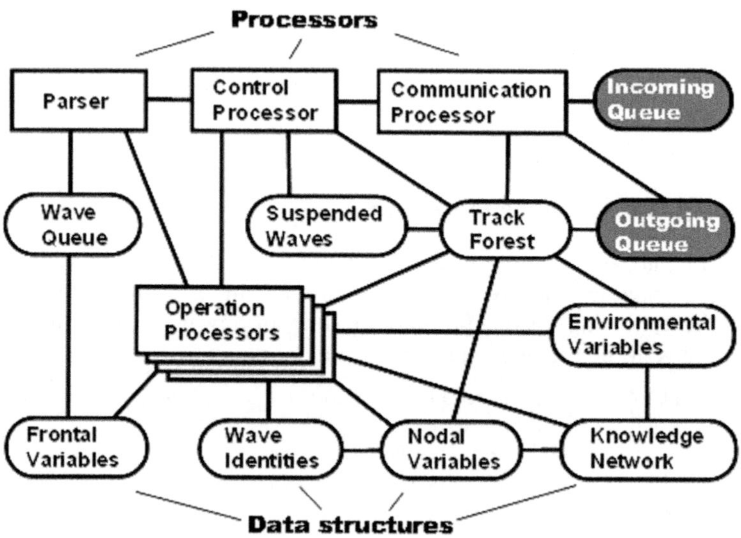

Fig. 10. The WPL interpreter architecture

The interpreter consists of a number of specialized modules working in parallel and handling and sharing specific data structures, which are supporting persistent virtual worlds and temporary hierarchical control mechanisms. The whole network of the interpreters can be mobile and open, changing the number of nodes and communication structure between them.

The heart of the distributed interpreter is its spatial track system enabling hierarchical command and control and remote data and code access, with high integrity of emerging parallel and distributed solutions. The interpreters can be embedded into any other systems, like mobile robots, allowing them to behave as integral teams, as shown in Fig. 11.

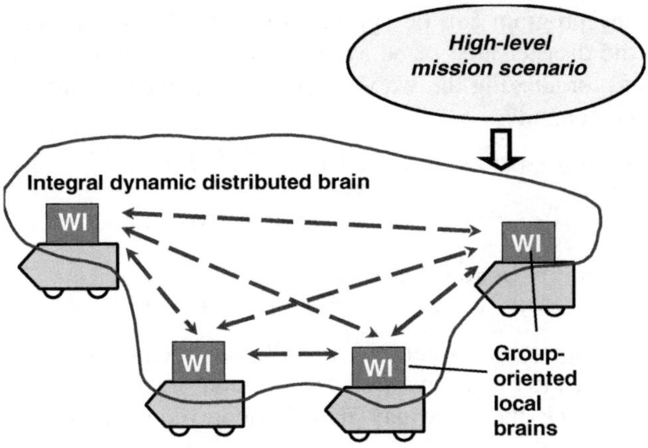

Fig. 11. WPL interpreters (WI) forming distributed robotic brain

6 Emergency Management

Emergency management, EM [5] due to the increased world dynamics, is becoming one of the hottest topics today. The emergency managers around the world are faced with new threats, new responsibilities, and new opportunities. Novel technologies, like the one of this paper, can alleviate consequences of natural or manmade disasters. They can allow law enforcement and intelligence investigators to identify potential terrorist plots and then mount preemptive strikes to stop their plans.

The technology described can help in solving many EM problems by using communicating interpreters (WI) embedded in different electronic devices, like laptops or mobile phones, with a symbolic disaster situation depicted in Fig 12.

A very simple example of doing something globally in this distributed world may be counting the total number of casualties in the disaster area, on all its affected regions.

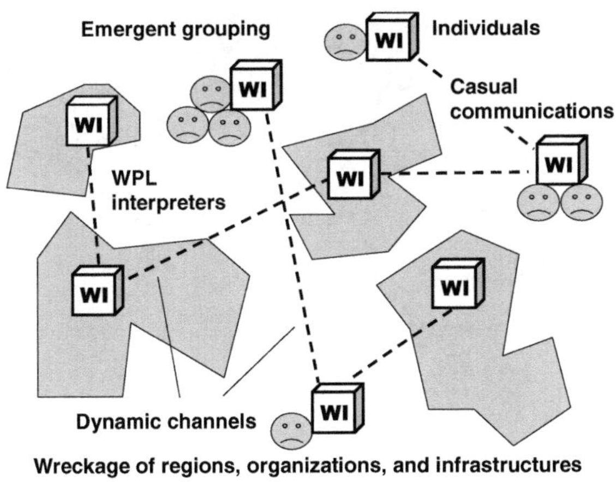

Fig. 12. A disaster area with WPL interpreters embedded

The following program can be applied from any WI as an entry one, which can reside within the disaster area or be away from it, and then can self-spread via local communications, organizing the whole region with embedded interpreters to work as a parallel spatial computer.

```
frontal Area = <disaster area definition>;
output sum (
  hop (directly, first come, nodes (Area));
  repeat(
    done (count casualties),
    hop (any links, first come, nodes (Area))))
```

More complex operations organized in WPL may include the delivery of relief aid, synchronized evacuation from the disaster area, and organization of (and cooperation with) the rescue teams (the latter may include robotic components).

7 Sensor Networks

Sensor networks are a sensing, computing and communication infrastructure that allows us to instrument, observe, and respond to phenomena in the natural environment, and in our physical and cyber infrastructure. The sensors themselves can range from small passive microsensors to larger scale, controllable platforms. Typical applications of wireless sensor networks (WSN) include monitoring, tracking, and controlling. Some of the specific applications are habitat monitoring, object tracking, nuclear reactor controlling, fire detection, traffic monitoring, etc. Any distributed problems can be solved by dynamic self-organized sensor networks working in WPL [6].

Starting from all transmitter nodes, the following program regularly (with interval of 20 sec.) covers stepwise, through local communications between sensors, the whole sensor network with a spanning forest, lifting information about observable events in each node reached, as shown in Fig. 13. Through this forest, by the internal

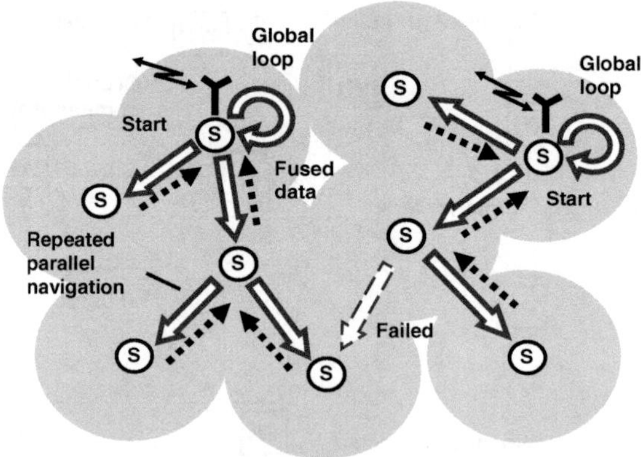

Fig. 13. Collecting data by a sensor network

interpretation infrastructure, the data lifted in nodes is moved and fused upwards the spanning trees, with final results collected in transmitter nodes and subsequently sent outside the system in parallel.

```
hop to all transmitters;
loop (
   sleep (20);
   IDENTITY = TIME;
   transmit (
     fuse (
        repeat (free (observe (events));
           hop to (directly reachable, first come)))))
```

Another program, which follows, provides for spanning tree coverage of some distributed phenomenon, with hierarchical collection, merging and fusing partial results from different sensors into the global picture. This picture will be forwarded to a nearest transmitter via the previously created infrastructure with links infra, as shown in Fig. 14.

```
hop to (random, all nodes, detected phenomenon).
loop (
   frontal Full = fuse (
      repeat (
        free (collect phenomenon),
        hop (directly reachable, first come,
                detected phenomenon)));
    repeat (hop by links -infra). Transmit Full)
```

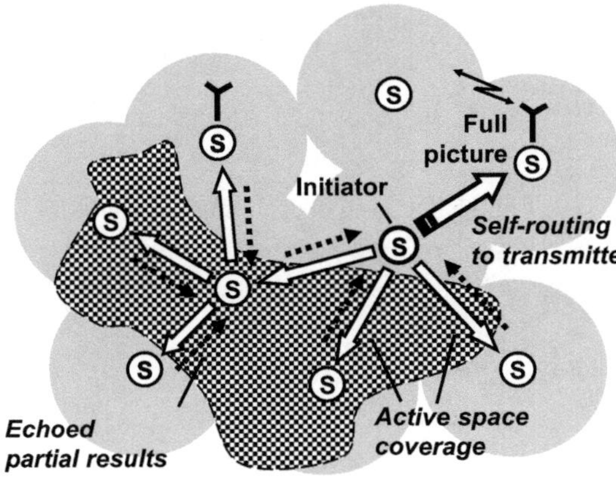

Fig. 14. Space coverage with hierarchical assembling of a distributed phenomenon

In more complex situations, we may have a number of simultaneously existing phenomena, which can intersect in a distributed space. We may also face a combined phenomenon integrating features of different phenomena. The phenomena (like flocks of birds, manned or unmanned groups or armies, spreading fire or flooding) covering certain regions may change in size and shape, they may also move as a whole, preserving internal organization. All these can be effectively managed in WPL.

8 Directed Energy Systems

Directed energy (DE) systems are of growing interest for broad applications in the nearest future, especially in infrastructure protection and defense. The DE-based systems will be able to operate under flexible command and control in WPL, restructuring and recovering in unpredictable environments [7].

An elementary DE-based system may consist of a control center, DE source, relay mirror (RM), and target. Using WPL, the system functionality can be set up dynamically by the following program:

```
sequence (
  parallel (
    (hop (DE); adjust (RM)),
    (hop (RM); adjust (DE, Target))),
  (hop (DE); activate (DE)))
```

Three snapshots of the system operation under this program are shown in Figs. 15-17.

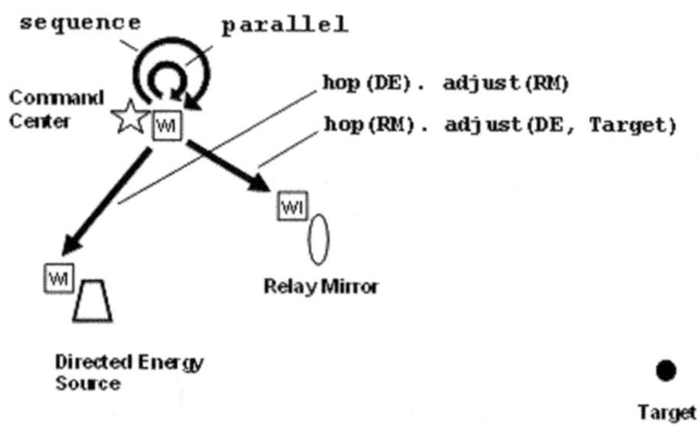

Fig. 15. DE system operation, Snapshot 1

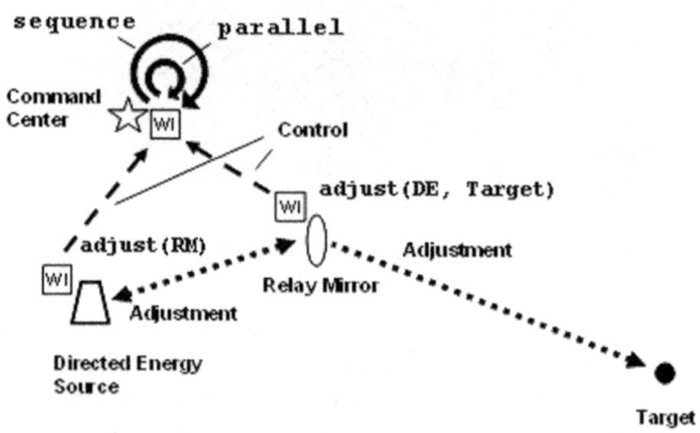

Fig. 16. DE system operation, Snapshot 2

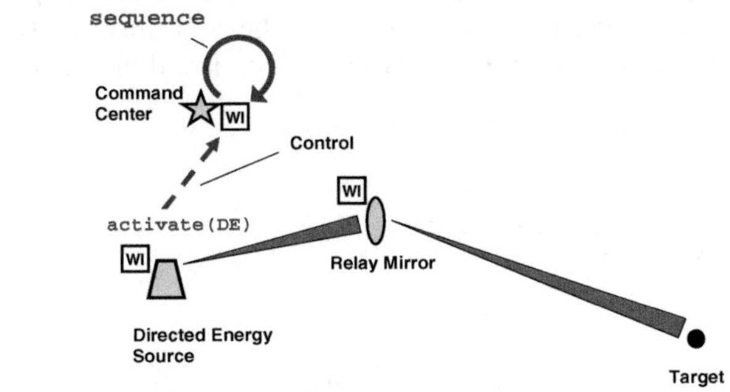

Fig. 17. DE system operation, Snapshot 3

Boeing's Advanced Relay Mirror System (ARMS) concept plans to entail a constellation of as many as two dozen orbiting mirrors that would allow a constant coverage of every corner of the globe. When activated, this would enable a directed energy response to critical trouble spots anywhere.

We will show here, be the program below, how the shortest path tree (SPT) starting from any DE source and covering the whole set of distributed mirrors can be created at runtime with the use of the technology presented. This will enable us to make optimal delivery of the directed energy to any point of the globe. The distributed SPT creation process is shown in Fig. 18.

```
nodal (Distance, Predecessor);
frontal (Length, Range = 400);
hop (DE); Distance = 0. Length = 0;
repeat (hop (Range, all);
        Length += between (WHERE, BACK);
        or (Distance == nil, Distance > Length);
        Distance = Length; Predecessor = BACK)
```

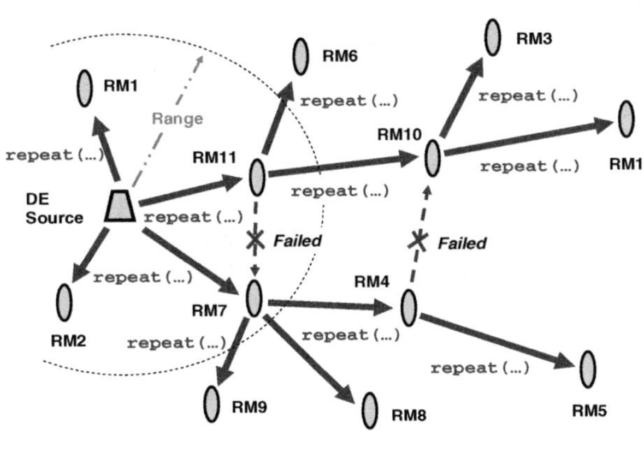

Fig. 18. Dynamic shortest path tree through all RMs

In case the target is defined, the following program forms a path from the DE source to the target via the relay mirrors, using the SPT formed, with a subsequent activation of the DE source to impact the target, as depicted in Fig. 19.

```
adjust (Seen (range), Predecessor);
repeat (
   hop (Predecessor, first);
   adjust (BACK, Predecessor));
activate (DE)
```

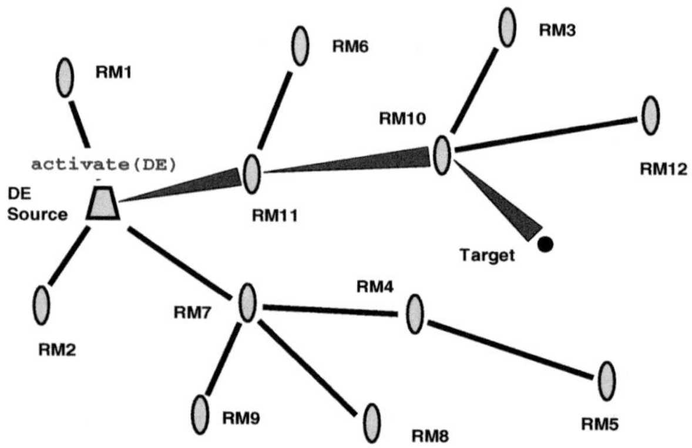

Fig. 19. Energy delivery via the path found

9 Electronic Warfare

Electronic warfare (EW) is becoming one of the main technological challenges of this century. All existing and being developed electronic support, attack, and protection measures usually have a very limited scope and effect if used alone. But taken together they may provide a capability for fulfilling the rapidly growing needs. Traditional communication and cooperation between these systems may not be sufficient. They should comprise altogether a much more integral system of systems with global situation awareness and "global will", which can be expressed and provided in WPL [8].

One of the typical EW tasks is fighting malicious intrusions and viruses in computer networks. Being itself a super-virus on the implementation level, the technology proposed, via the embedded network of WPL interpreters, can simultaneously discover and analyze electronic viruses, with blocking their spread and inferring attack sources. For example, the following scenario can find all virus sources in parallel, as shown in Fig 20:

```
nodal (Trace, Predecessor);
sequence (
  (hop (all nodes);
   nonempty (check general (viruses));
   repeat (
```

```
      increment (Trace);
       nonempty (Predecessor = check special (viruses));
       hop (Predecessor))),
   output (
    sort (
      hop (all nodes); empty (Predecessor);
      nonempty (Trace); Trace & ADDRESS)))
```

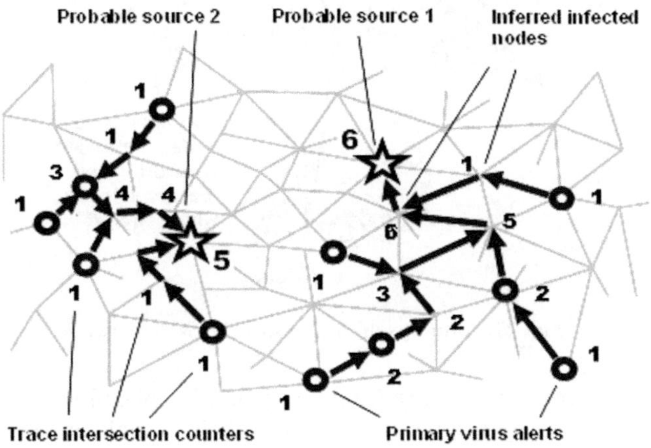

Fig. 20. Finding virus sources in parallel

10 Avionics

Avionics, or aviation electronics, represents a substantial share of the cost of any modern flying devices.

- Any avionics system, whether for a single aircraft or a group of them with manned or unmanned units, may be considered as a complex organization consisting of numerous components properly interacting with each other to pursue global goals. This organization can be effectively expressed in WPL on a variety of levels.
- This organization can be made flexible enough to recover from indiscriminate damages and restructure at runtime.
- The WP approach may offer real possibilities for a runtime recovery after damages, including reassembling of the whole system (or what remains of it) from any point.

Implanting communicating WPL interpreters into main components of an aircraft, as universal control modules U (see Fig. 21), may allow us to convert the whole distributed object into a parallel computer capable of solving a variety of complex problems at runtime, including aircraft's safety and recovery [9].

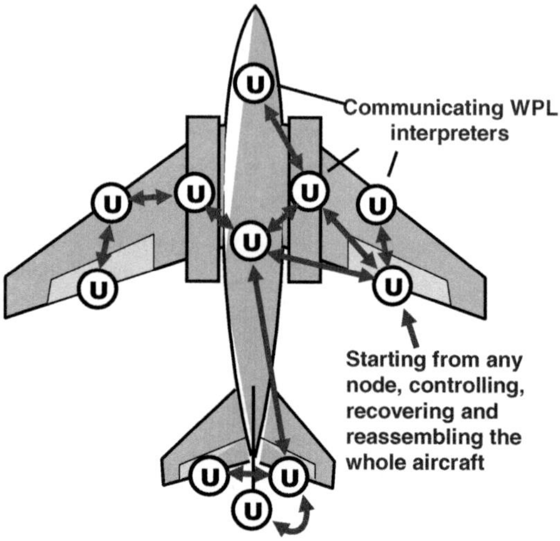

Fig. 21. Aircraft self-analysis by the WPL network

The following program, starting from any point, is collecting availability of vital mechanisms of a damaged aircraft, analyzing their completeness to operate as a system, with making proper decisions (which may include the alarm with emergent evacuation of the crew).

```
nodal Available_Set =
  repeat (
    free (if CONTENT belongs_to
           (left_aileron, right_aileron, left_elevator,
            right_elevator, rudder, left_engine,
            right_engine, left_chassis, right_chassis,…)
               then CONTENT),
    hop_first to all_neighbors);
if sufficient Available_Set
          then control_with Available_Set
                 otherwise alarm
```

11 Distributed Objects Tracking

Tracking mobile objects in distributed environments is an important task in a number of areas like air and road traffic, infrastructure protection, national and international crime, or missile defense. The example here relates to tracking aerial objects by a dynamic network of unmanned aerial vehicles, UAVs [9], with the following features to be taken into account.

- Each UAV can observe only a limited part of space (as in Fig. 22).
- To keep the whole observation continuous, the object discovered should be handed over between neighboring UAVs during its movement, along with the data accumulated about it.

- The model can catch each object and accompany it individually by the mobile intelligence, while propagating between WPL interpreters in UAVs.
- Many such objects can be picked up and chased in parallel by a dynamic UAV network.

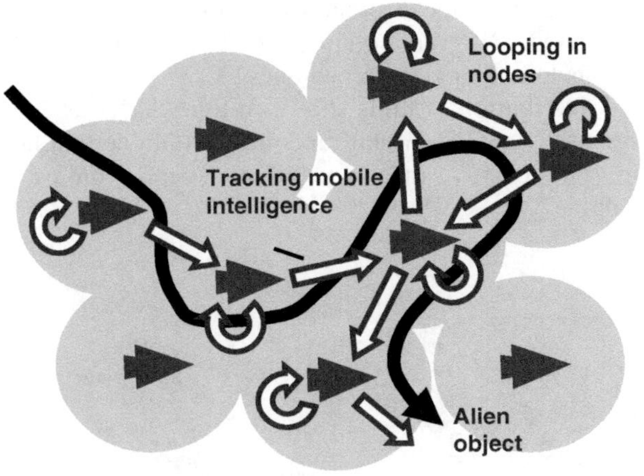

Fig. 22. Collective tracking of a mobile object

The following program, starting in all units, catches the object it sees and follows it wherever it goes, if it is not seen from this point any more (its visibility becomes lower than a given threshold).

```
hop all_nodes; Frontal Threshold = 0.1;
frontal Object =
   select_max_visible (aerial, Threshold);
repeat (
   loop (visibility (Object) > Threshold);
   choose_destination_with_max_value (
      hop to all_neighbors.
      visibility (Object) > Threshold))
```

A snapshot of a possible situation in a distribute space is shown in Fig. 22. The information about the tracked objects can be accumulated by individual mobile intelligences [10], which can cooperate with each other, making individual or collective decisions about the further fate of the objects (e.g. classifying them as friendly or hostile).

12 Collective Behavior

The higher-level, semantic WPL scenarios are well understandable by humans who can perform jobs written in the language and delegate other jobs to other group members, establishing runtime relations with each other. These scenarios also represent formal descriptions that can be effectively interpreted by robots and their groups automatically.

Both human and robotic suitability allow for a unified approach to organization of teams that can range from purely human to purely robotic. These teams can be open and emergent, and can operate in unpredictable environments, where team members can indiscriminately fail at any time but the mission scenario, collectively interpreted by the distributed group, can survive and fulfill objectives. The collective team behavior can be based on a loose organization like swarms, or can be strictly or hierarchically controlled. Different solutions in WPL throughout this organizational range are possible, including any combined ones [2].

With the initial distribution of units shown in Fig. 23, let us consider a collective swarm-like movement, where each unit randomly, within certain hop limits, defining general direction, tries to move in new positions, keeping the established threshold distance to other units.

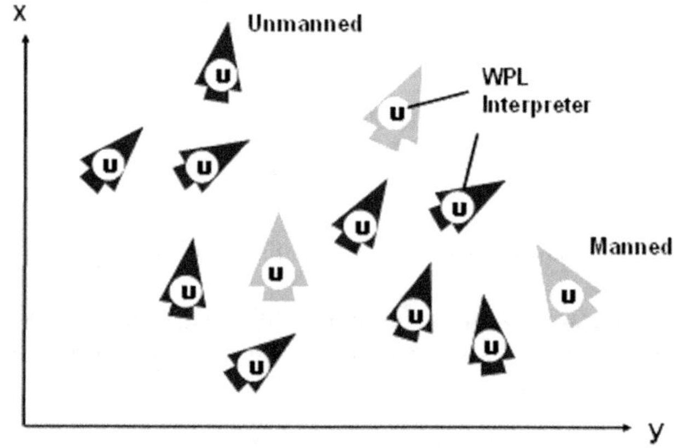

Fig. 23. Initial distribution of units

This can be done by the following program, which can start from any unit, manned or unmanned.

```
nodal (Limits, Range, Shift);
hop all_nodes;
Limits = (dx (0, 8), dy (- 2, 5)); Range = 5;
repeat (
   Shift = random (Limits );
   if empty hop (Shift, Range) then move Shift)
```

A snapshot of the group movement by this spatial program is depicted in Fig. 24.

To have more coordinated actions of the group, we may set up a distributed hierarchical infrastructure over it, to be used in command and control and in maintaining global awareness. As the group is distributed in space and distances between units can change, such an infrastructure should be preferably based on the current physical position of the units, with top of the hierarchy to be close to the group's center, in order to optimize global coordination. We will consider here how the topologically central unit can be found at runtime, during the movement within a swarm, and how the C2 hierarchy can be formed starting from this central unit.

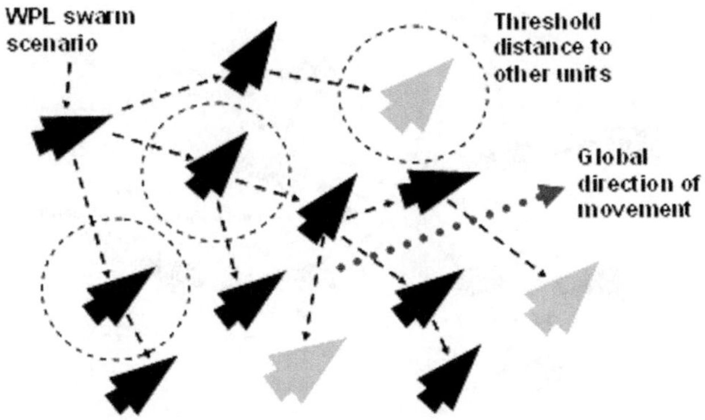

Fig. 24. A swarm movement snapshot

The following distributed program, starting from any unit, finds topologically central unit of the distributed swarm, which is shown in Fig. 25.

```
frontal Aver =
   average (hop all_nodes;  WHERE);
nodal Center =
   element (
      min (
         hop all_nodes;
         distance (Aver, WHERE) & ADDRESS), 2)
```

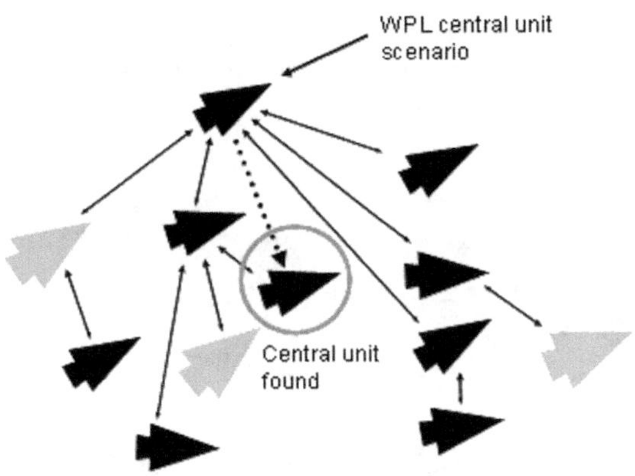

Fig. 25. Finding topologically central unit

Starting from the central unit found, the next program creates runtime hierarchical infrastructure with oriented links infra, as shown in Fig. 26.

```
frontal Range = 20.
   repeat (
      create_links (
          + infra, first_come, nodes (Range)))
```

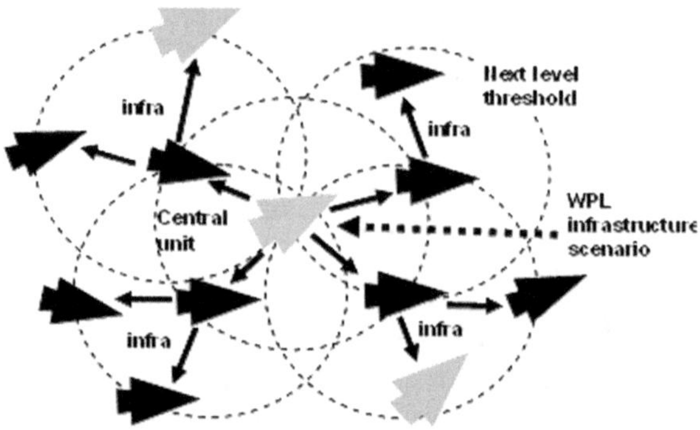

Fig. 26. Hierarchical infrastructure built

This runtime hierarchy created may be effectively used for maintaining global awareness in the distributed space, collection and fusion of targets seen by individual units, spreading the set of collected targets back to all units, which may select the most suitable ones for an individual impact. The following program, navigating the infrastructure created, follows this scenario, as shown in Fig. 27.

```
repeat (
   if nonempty (
      frontal Seen = Repeat (
                     Free (detect targets),
                     Hop_links + infra))   then
      repeat (
         free (if TYPE == UAV   then
                     select_move_shoot Seen),
         hop_links + infra))
```

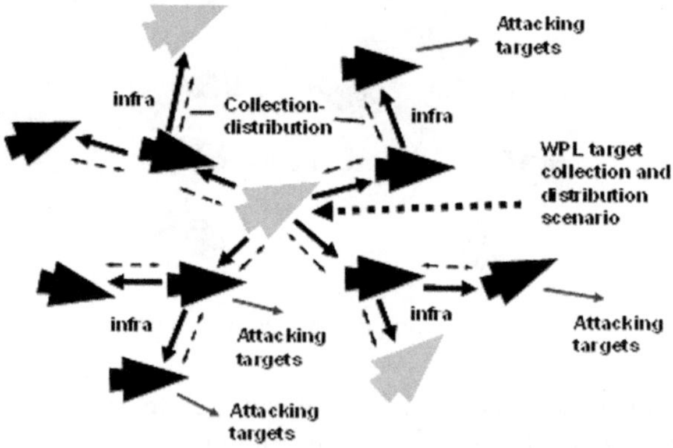

Fig. 27. Hierarchical fusion and distribution of targets

As the whole group moves and distances between separate units may change in the swarm, the programs of finding the center and hierarchical infrastructure may be repeated regularly, which will help to maintain the group's optimal spatial organization in a distributed environment. Any position in this dynamic hierarchy (the top one including) may happen to be occupied by any unit at any moment of time, regardless of whether it is manned or unmanned.

Many more applications of this world processing paradigm (previously known as WAVE) can be found in [1, 2, 3, 10].

13 Conclusions

We have touched only a few application areas currently in investigation for the WP technology. The experience obtained allows us to conclude the following.

- The technology converts any distributed system into a universal spatial computer capable of solving complex problems on itself and on the surrounding environment.
- This system, for example, can be represented by a single unit or by a group (army) of them, with individual units bring manned or unmanned.
- The whole system is driven by high-level scenarios setting how to behave and what to do as a whole, while omitting traditional implementation details, which are effectively delegated to intelligent distributed interpretation system.
- The system scenarios in the World Processing Language are very compact and can be created at runtime, on the fly, swiftly reacting on rapidly changing environment and mission goals.
- Any scenario can start from any available component and can cover the system at runtime, during its evolution.
- The approach may offer real possibilities for a runtime recovery after indiscriminate damages, including reassembling of the whole system (or what remains of it) at any stage and from any system point.
- The technology can help dominate over other distributed system organizations, especially over those explicitly based on communicating and interacting parts (or agents).

References

1. Sapaty, P.S.: Over-Operability in Distributed Simulation and Control. The MSIAC's M&S Journal Online 4(2) (Winter Issue, 2002)
2. Sapaty, P.S.: Ruling Distributed Dynamic Worlds. John Wiley & Sons, New York (2005)
3. Sapaty, P.S.: Mobile Processing in Distributed and Open Environments. John Wiley & Sons, New York (1999)
4. Sapaty, P.S.: A distributed Processing System. European Patent No. 0389655, Publ. 10.11.93, European Patent Office (1993)
5. Sapaty, P., Sugisaka, M., Finkelstein, R., Delgado-Frias, J., Mirenkov, N.: Advanced IT Support of Crisis Relief Missions. Journal of Emergency Management 4(4) (July/August 2006)

6. Sapaty, P.: Intelligent Management of Distributed Sensor Networks. In: Sensors, and Command, Control, Communications, and Intelligence (C3I) Technologies for Homeland Security and Homeland Defense VI. In: Carapezza, E.M. (ed.) Proc. SPIE, vol. 6538 (2007)
7. Sapaty, P., Morozov, A., Sugisaka, M.: DEW in a Network Enabled Environment. In: Proc. International Conference Directed Energy Weapons 2007, Le Meridien Piccadilly, London, UK (2007)
8. Sapaty, P.: Global Management of Distributed EW-related Systems. In: Proc. Electronic Warfare: Operations & Systems 2007, Thistle Selfridge, London, UK (2007)
9. Sapaty, P.: Grasping the Whole by Spatial Intelligence: A higher Level for Distributed Avionics. In: Proc. international conference Military Avionics 2008, Café Royal, London, UK (2008)
10. Sapaty, P.S., Corbin, M.J., Seidensticker, S.: Mobile Intelligence in Distributed Simulations. In: Proc. 14th Workshop on Standards for the Interoperability of Distributed Simulations, IST UCF, Orlando, FL (1995)

Dealing with Uncertainty in the Hybrid World*

Luís Pina and Miguel Ayala Botto

Department of Mechanical Engineering, IDMEC, Instituto Superior Técnico
Technical University of Lisbon, Portugal
luispina@dem.ist.utl.pt, ayalabotto@ist.utl.pt

Abstract. This paper presents an efficient hybrid estimation algorithm. The proposed algorithm is based on a least-squares Interacting Multiple-Model setup that simultaneously estimates the continuous state and the discrete mode of an hybrid system, providing also the associated uncertainties. The efficiency of the algorithm is achieved by excluding as many discrete mode sequences as possible with little computational effort. This is done by rapidly computing good estimates, separating the constrained and unconstrained estimates, and using some auxiliary coefficients computed off-line. The success of this state estimation algorithm is shown while compared with the Moving Horizon Estimation algorithm in a fault detection problem of the benchmark AMIRA DTS200 three-tanks system experimental setup.

1 Introduction

In the last decade hybrid systems have become a major research topic in Control Engineering [4]. Hybrid systems are dynamical systems composed by both discrete valued and continuous valued states. The dynamics of a hybrid system is governed by a mode selector that determines, at each time instant, which discrete mode is active from endogenous and/or exogenous variables. The continuous state is then updated through a dynamic relation that is selected from a set of possible dynamics according to the value of the active discrete mode. In fact, the presence of physical components such as on/off switches or valves, gears or speed selectors, or behaviors dependent on if-then-else rules imply explicitly or implicitly the discrete/continuous interaction. This interaction can be found in many real world applications such as automotive control, urban and air traffic control, communications networks, embedded control systems, and in the control of complex industrial systems via the combination of classical continuous control laws with supervisory switching logic.

The hybrid nature has attracted the interest of mathematicians, control engineers and computer scientists, therefore leading to different modelling languages and paradigms that influenced the line of research on hybrid systems in several different ways. For instance, the computer science research community is more focused on systems whose

* This work was supported by project PTDC/EME-CRO/69117/2006 co-sponsored by FEDER, Programa Operacional Ciência e Inovação 2010, and by the grant SFRH/BPD/41496/2007, from FCT, Portugal.

J.A. Cetto et al. (eds.): Informatics in Control, Automation and Robotics, LNEE 37, pp. 25–45.
springerlink.com

variables take values in a finite set, so adopted the discrete events modelling formalism to model hybrid systems, using finite state machines, Petri nets, temporal logic, etc. On the other hand, the control systems community typically considers a continuous valued world, where time is continuously changing, thus considering a hybrid system as described by a differential (or difference) equation with some switching mechanism. Examples of such hybrid models include Piece-Wise Affine (PWA) [19] and Mixed Logical Dynamical (MLD) [7] models. A PWA model is the most intuitive representation of a hybrid system since it provides a direct relation to linear systems while still capturing very complex dynamical behaviors. However, a MLD representation is most adequate to be used in optimization problems since it is able to embed both propositional logic statements (if-then-else rules) and operating constraints in a state linear dynamics equation by transforming them to mixed-integer linear inequalities. Despite these differences, PWA and MLD are equivalent models of hybrid systems in respect to well-posedness and boundness of input, state, output or auxiliary variables [13]. This fact allows to interchange analysis and synthesis tools between them.

Research on hybrid systems spans to a wide range of topics (and approaches), from modelling to stability analysis, reachability analysis and verification, study of the observability and controllability properties, methods of state estimation and fault detection, identification techniques, and control methodologies. Typically, hybrid tools rely on the solution of optimization problems. However, due to the different nature of the optimization variables involved (integer and continuous) the main source of complexity becomes the combinatorial (yet finite) number of possible switching sequences that have to be considered. A hybrid optimal solution thus requires solving mixed-integer non-convex optimization algorithms with NP-complete complexity [20].

Analysis and synthesis procedures for hybrid systems when disturbances are present either on the continuous dynamics or on the discrete mode of the hybrid system, is still an open research topic that has being tackled by several authors using distinct approaches. In the state estimation problem two distinct approaches are usually followed, the main difference being the knowledge of the active mode: some approaches consider only continuous state uncertainty with known discrete mode, while others assume that both the discrete mode and the continuous state are unknown. The combination of both uncertainties (state and mode) on the estimation process of a hybrid system presents a very difficult problem for which a global solution is not yet found. When the discrete mode is known in advance, the problem is greatly simplified and the state estimation methodologies for linear systems can be applied with very little modifications. For example in [9] a bank of Kalman filters is used and in [1] an LMI based algorithm computes the stabilizing gains for a set of Luenberger observers. If, on the other hand, the discrete mode must also be estimated the estimation problem becomes much more complex and every discrete mode sequence (*dms*) must be checked to choose the one that provides the best fit for the observed data. The continuous state estimates are then computed for the estimated *dms*. Several works address this problem, see [6] where a location observer is used to estimate the discrete mode and a Luenberger observer is then used to estimate the continuous state. In [11] and [16] a Moving Horizon Estimation (MHE) scheme simultaneously estimates the discrete mode and the continuous state, differing in the fact that the latter can also estimate the input disturbances.

The derivation of the truly optimal filter for systems with switching parameters was first presented in [5]. The objective was to perform simultaneous system identification and state estimation for linear systems but the derivation is quite general and is directly applicable to the hybrid state estimation problem. This method requires the consideration of all admissible *dms* starting from the initial time instant, being obviously unpractical since the number of *dms* grows exponentially in time, and so, suboptimal methods were developed. From the various possibilities, setting a time window and considering all the admissible *dms* of that given length is a very popular methodology to bound the complexity and will be used to develop the proposed algorithm. Several suboptimal multiple model estimation schemes where developed and applied for tracking maneuvering vehicles, as surveyed in [15], and systems with Markovian switching coefficients, [8], proving their efficiency for state estimation in multiple model systems. Multiple model estimation algorithms use a set of filters, one for each possible dynamic of the system. Here a least-squares filter will be designed for each sequence of discrete modes with the predefined length, and were chosen due to their applicability under very slim assumptions, optimality in special setups and due to the possibility of considering constraints.

In this paper an efficient state estimation algorithm for stochastic hybrid systems, based on the Interacting Multiple-Model (IMM) estimation algorithm, is proposed. The method is applicable to most of the existing models of hybrid systems subject to disturbances with explicitly known probability density function, so being rather general. This estimation method will be further compared to the Moving Horizon Estimation (MHE) algorithm and tested in the benchmark AMIRA DTS200 three-tanks system experimental setup. The MHE algorithm is one of the best existing solutions to the hybrid estimation problem with ensured convergence properties, however, it relies on brute force optimization of mixed-integer programs and is not efficient enough for most online applications.

The paper is organized as follows. Section 2 provides a description of the considered PWA model and in section 3 the proposed Interacting Multiple-Model estimation algorithm is presented, with reference to the composing parts, computational complexity and accuracy of the estimates. Section 4 presents an experimental application of the proposed algorithms to the AMIRA DTS200 three-tanks system experimental setup. First the experimental setup is presented and modelled, including a full characterization of all uncertainties. Then the proposed algorithms are tested and their performance is compared. Finally, in section 5 some conclusions are drawn along with some possible future developments.

2 System Description

The proposed estimation algorithm is developed for PWA systems which were introduced in [19]. The following stochastic PWA model will be considered:

$$x(k+1) = A_{i(k)}x(k)+B_{i(k)}u(k)+f_{i(k)}+W_{i(k)}w(k) \tag{1a}$$

$$y(k) = C_{i(k)}x(k)+D_{i(k)}u(k)+g_{i(k)}+v(k) \tag{1b}$$

$$\text{iff} \begin{bmatrix} x(k) \\ u(k) \\ w(k) \end{bmatrix} \in \Omega_{i(k)} \tag{1c}$$

where k is the discrete time, $x(k) \in \mathbb{X} \subset \mathbb{R}^{n_x}$ is the continuous state, $u(k) \in \mathbb{U} \subset \mathbb{R}^{n_u}$ is the input, $y(k) \in \mathbb{R}^{n_y}$ is the output, $i(k) \in \mathcal{I} = \{1, \ldots, s\}$ is the discrete mode, and s is the total number of discrete modes. The matrices and vectors A_i, B_i, f_i, L_i, C_i, D_i, g_i depend on the discrete mode $i(k)$ and have appropriate dimensions. The input disturbance $w(k)$ and the measurement noise $v(k)$ are modelled as independent identically distributed random variables, belonging to the sets $\mathbb{W}_i$ and $\mathbb{V}_i$, with expected values $E\{w(k)\} = 0$, $E\{v(k)\} = 0$ and covariances Σ_{w_i} and Σ_{v_i}, respectively. These conditions are not restrictive at all since the zero mean can be imposed by summing a constant vector to the disturbances and compensated in the affine term of the system dynamics (1) and, the sets $\mathbb{W}_i$ and $\mathbb{V}_i$ can be considered large enough to contain all possible disturbances relevant for practical applications, for instance 99.99% of all admissible values. Notice that the input disturbance and measurement noise *pdf*s may depend on the actual mode of the system $i(k)$. The sets $\mathbb{W}_i$ and $\mathbb{V}_i$ are respectively defined for each mode $i(k)$ by a set of linear inequalities:

$$H_{\mathbb{W}_{i(k)}}\, w(k) \leq h_{\mathbb{W}_{i(k)}}, \quad \forall_{k \in \mathbb{N}_0} \tag{2}$$

$$H_{\mathbb{V}_{i(k)}}\, v(k) \leq h_{\mathbb{V}_{i(k)}}, \quad \forall_{k \in \mathbb{N}_0} \tag{3}$$

The discrete mode $i(k)$ is a piecewise constant function of the state, input and input disturbance of the system whose value is defined by the regions Ω_i:

$$\Omega_i \quad : \quad S_i\, x(k) + R_i\, u(k) + Q_i\, w(k) \leq T_i \tag{4}$$

Some helpful notation regarding the time-compressed representation of [14] for system (1) will now be introduced. The time-compressed representation of a system defines the dynamics of the system over a sequence of time instants in opposition to the single time step state-space representation. Consider the time interval $[k, k+T-1]$, the sequence of discrete modes over this interval is represented as $\mathbf{i}_T \equiv \mathbf{i}_T(k) \triangleq \{i(k), \ldots, i(k+T-1)\}$. To simplify the notation, the time index k is removed from the discrete mode sequence (*dms*) whenever it is obvious from the other elements in the equations. In view of this, the state and output sequences over the same interval can be respectively computed by:

$$X_T(k) = \mathbf{A}_{\mathbf{i}_T} x(k) + \mathbf{B}_{\mathbf{i}_T} U_T(k) + \mathbf{f}_{\mathbf{i}_T} + \mathbf{W}_{\mathbf{i}_T} W_T(k) \tag{5}$$

$$Y_T(k) = \mathbf{C}_{\mathbf{i}_T} x(k) + \mathbf{D}_{\mathbf{i}_T} U_T(k) + \mathbf{g}_{\mathbf{i}_T} + \mathbf{L}_{\mathbf{i}_T} W_T(k) + V_T(k) \tag{6}$$

where the input, input disturbance, output and measurement noise sequences $U_T(k)$, $W_T(k)$, $V_T(k)$ and $V_T(k)$ respectively are defined in the same way as the output sequence $X_T(k) \triangleq [x(k)^{\mathrm{T}}, \ldots, x(k+T-1)^{\mathrm{T}}]^{\mathrm{T}}$. The matrices and vectors $\mathbf{A}_{\mathbf{i}_T}$, $\mathbf{B}_{\mathbf{i}_T}$, $\mathbf{f}_{\mathbf{i}_T}$, $\mathbf{W}_{\mathbf{i}_T}$, $\mathbf{C}_{\mathbf{i}_T}$, $\mathbf{D}_{\mathbf{i}_T}$, $\mathbf{g}_{\mathbf{i}_T}$ and $\mathbf{L}_{\mathbf{i}_T}$ are computed from the system dynamics (1a-1b) according to what is presented in (Kamen, E.W., 1992):

$$\mathbf{A_{i_T}} = \begin{bmatrix} I_{n_x} \\ A_{i_0} \\ A_{i_1}A_{i_0} \\ \vdots \\ A_{i_{T-2}}\dots A_{i_1}A_{i_0} \end{bmatrix} \tag{7}$$

$$\mathbf{B_{i_T}} = \begin{bmatrix} 0 & \cdots & 0 & 0 \\ B_{i_0} & \cdots & \vdots & 0 \\ A_{i_1}B_{i_0} & \cdots & 0 & \vdots \\ \vdots & \ddots & 0 & 0 \\ A_{i_{T-2}}\dots A_{i_1}B_{i_0} & \cdots & B_{i_{T-2}} & 0 \end{bmatrix} \tag{8}$$

$$\mathbf{f_{i_T}} = \begin{bmatrix} 0 \\ f_{i_0} \\ A_{i_1}f_{i_0} + f_{i_1} \\ \vdots \quad\quad \ddots \\ A_{i_{T-2}}\dots A_{i_1}f_{i_0} + \cdots + f_{i_{T-2}} \end{bmatrix} \tag{9}$$

$$\mathbf{W_{i_T}} = \begin{bmatrix} 0 & \cdots & 0 & 0 \\ W_{i_0} & \cdots & 0 & 0 \\ A_{i_1}W_{i_0} & \cdots & 0 & 0 \\ \vdots & \ddots & \ddots & \vdots \\ A_{i_{T-2}}\dots A_{i_1}W_{i_0} & \cdots & W_{i_{T-2}} & 0 \end{bmatrix} \tag{10}$$

$$\mathbf{C_{i_T}} = \begin{bmatrix} C_{i_0} \\ C_{i_1}A_{i_0} \\ C_{i_2}A_{i_1}A_{i_0} \\ \vdots \\ C_{i_{T-1}}A_{i_{T-2}}\dots A_{i_1}A_{i_0} \end{bmatrix} \tag{11}$$

$$\mathbf{D_{i_T}} = \begin{bmatrix} D_{i_1} & \cdots & 0 & 0 \\ C_{i_1}B_{i_0} & \cdots & \vdots & 0 \\ C_{i_2}A_{i_1}B_{i_0} & \cdots & 0 & \vdots \\ \vdots & \ddots & D_{i_{T-2}} & 0 \\ C_{i_{T-1}}A_{i_{T-2}}\dots A_{i_1}B_{i_0} & \cdots & C_{i_{T-1}}B_{i_{T-2}} & D_{i_{T-1}} \end{bmatrix} \tag{12}$$

$$\mathbf{g_{i_T}} = \begin{bmatrix} g_{i_0} \\ C_{i_1}f_{i_0} + g_{i_1} \\ C_{i_2}A_{i_1}f_{i_0} + C_{i_2}f_{i_1} + g_{i_2} \\ \vdots \quad\quad \ddots \\ C_{i_{T-1}}A_{i_{T-2}}\dots A_{i_1}f_{i_0} + \cdots + C_{i_{T-1}}f_{i_{T-2}} + g_{i_{T-1}} \end{bmatrix} \tag{13}$$

$$\mathbf{L}_{\mathbf{i_T}} = \begin{bmatrix} 0 & \cdots & 0 & 0 \\ C_{i_1} W_{i_0} & \cdots & 0 & 0 \\ C_{i_2} A_{i_1} W_{i_0} & \cdots & 0 & 0 \\ \vdots & \ddots & \ddots & \vdots \\ C_{i_{T-1}} A_{i_{T-2}} \dots A_{i_1} W_{i_0} & \cdots & C_{i_{T-1}} W_{i_{T-2}} & 0 \end{bmatrix} \tag{14}$$

The same reasoning can be applied to the constraints $\Omega_{\mathbf{i}_T}$:

$$\boldsymbol{\Omega}_{\mathbf{i}_T} \quad : \quad \mathbf{S}_{\mathbf{i}_T}\, x(k) + \mathbf{R}_{\mathbf{i}_T}\, U_T(k) + \mathbf{Q}_{\mathbf{i}_T}\, W_T(k) \le \mathbf{T}_{\mathbf{i}_T} \tag{15}$$

where the matrices $\mathbf{S}_{\mathbf{i}_T}$, $\mathbf{R}_{\mathbf{i}_T}$, $\mathbf{Q}_{\mathbf{i}_T}$ and $\mathbf{T}_{\mathbf{i}_T}$ can be computed from the system dynamics (1a) and partitions (4):

$$\mathbf{S}_{\mathbf{i_T}} = \begin{bmatrix} S_{i_0} \\ S_{i_1} A_{i_0} \\ S_{i_2} A_{i_1} A_{i_0} \\ \vdots \\ S_{i_{T-1}} A_{i_{T-2}} \dots A_{i_1} A_{i_0} \end{bmatrix} \tag{16}$$

$$\mathbf{R}_{\mathbf{i_T}} = \begin{bmatrix} R_{i_1} & \cdots & 0 & 0 \\ S_{i_1} B_{i_0} & \cdots & \vdots & 0 \\ S_{i_2} A_{i_1} B_{i_0} & \cdots & 0 & \vdots \\ \vdots & \ddots & R_{i_{T-2}} & 0 \\ S_{i_{T-1}} A_{i_{T-2}} \dots A_{i_1} B_{i_0} & \cdots & S_{i_{T-1}} B_{i_{T-2}} & R_{i_{T-1}} \end{bmatrix} \tag{17}$$

$$\mathbf{Q}_{\mathbf{i_T}} = \begin{bmatrix} Q_{i_1} & \cdots & 0 & 0 \\ S_{i_1} W_{i_0} & \cdots & \vdots & 0 \\ S_{i_2} A_{i_1} W_{i_0} & \cdots & 0 & \vdots \\ \vdots & \ddots & Q_{i_{T-2}} & 0 \\ S_{i_{T-1}} A_{i_{T-2}} \dots A_{i_1} W_{i_0} & \cdots & S_{i_{T-1}} W_{i_{T-2}} & Q_{i_{T-1}} \end{bmatrix} \tag{18}$$

$$\mathbf{T}_{\mathbf{i_T}} = \begin{bmatrix} T_{i_0} \\ T_{i_1} - S_{i_1} f_{i_0} \\ T_{i_2} - S_{i_2} A_{i_1} f_{i_0} - S_{i_2} f_{i_1} \\ \vdots \quad \ddots \\ T_{i_{T-1}} - S_{i_{T-1}} A_{i_{T-2}} \dots A_{i_1} f_{i_0} - \cdots - S_{i_{T-1}} f_{i_{T-2}} \end{bmatrix} \tag{19}$$

The inequalities that define the disturbance and noise sets over a *dms* $\mathbf{i}_T$, $\mathbb{W}_{\mathbf{i}_T}$ and $\mathbb{V}_{\mathbf{i}_T}$ respectively, can also be easily found from equations (2) and (3):

$$\mathbf{H}_{\mathbb{W}_{\mathbf{i}_T}} W_T(k) \leq \mathbf{h}_{\mathbb{W}_{\mathbf{i}_T}} \tag{20}$$

$$\mathbf{H}_{\mathbb{V}_{\mathbf{i}_T}} V_T(k) \leq \mathbf{h}_{\mathbb{V}_{\mathbf{i}_T}} \tag{21}$$

3 Interacting Multiple Model Estimation

The proposed Interacting Multiple-Model (IMM) Estimation algorithm is composed of three parts; the Unconstrained Filter Bank (UFB), the Constrained Filter Bank (CFB) and, the Discrete Mode Sequence Estimator (DMSE). A schematic representation is presented in figure 1.

The estimation algorithm works as follows: first the continuous state estimates are computed in the UFB without considering the constraints. Then, the DMSE computes the squared errors of these estimates and ranks them. Finally, starting with the estimate with the lowest squared error, the estimates are recomputed in the CFB considering the

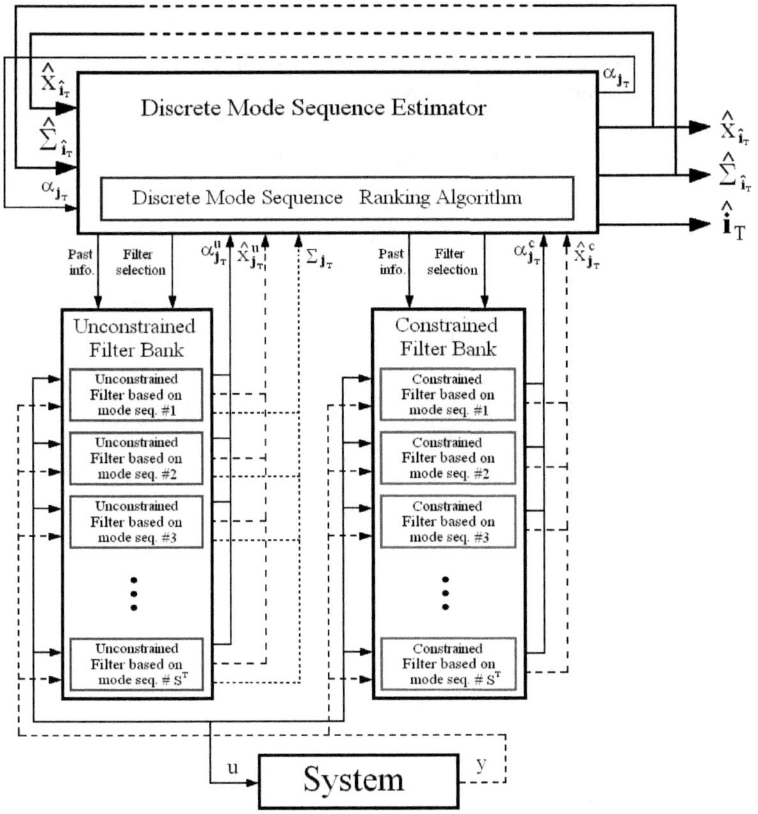

Fig. 1. Interacting Multiple-Model Estimation Algorithm

existence of constraints. When the estimate with the least squared error is already a constrained estimate the whole process stops since the most accurate estimate has been found.

As the estimation is based on sequences of measurements $Y_T(k)$ and discrete modes $\mathbf{i}_T(k)$, two distinct time instants must be considered: the time instant at the beginning of the sequences, k, and the time instant at the end of these sequences, which is the present time instant $t = k+T-1$. The state estimates $\hat{x}(k|t)$ at time instant k are computed at the present time instant t, and can be propagated to the present time according to the dynamics (5) defined by the discrete mode sequence estimate $\hat{\mathbf{i}}_T(k|t)$.

3.1 Unconstrained Filter Bank

The UFB computes the unconstrained state estimates. It is composed by a set of unconstrained least-squares filters, one for each possible *dms* $\mathbf{j}_T$:

$$\hat{x}^u_{\mathbf{j}_T}(k|t) = \hat{x}_{\mathbf{j}_T}(k|t-1)+ \tag{22}$$

$$\mathbf{K}_{\mathbf{j}_T}(k|t-1)\left[\,(Y_T(k)-\mathbf{D}_{\mathbf{j}_T}U_T(k)-\mathbf{g}_{\mathbf{j}_T})-\mathbf{C}_{\mathbf{j}_T}\hat{x}_{\mathbf{j}_T}(k|t-1)\right] \tag{23}$$

where $\hat{x}_{\mathbf{j}_T}(k|t-1)$ is the *a priori* continuous state estimate for mode sequence $\mathbf{j}_T$ using measurements up to time instant $t-1$. $\mathbf{K}_{\mathbf{j}_T}(k|t-1)$ is the filter gain:

$$\mathbf{K}_{\mathbf{j}_T}(k|t-1)=\left(\boldsymbol{\Sigma}^{-1}_{x_{\mathbf{j}_T}}(k|t-1)+\mathbf{C}^{\mathrm{T}}_{\mathbf{j}_T}\boldsymbol{\Sigma}^{-1}_{Y_{\mathbf{j}_T}}\mathbf{C}_{\mathbf{j}_T}\right)^{-1}\mathbf{C}^{\mathrm{T}}_{\mathbf{j}_T}\boldsymbol{\Sigma}^{-1}_{Y_{\mathbf{j}_T}} \tag{24}$$

$$\boldsymbol{\Sigma}_{Y_{\mathbf{j}_T}} = \left[\mathbf{L}_{\mathbf{j}_T}\ I_{T.n_y}\right]\begin{bmatrix}\boldsymbol{\Sigma}_{W_{\mathbf{j}_T}} & 0\\ 0 & \boldsymbol{\Sigma}_{V_{\mathbf{j}_T}}\end{bmatrix}\left[\mathbf{L}_{\mathbf{j}_T}\ I_{T.n_y}\right]^{\mathrm{T}} \tag{25}$$

The covariance of the obtained unconstrained estimate can also be computed:

$$\boldsymbol{\Sigma}_{x_{\mathbf{j}_T}}(k|t) = \left(\boldsymbol{\Sigma}^{-1}_{x_{\mathbf{j}_T}}(k|t-1) + \mathbf{C}^{\mathrm{T}}_{\mathbf{j}_T}\boldsymbol{\Sigma}^{-1}_{Y_{\mathbf{j}_T}}\mathbf{C}_{\mathbf{j}_T}\right)^{-1} \tag{26}$$

This covariance matrix not only provides some insight on the accuracy of the continuous state estimate $\hat{x}^u_{\mathbf{j}_T}(k|t)$, but also defines the confidence on the past information at the subsequent time instant $\hat{x}_{\mathbf{j}_T}(k+1|t)$:

$$\boldsymbol{\Sigma}_{x_{\mathbf{j}_T}}(k+1|t)=A_{j(k)}\boldsymbol{\Sigma}_{x_{\mathbf{j}_T}}(k|t)A^{\mathrm{T}}_{j(k)}+L_{j(k)}\boldsymbol{\Sigma}_{wj(k)}L^{\mathrm{T}}_{j(k)} \tag{27}$$

When computing the unconstrained state estimate, no *a priori* information may be available or one may be interested in discarding it, then $\boldsymbol{\Sigma}^{-1}_{x_{\mathbf{j}_T}}(k|t-1)$ should be set to 0. The corresponding unconstrained state estimate is referred to as $\hat{x}^{u*}_{\mathbf{j}_T}(k|t)$.

3.2 Constrained Filter Bank

The CFB will recompute the state estimates but now considering the constraints (15), (20) and (21). The constrained least-squares filter is somehow more complicated. First the least-squares state vector must be augmented to include both the input disturbance and measurement noise vectors, since there exist explicit constraints on these variables:

$$\begin{bmatrix} x_{\mathbf{j}_T}(k) \\ W_{\mathbf{j}_T}(k) \\ V_{\mathbf{j}_T}(k) \end{bmatrix} \tag{28}$$

Notice that by explicitly considering the input disturbance and measurement noise sequences, all the uncertainty is removed from the observation equation (6) and it becomes an equality constraint:

$$\mathbf{H_e} \cdot \begin{bmatrix} x_{\mathbf{j}_T}(k) \\ W_{\mathbf{j}_T}(k) \\ V_{\mathbf{j}_T}(k) \end{bmatrix} = \mathbf{h_e} \quad \Leftrightarrow \tag{29}$$

$$\Leftrightarrow \quad \begin{bmatrix} \mathbf{C}_{\mathbf{j}_T} & \mathbf{L}_{\mathbf{j}_T} & I_{nY} \end{bmatrix} \cdot \begin{bmatrix} x_{\mathbf{j}_T}(k) \\ W_{\mathbf{j}_T}(k) \\ V_{\mathbf{j}_T}(k) \end{bmatrix} = \begin{bmatrix} Y_T(k) - \mathbf{D}_{\mathbf{j}_T} U_T(k) - \mathbf{g}_{\mathbf{j}_T} \end{bmatrix}$$

The constraints defined by the region $\mathbf{\Omega}_{\mathbf{j}_T}$ (15) and the bounds on the input disturbance and measurement noise vectors defined by the sets $\mathbb{W}_{\mathbf{j}_T}$ and $\mathbb{V}_{\mathbf{j}_T}$ described by equations (20) and (21) compose the inequality constraints of the least-squares problem, according to:

$$\mathbf{H_i} \begin{bmatrix} x_{\mathbf{j}_T}(k) \\ W_{\mathbf{j}_T}(k) \\ V_{\mathbf{j}_T}(k) \end{bmatrix} \leq \mathbf{h_i} \Leftrightarrow \begin{bmatrix} \mathbf{S}_{\mathbf{j}_T} & \mathbf{Q}_{\mathbf{j}_T} & 0 \\ 0 & \mathbf{H}_{\mathbb{W}_{\mathbf{j}_T}} & 0 \\ 0 & 0 & \mathbf{H}_{\mathbb{V}_{\mathbf{j}_T}} \end{bmatrix} \cdot \begin{bmatrix} x_{\mathbf{j}_T}(k) \\ W_{\mathbf{j}_T}(k) \\ V_{\mathbf{j}_T}(k) \end{bmatrix} \leq \begin{bmatrix} \mathbf{T}_{\mathbf{j}_T} - \mathbf{R}_{\mathbf{j}_T} U_T(k) \\ \mathbf{h}_{\mathbb{W}_{\mathbf{j}_T}} \\ \mathbf{h}_{\mathbb{V}_{\mathbf{j}_T}} \end{bmatrix} \tag{30}$$

Having defined the constraints matrices, the constrained least-squares filter corresponding to the mode sequence $\mathbf{j}_T$ is given by:

$$\begin{bmatrix} \hat{x}_{\mathbf{j}_T}(k|t) \\ \hat{W}_{\mathbf{j}_T}(k|t) \\ \hat{V}_{\mathbf{j}_T}(k|t) \end{bmatrix} = \begin{bmatrix} \hat{x}_{\mathbf{j}_T}(z, k|t-1) \\ \hat{W}_{\mathbf{j}_T}(k|t-1) \\ \hat{V}_{\mathbf{j}_T}(k|t-1) \end{bmatrix} + \mathbf{K}_{\mathbf{j}_T}(k|t) \left(\begin{bmatrix} \mathbf{h_e} \\ \mathbf{h_i} \end{bmatrix} - \begin{bmatrix} \mathbf{H_e} \\ \mathbf{H_i} \end{bmatrix} \cdot \begin{bmatrix} \hat{x}_{\mathbf{j}_T}(k|t-1) \\ \hat{W}_{\mathbf{j}_T}(k|t-1) \\ \hat{V}_{\mathbf{j}_T}(k|t-1) \end{bmatrix} \right) \tag{31}$$

The constrained least-squares filter gain is defined as:

$$\mathbf{K}_{\mathbf{j}_T}(k|t) = \left(\begin{bmatrix} \mathbf{\Sigma}_{x_{\mathbf{j}_T}}(k|t-1) & 0 & 0 \\ 0 & \mathbf{\Sigma}_{W_{\mathbf{j}_T}} & 0 \\ 0 & 0 & \mathbf{\Sigma}_{V_{\mathbf{j}_T}} \end{bmatrix}^{-1} + \begin{bmatrix} \mathbf{H_e} \\ \mathbf{H_i} \end{bmatrix}^{\mathrm{T}} \mathbf{Z}_{\mathbf{j}_T}(k|t) \begin{bmatrix} \mathbf{H_e} \\ \mathbf{H_i} \end{bmatrix} \right)^{-1} \begin{bmatrix} \mathbf{H_e} \\ \mathbf{H_i} \end{bmatrix}^{\mathrm{T}} \mathbf{Z}_{\mathbf{j}_T}(k|t) \tag{32}$$

where $\mathbf{\Sigma}_{x_{\mathbf{j}_T}}(k|t-1)$ is the covariance matrix associated with the *a priori* state estimate $\hat{x}_{\mathbf{j}_T}(k|t-1)$. $\mathbf{Z}_{\mathbf{j}_T}(k|t)$ is the diagonal matrix that defines the active constraints.

There are several methods, most of them iterative, for determining the matrix $\mathbf{Z}_{\mathbf{j}_T}(k|t)$, or equivalently the set of active constraints. Here, the active set method presented in (Fletcher, R., 1987) will be used.

As in the unconstrained case, *a priori* information may be discarded by setting $\mathbf{\Sigma}^{-1}_{x_{\mathbf{j}_T}}(k|t{-}1)$ to 0. The corresponding constrained state estimate is referred to as $\hat{x}^{c*}_{\mathbf{j}_T}(k|t)$.

3.3 Discrete Mode Sequence Estimator

The DMSE deals with the estimation of the discrete mode sequence and, consequently, selects the filter which will provide the final continuous state estimate.

According to the least-squares philosophy, an approximation of the measured output sequence is computed for every possible *dms* and then, the one providing the smallest squared error should be selected as the least-squares estimate.

The *dms* estimate is then selected as the one that presents the lowest constrained squared error, $\alpha^c_{\mathbf{j}_T}$:

$$\hat{\mathbf{i}}_T(k|t) = \arg \min_{\mathbf{j}_T} \alpha^c_{\mathbf{j}_T}(k|t) \tag{33}$$

The squared error associated with the *dms* $\mathbf{j}_T$ is given by:

$$\begin{aligned}\alpha_{\mathbf{j}_T}(k|t) &= \left\| \hat{Y}^*_{\mathbf{j}_T}(k|t) - Y_T(k) \right\|^2_{\mathbf{\Sigma}^{-1}_{Y_{\mathbf{j}_T}}} = \\ &= \left[\hat{Y}^*_{\mathbf{j}_T}(k|t) - Y_T(k) \right]^{\mathrm{T}} \mathbf{\Sigma}^{-1}_{Y_{\mathbf{j}_T}} \left[\hat{Y}^*_{\mathbf{j}_T}(k|t) - Y_T(k) \right]\end{aligned} \tag{34}$$

where:

$$\hat{Y}^*_{\mathbf{j}_T}(k|t) = \mathbf{C}_{\mathbf{j}_T} \hat{x}^*_{\mathbf{j}_T}(k|t) + \mathbf{D}_{\mathbf{j}_T} U_T(k) + \mathbf{g}_{\mathbf{j}_T} \tag{35}$$

and $\hat{x}^*_{\mathbf{j}_T}(k|t)$ is the estimated state of the *dms* $\mathbf{j}_T$ when all past information is discarded, $\left(\mathbf{\Sigma}^{-1}_{x_{\mathbf{j}_T}}(k|t-1) = 0\right)$.

The squared errors computed by equation and (34) are useful when comparing continuous state estimates from the same *dms*. However, when the covariance matrices are different, an additional factor, $\bar{\alpha}_{\mathbf{j}_T}$, must be considered to allow a meaningful comparison between squared errors. Recalling the relation between least-squares and the maximization of the Gaussian likelihood function (or its logarithm), the value of $\bar{\alpha}_{\mathbf{j}_T}$ should be defined as:

$$\bar{\alpha}_{\mathbf{j}_T} = -\frac{1}{2} \ln \left((2\pi)^{n_Y} \det \left(\mathbf{\Sigma}_{Y_{\mathbf{j}_T}} \right) \right) \tag{36}$$

And equation (34) should be modified to:

$$\alpha_{\mathbf{j}_T}(k|t) = \bar{\alpha}_{\mathbf{j}_T} + \left\| \hat{Y}^*_{\mathbf{j}_T}(k|t) - Y_T(k) \right\|^2_{\mathbf{\Sigma}^{-1}_{Y_{\mathbf{j}_T}}} \tag{37}$$

Equation (37) can be used to compute the squared errors of both the unconstrained estimates,$\alpha^u_{\mathbf{j}_T}(k|t)$, and the constrained estimates, $\alpha^c_{\mathbf{j}_T}(k|t)$, using $\hat{x}^{u*}_{\mathbf{j}}(k|t)$ and $\hat{x}^{c*}_{\mathbf{j}}(k|t)$, respectively.

The option of discarding all past information to compute the state estimates was chosen to improve the computational efficiency of the algorithm and to allow a precise determination of the uncertainty associated with the *dms* estimates.

3.4 Computational Issues

Regarding the computational requirements of the proposed algorithm, it is noticed that there can be as many as n_s^T *dms*, which becomes an extremely large number even for relatively small n_s and T. So, computationally demanding calculations should be preformed for the minimum number of *dms* possible.

Analyzing the required computations one concludes that $\hat{x}^{u*}_{\mathbf{j}_T}(k|t)$ can be determined by simple matrix sums and multiplications if the filter gain $\mathbf{K}_{\mathbf{j}_T}(k|t\text{–}1)$ is computed off-line, since there are no varying terms as can be seen in equation (22). The corresponding squared error $\alpha^u_{\mathbf{j}_T}(k|t)$, computed through equation (37), can also be determined using simple matrix sums and multiplications from $\hat{x}^{u*}_{\mathbf{j}_T}(k|t)$. The continuous state estimate $\hat{x}^u_{\mathbf{j}_T}(k|t)$ on the other hand, requires a matrix inversion to determine the corresponding filter gain using equation (24) since the matrix $\Sigma^{-1}_{x_{\mathbf{j}_T}}(k|t-1)$ is not known in advance, which can not be done in a computational efficient way for all *dms*.

The constrained estimates require much more complex computations in the solution of the inequality constrained least-squares problem. An iterative algorithm has to be preformed online, and involves one matrix inversion at each iteration which is computationally heavy. There is the possibility that the solution corresponding to the true *dms* is the same as the unconstrained solution and the iterative algorithm stops at the first iteration. In general, however, this will not be the case. So, the computation of constrained solutions should only be done in cases of absolute necessity. The squared error of the constrained estimates $\alpha^c_{\mathbf{j}_T}(k|t)$ can be determined using simple matrix sums and multiplications from $\hat{x}^{c*}_{\mathbf{j}}(k|t)$.

The proposed algorithm should take this knowledge into account and arrive at the final estimates in the most efficient way possible.

To avoid the computation of the constrained least-squares estimates from all discrete mode sequences, the following relation between the constrained and unconstrained squared errors for a given discrete mode sequence is used:

$$\alpha^u_{\mathbf{j}_T}(k|t) \leq \alpha^c_{\mathbf{j}_T}(k|t) \tag{38}$$

An efficient reduction on the number of constrained estimates that have to be computed can be achieved by computing all unconstrained estimates $\hat{x}^{u*}_{\mathbf{j}}(k|t)$ and the corresponding squared errors $\alpha^u_{\mathbf{j}_T}(k|t)$ and then, start replacing the unconstrained solutions with the corresponding constrained ones, from the lower values of the squared error. Whenever the lowest squared error corresponds to a constrained solution, the algorithm stops since no further reduction of the squared error can be done. The discrete mode sequence and continuous state estimates are the ones corresponding to that lowest squared error.

This algorithmic procedure may provide a substantial reduction in the number of inequality constrained least-squares problems to be solved since the increase in the squared error should be small, or even zero, for the true *dms*. However, the unconstrained solutions of incorrect *dms* may have low squared errors, which rise substantially only when the respective constrained solutions are computed. An efficient procedure to detect these incorrect *dms* before computing the respective constrained estimates would reduce the computational requirements even more.

To further improve the algorithm, the following $\mathcal{B}$ matrix must be introduced. Each coefficient $\beta_{\mathbf{i}_T,\mathbf{j}_T}$ of the matrix $\mathcal{B}$ is defined as the maximum value of $\alpha^c_{\mathbf{i}_T}$ under which $\alpha^c_{\mathbf{i}_T}$ is always smaller than $\alpha^c_{\mathbf{j}_T}$, or in an even more restrictive way, under which $\mathbf{j}_T$ is never the estimated sequence. The coefficients $\beta_{\mathbf{i}_T,\mathbf{j}_T}$ can be computed off-line by the following optimization problem, that belongs the general class of Second-Order Cone Programs for which efficient solvers have already been developed, for instance, by [2]:

$$\begin{aligned} \beta_{\mathbf{i}_T,\mathbf{j}_T} &= \min_{Y_T,U_T} \quad \alpha^c_{\mathbf{i}_T}(Y_T,U_T) \\ \text{subject to :} & \\ U_T &\in \mathbb{U}^T \\ \hat{\mathbf{i}}_T &= \mathbf{j}_T \end{aligned} \tag{39}$$

By this definition of $\beta_{\mathbf{i}_T,\mathbf{j}_T}$, when the constrained solution of a *dms* $\mathbf{i}_T$ is computed, all *dms* $\mathbf{j}_T$ such that $\beta_{\mathbf{i}_T,\mathbf{j}_T}$ is greater than $\alpha^c_{\mathbf{i}_T}(k|t)$ can be discarded. This algorithmic procedure, despite requiring substancial memory to store the $\mathcal{B}$ matrix, provides an even greater reduction on the number of constrained problems to be solved. Notice that this procedure does not even require the computation of the unconstrained solutions of the *dms* to be discarded.

Both previous modifications to the algorithm require the existence of one constrained solution to discard any other *dms*. Furthermore, the number of discarded *dms* depends on the quality of the constrained solution. In the following, some attention will be given to the recursiveness of the DMSE and the methodology to determine the *dms* that will most likely provide good constrained estimates.

At a given time instant $t+1$ the following quantities have been computed at the previous time instant: the discrete mode sequence estimate, $\hat{\mathbf{i}}_T(k|t)$, the squared errors (or lower bounds) of all *dms*, $\alpha^c_{\mathbf{j}_T}(k|t)$ and, the continuous state estimates $\hat{x}^{c*}_{\mathbf{j}_T}(k|t)$ and the values of the estimated input disturbances $\hat{W}_{\mathbf{j}_T}(k|t)$ for the *dms* whose squared errors have been computed, including the *dms* estimate. These quantities allow the computation of the *a priori* continuous state estimate corresponding to the discrete mode sequence estimate at the following time instant:

$$\begin{aligned} \hat{x}^*_{\mathbf{j}_T}(t+1|t) = &\left(A_{j(t)} \,\cdots\, A_{j(k)}\right)\hat{x}^*_{\mathbf{j}_T}(k|t) + \\ &\left[A_{j(t)}\cdots A_{j(k+1)}B_{j(k)},\ldots,B_{j(t)}\right]U_T(k) + \\ &\left[A_{j(t)}\cdots A_{j(k+1)}W_{j(k)}\;,\ldots,\;W_{j(t)}\right]\hat{W}_{\mathbf{j}_T}(k|t) + \\ &\left(A_{j(t)}\cdots A_{j(k+1)}f_{j(k)} + \ldots + f_{j(t)}\right) \end{aligned} \tag{40}$$

This estimate can be used to obtain some insight on the likelihood of the discrete mode at the next time instant $j(t+1)$, by verifying if it does not violate the constraints. The discrete modes $j(t+1)$ can be sorted by ascending values of:

$$\begin{aligned} &\gamma_{\mathbf{j}_T,j}(t+1|t) = \\ &\max\left(S_j\,\hat{x}^*_{\mathbf{j}_T}(t+1|t) + R_j u(t+1) + Q_j\hat{w}(t+1|t) - T_j\right) \end{aligned} \tag{41}$$

The value of $\hat{w}(t+1)$ should be set to $\mathrm{E}\{w_j\}$.

The discrete modes $j(t{+}1)$ that provide the lower values of $\gamma_{\mathbf{j}_T,j}(t{+}1|t)$ correspond the discrete mode sequences $\mathbf{j}_T = \{j(k{+}1),\ldots,j(t),j(t{+}1)\}$ at time instant $t+1$ most likely to succeed to $\mathbf{j}_T$ at time instant t.

Applying this methodology to the discrete mode sequence estimate at the previous time instant, $\hat{\mathbf{i}}_T(k|t)$, should provide *dms* with very low squared errors that discard most of the other candidate *dms*. The same reasoning should be applied to all other discrete mode sequences of the previous time instant that have not been discarded yet, starting from the ones that present lower squared errors and then the ones with the lower bounds.

3.5 Uncertainty Analysis

Both the continuous state and discrete mode estimates have an associated uncertainty, which must be determined exactly or at least conservatively approximated to provide a notion of their accuracy. For an exact description of the uncertainty associated with the computed estimates one should determine their probability density function, however, this may be too complex, computationally demanding, or impractical for most applications. The uncertainty associated with the continuous state estimates will be approximated by a covariance matrix, while a probability of correctly estimating the *dms* will be used for the discrete mode.

The covariance associated with the unconstrained state estimates is given be eq. (26). The covariance of the constrained state estimates is much more difficult to determine and the unconstrained covariance will be used as a conservative approximation, according to the following proposition:

Proposition 1. *Reduction of Covariance through Truncation (Chow, Y.S., Studden, W.J., 1969):*
Given a random variable $\mathbf{x}$ *and a polytope* Ω, *it holds:*

$$\begin{aligned}&\mathrm{E}\Big\{\big(\mathbf{x}-\mathrm{E}\{\mathbf{x}\}\big)\big(\mathbf{x}-\mathrm{E}\{\mathbf{x}\}\big)^{\mathrm{T}}\Big\} \geq \\ &\geq \mathrm{E}\Big\{\big(\mathbf{x}-\mathrm{E}\{\mathbf{x}|\mathbf{x}{\in}\Omega\}\big)\big(\mathbf{x}-\mathrm{E}\{\mathbf{x}|\mathbf{x}{\in}\Omega\}\big)^{\mathrm{T}}\big|\mathbf{x}{\in}\Omega\Big\}\end{aligned} \tag{42}$$

If all past information is disregarded when computing the continuous state estimates, these covariance matrices can even be determined off-line. To determine the uncertainty associated with the discrete mode estimates, the off-line numerical integration method presented in [17] can be used. This method numerically integrates eq. (43) to compute the probability of correctly estimating the *dms* for every hybrid trajectory $(x, U_T, \mathbf{i}_T)$ and arrive at the maps of probability of correct mode estimation.

$$\begin{aligned}&\mathcal{P}^{a}_{\mathbf{i}_T,\{\mathbf{i}_T,\mathbf{j}_T\}}\Big((x,U_T,\mathbf{i}_T)\Big) = \\ &= \Pr\Big(\hat{\mathbf{i}}^{a}_T\big(Y(x,U_T,W_T,V_T,\mathbf{i}_T),U_T,\{\mathbf{i}_T,\mathbf{j}_T\}\big) = \mathbf{i}_T\big|(x,U_T,\mathbf{i}_T)\Big) = \\ &= \int_{\mathbb{R}^{T.n_w}}\int_{\mathbb{R}^{T.n_v}} \Big[\hat{\mathbf{i}}^{a}_T\big(Y(x,U_T,W_T,V_T,\mathbf{i}_T),U_T,\{\mathbf{i}_T,\mathbf{j}_T\}\big) = \mathbf{i}_T\Big]. \\ &\qquad . \Pr\Big(W_T,V_T\big|(x,U_T,\mathbf{i}_T)\Big)\; dV_T\, dW_T\end{aligned} \tag{43}$$

The following function is used in eq. (43) and compares the *dms* estimate with the desired *dms*:

$$\left[\hat{\mathbf{i}}_T^a(Y_T, U_T, \mathcal{J}_T) = \mathbf{j}_T\right] = \begin{cases} 1 \text{ iff } \hat{\mathbf{i}}_T^a = \mathbf{j}_T \\ 0 \text{ iff } \hat{\mathbf{i}}_T^a \neq \mathbf{j}_T \end{cases}, \quad \hat{\mathbf{i}}_T^a(\cdot) \in \{\mathcal{J}_T, \mathbf{0}\} \tag{44}$$

where the estimate is chosen from the candidate discrete mode sequence set $\{\mathcal{J}_T, \mathbf{0}\}$ that also includes the null sequence $\mathbf{0}$ that is issued if non of the candidate *dms* is chosen.

These maps clearly show the hybrid trajectories where the *dms* can be correctly estimated with probability close to 1 and those other trajectories that will most likely produce incorrect *dms* estimates. The following experimental application will further show the relevance of these maps of probability of correct mode estimation.

4 Experimental Application

To demonstrate the applicability of the hybrid estimation algorithms, the laboratory setup of the DTS200 three-tanks system from AMIRA® [3] will be used to simulate different situations common in hybrid estimation. A photo of the three-tanks system is presented in figure 2 showing the different components of the experimental setup.

Fig. 2. The three-tanks setup

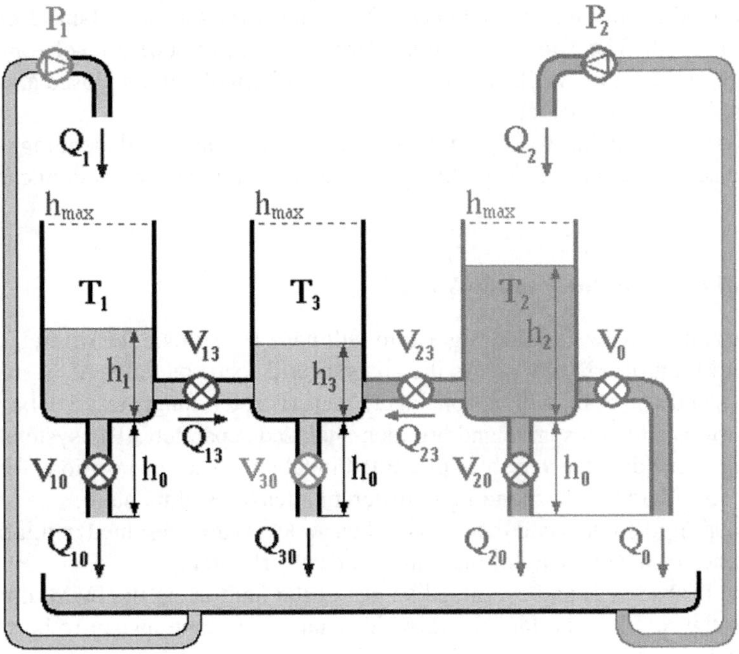

Fig. 3. Final model of the three-tanks system

The plant consists of three plexiglas cylinders or tanks, T_1, T_2 and T_3 with similar cross section. These are connected in series with each other by cylindrical pipes with cross section S_n. Located at T_2 is the single so called nominal outflow valve V_0 which also has a circular cross section S_n. The outflowing liquid (colored distilled water) is collected in a reservoir, which supplies the pumps P_1 and P_2. Here the water circuit is closed. h_{max} denotes the highest possible liquid level in any of the tanks. In case the liquid level of T_1 or T_2 exceeds this limit the corresponding pump will be switched off automatically. Q_1 and Q_2 are the flow rates from pumps P_1 and P_2, respectively.

The pump flow rates Q_1 and Q_2 and the position of the valves V_{13}, V_{23}, V_0, V_{10}, V_{20}, V_{30}, denote the controllable variables, while the liquid levels of h_1, h_2 and h_3 are the output variables. The necessary level measurements are carried out by piezo-resistive difference pressure sensors. There are also potentiometric sensors that measure the position of each valve. The sensor signals are preprocessed to the interval $[0\,;1]$ and so need to be adjusted to $[0\,;h_{max}]$ for the water levels. For the remainder of this section the three-tanks system will be adapted so that more realistic hybrid estimation problems can be studied while simultaneously simplifying the presentation of results. The new model is present in figure 3 where the elements in grey are assumed to be nonexistent, the elements in green are fully operational and the elements in red may be subject to faults and will be used to model input disturbances.

Pump P_1 is considered to be a fully operational on/off valve. Valve V_{13} will have two nominal values "on" and "off", while Valve V_{10} will remain closed. Both these valves are subject to a possible fault resulting in an unmeasurable flow to cross them and described as an input disturbance. The water level sensor of tank 3 can also be subject to a fault. The Valve V_{30} is considered to be a fully operational "on/off" valve with no possible faults, while Valves V_{20}, V_{23} and V_0 will remain closed and so can be considered to be nonexistent.

The system can exhibit a large number of different dynamics, depending on the state of each discrete variable. The full hybrid model description of the system can be found in [18].

4.1 Estimation of the Fault in Valve $\mathbf{V_{13}}$

In this example, the estimation algorithm will have to estimate the discrete mode that indicates a fault on valve $\mathbf{V_{13}}$. As the analysis will focus on valve $\mathbf{V_{13}}$, the faults on valve $\mathbf{V_{10}}$ and sensor $\mathbf{h_3}$ will be considered nonexistent. A single test will be performed where various situations arise and are then analyzed separately. The system is excited according to the discrete variables presented in table 1. Various positions for the valve $\mathbf{V_{13}}$ are considered, corresponding to different intensities of the fault.

The measured outputs and the estimated water levels are presented in figure 4, where the influence of the intensity of the fault can be clearly seen.

The real (observed) and estimated values of the fault using the IMM algorithm are shown in figure 5. As the fault in valve $\mathbf{V_{13}}$ takes one time instant to be reflected in the water level measurements, only the value of $f_{V_{13}}(k-1|k)$ is relevant. Note that $f_{V_{13}}(k-1|k)$ is a discrete variable that takes value 1 when a leak occurs, and value 0 when there is no fault.

The corresponding estimated continuous input disturbances by both algorithms are shown in figure 6. As the fault in valve $\mathbf{V_{13}}$ takes one time instant to be reflected in the water level measurements, only the value of $w_{V_{13}}(k-1|k)$ is estimated. The variable $w_{V_{13}}$ determines the leaking flow and is considered to be a uniformly distributed random variable defined in the interval $[-0.3\,;0.3]$ cm, with zero mean and variance $\frac{0.6^2}{12}$ cm^2 for all k, where 0.6 is the maximum water level change when the valve $\mathbf{V_{13}}$ is fully open.

The difference observed in both algorithms for the estimation of the disturbance $w_{V_{13}}(k-1|k)$ shows that the MHE algorithm is not able to weight the disturbance with any prior value so allowing it to change freely, which increases the variation of the input disturbance estimates.

The estimation results presented in figures 4 and 5 will now be analyzed independently for the 3 considered valve $\mathbf{V_{13}}$ fault intensities.

Table 1. Evolution of the discrete variables

Time(s)	0-49	50-99	100-149	150-199	200-249	250-300
$\mathbf{V_{13}}$	"faulty" *max*	"ok"	"faulty" *med*	"faulty" *med*	"ok"	"faulty" *max*
$\mathbf{V_{10}}$	"ok"	"ok"	"ok"	"ok"	"ok"	"ok"
$\mathbf{h_3}$	"ok"	"ok"	"ok"	"ok"	"ok"	"ok"
$\mathbf{P_1}$	"on"	"off"	"on"	"off"	"on"	"off"
$\mathbf{V_{13}}$	"open"	"open"	"open"	"open"	"open"	"open"
$\mathbf{V_{30}}$	"open"	"open"	"open"	"open"	"open"	"open"

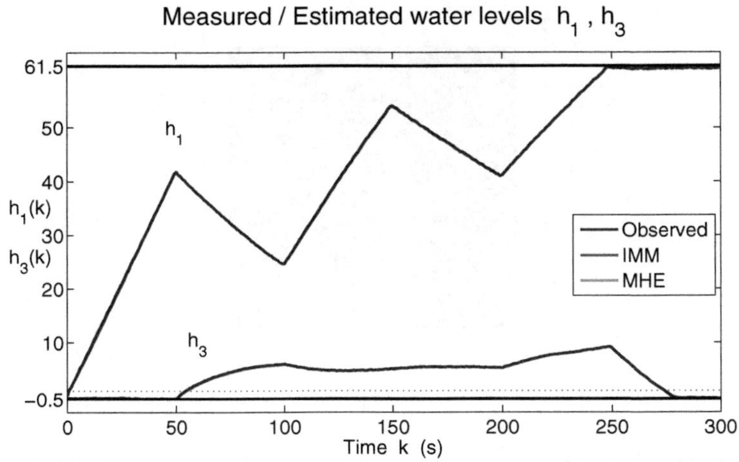

Fig. 4. Water levels estimation using the IMM and MHE estimation algorithms

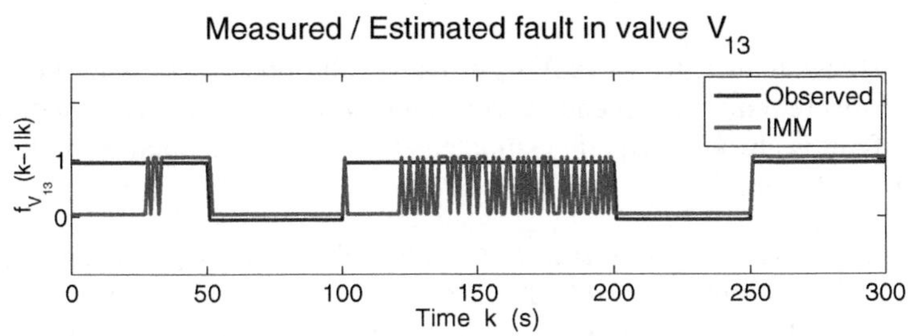

Fig. 5. Estimation of the discrete mode sequence relative to the fault in valve $\mathbf{V_{13}}$

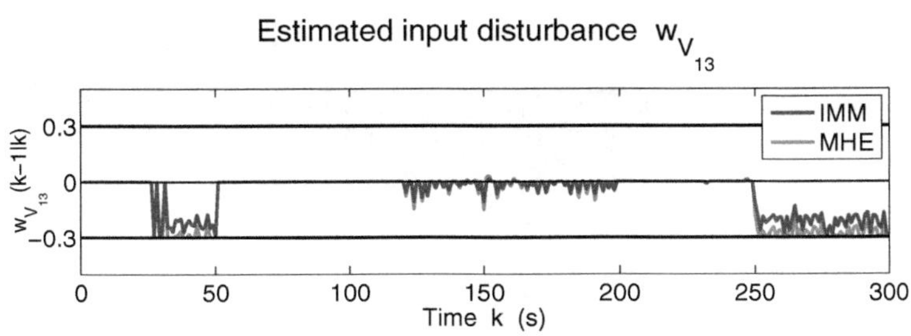

Fig. 6. Estimation of the input disturbance $w_{V_{13}}(k-1|k)$ corresponding to the fault in valve $\mathbf{V_{13}}$

Case 1 - Fault Inactive. For time intervals $[\,50\ ;100\,[$ s and $[\,200\ ;250\,]$ s valve $\mathbf{V_{13}}$ remained closed and the fault is considered inactive. Despite being inactive, there is still a possibility of a wrong estimate reflected on the value of the discrete variable $f_{V_{13}}$. However, as shown in figure 7, the valve's true state was correctly estimated during these time periods.

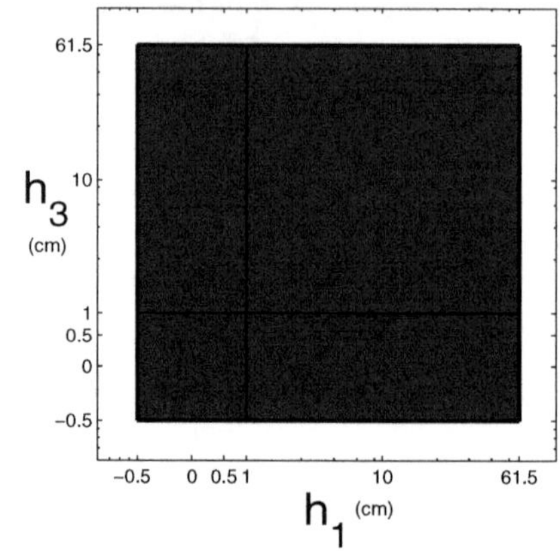

Fig. 7. Map of probability of correct mode estimation, with 1s delay, when valve $\mathbf{V_{10}}$ is fully closed.(Red - probability of correct mode estimation 0, Blue - probability of correct mode estimation 1).

Figure 7 shows that if the valve $\mathbf{V_{13}}$ remains fully open there is no possibility of estimating a discrete mode sequence corresponding to a closed valve condition. Thus the inactive fault is always correctly estimated. This result was obtained independently of the position of pump $\mathbf{P_1}$.

Case 2 - Fault Active with Intermediate Intensity. The valve $\mathbf{V_{13}}$ has an intermediate open position during time interval $[\,100\ ;200\,[$ s allowing an unmeasured flow to cross it. In this case, a fully opened valve was estimated by the IMM algorithm in several time instants. These wrong estimates are understandable since the effects on the water levels of tanks 1 and 3 is not too drastic and can be mistaken by any other source of uncertainty, like

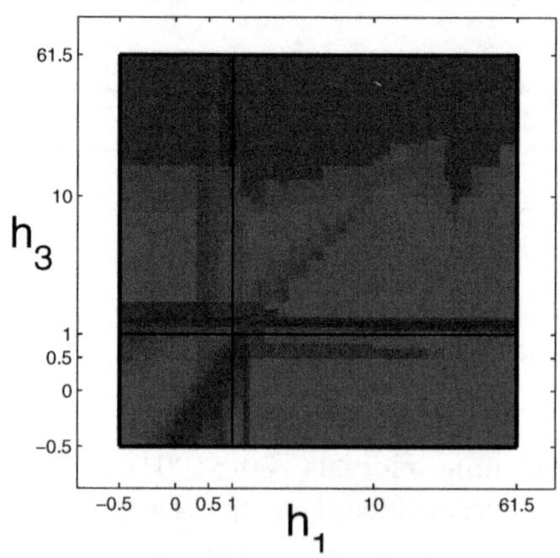

Fig. 8. Map of probability of correct mode estimation, with 1s delay, when fault $f_{V_{13}}$ is active and pump $\mathbf{P_1}$ is "on". (Red - probability of correct mode estimation 0, Blue - probability of correct mode estimation 1).

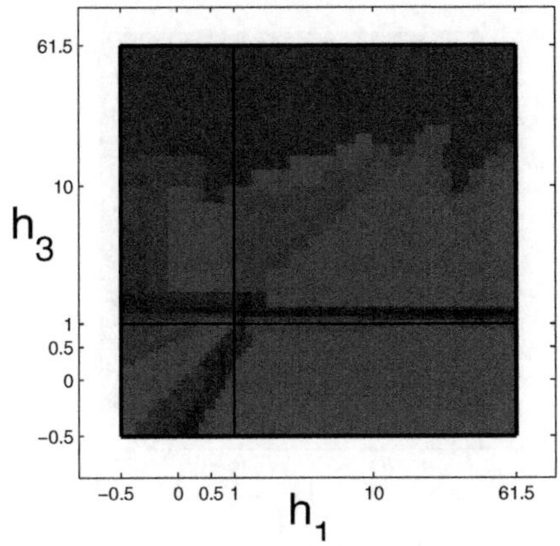

Fig. 9. Map of probability of correct mode estimation, with 1s delay, when fault $f_{V_{13}}$ is active and pump $\mathbf{P_1}$ is "off". (Red - probability of correct mode estimation 0, Blue - probability of correct mode estimation 1).

measurement noise for instance. This difficulty in discerning whether the valve is slightly open or fully closed is patent in the map of probability of correct mode estimation shown in figures 8 and 9. It can also be concluded that the probability of an incorrect estimation of the valve's condition increases as the water level of tank 3 becomes lower.

It can be seen that there is always the possibility of incorrectly determining the condition of the valve. The map of probability of correct mode estimation is not able to show the existing dependence between the probability of correctly determining the valve's condition and its real position. However, as it is understandable, the probability of making a mistake increases as the water levels become closer, since the water flow between the two tanks becomes very reduced. This dependence of the probability of correctly estimating the discrete mode with the proximity of the water levels is patent in figure 5 where the estimate for the condition of valve $\mathbf{V_{13}}$ is systematically incorrect when the water levels are not separated enough.

Case 3 - Fault Active with Maximum Intensity. If valve $\mathbf{V_{13}}$ is fully closed it becomes much easier to determine its position, thus allowing the IMM algorithm to provide correct estimates for the discrete mode sequence during time intervals $[0\ ;50[$ s and $[250\ ;300[$ s. This is quite obvious since the flow between the tanks is completely stopped and the very small disturbances are never assumed to come from a leak in the valve, independently of the position of pump $\mathbf{P_1}$, as depicted in figure 10.

This map of probability of correct mode estimation was computed considering an hypothetical model for the system where valve $\mathbf{V_{13}}$ can only be fully open or fully closed.

From the maps of figure 10, the fault $f_{V_{13}}$ should have always been correctly estimated. However, these maps were computed considering $w_{v_{13}} = -0.3$ which can not happen in the real system except when tank 1 is full and tank 3 is empty. For closer water levels in the tanks, the maximum value for the disturbance $w_{v_{13}}$ should be considered much smaller. When both tanks have similar water levels as in the initial instants of

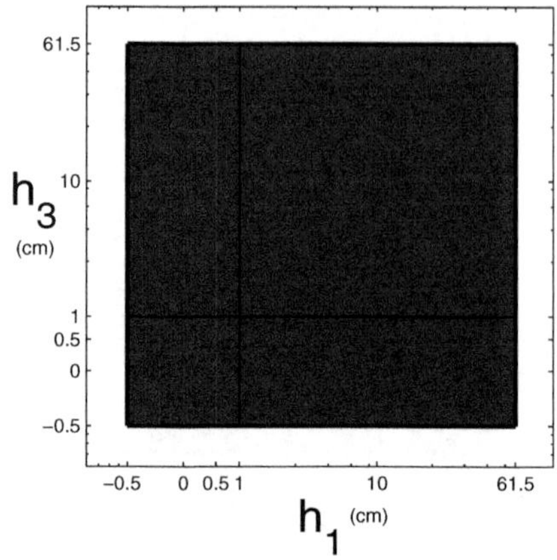

Fig. 10. Map of probability of correct mode estimation, with 1s delay, considering that fault $f_{V_{13}}$ has maximum intensity, $w_{v_{13}} = -0.3$. (Red - probability of correct mode estimation 0, Blue - probability of correct mode estimation 1).

the experiment, the flow across valve $\mathbf{V_{13}}$ would be very small, and small disturbances can be mistaken for water flowing across the valve, leading to the incorrect estimates obtained in that interval.

5 Conclusions

This paper presented an efficient state estimation algorithm for hybrid systems based on an IMM setup composed by a set of least-squares filters. The computational efficiency is obtained though some algorithmic procedures that discard as many candidate *dms* while avoiding heavy computations. These procedures rely on the early determination of good estimates, on the separation of constrained and unconstrained estimates and on some bounding parameters for the squared errors that are computed off-line.

The proposed algorithm was able to provide accurate online estimates of both continuous states and discrete modes when applied to the hybrid model of the benchmark AMIRA DTS200 three-tanks system experimental setup. The potential of the IMM based algorithm was demonstrated when comparing its computational efficiency with the MHE algorithm in this fault detection problem.

Most hybrid methodologies must deal with the high number of discrete modes sequences typically involved when dealing with a medium size hybrid system model. This fact eventually turns most of the problems untractable. For the case of the three-tanks system experimental setup, it was noticed that the consideration of all three tanks in the same hybrid would require huge computational resources due to the combination of the modes from the three tanks. Thus, authors believe that subdividing the model, for instance, using a multi-agent modelling architecture, could significantly simplify the model complexity while being able to retain its full hybrid dynamical flavor. As the size of the problems to be solved with hybrid systems grows exponentially with the number of discrete modes involved, multi-agent architectures may keep the combinations of discrete modes and the complexity at tractable levels, thus being a very interesting and possibly fruitful research topic.

References

1. Alessandri, A., Coletta,P.: Design of observers for switched discrete-time linear systems. In: Proc. American Control Conference, Denver, Colorado, pp. 2785–2790 (2003)
2. Alizadeh, F., Goldfarb, D.: Second-order Cone Programming. Technical Report RRR Report number 51-2001, RUTCOR, Rutgers University, Piscataway, New Jersey (2001)
3. Amira: DTS200 - Laboratory Setup Three-tank-system. Amira, Duisburg, Germany (2002)
4. Antsaklis, P.J.: A brief introduction to the theory and applications of hybrid systems. Proc. IEEE, Special Issue on Hybrid Systems: Theory and Applications 88(7), 879–886 (2000)
5. Athans, M., Chang, C.B.: Adaptive estimation and parameter identification using multiple model estimation algorithm. Technical Report 28, M.I.T. - Lincon Laboratory, Lexington, Massachusetts (1976)
6. Balluchi, A., Benvenuti, L., Di Benedetto, M.D., Sangiovanni-Vincentelli, A.L.: Design of observers for hybrid systems. In: Tomlin, C.J., Greenstreet, M.R. (eds.) HSCC 2002. LNCS, vol. 2289, pp. 76–89. Springer, Heidelberg (2002)
7. Bemporad, A., Morari, M.: Control of systems integrating logic, dynamics, and constraints. Automatica 35(3), 407–427 (1999)
8. Blom, H.A.P., Bar-Shalom, Y.: The interactive multiple model algorithm for systems with markovian switching coefficients. IEEE Trans. on Automatic Control 33(8), 780–783 (1988)
9. Böker, G., Lunze, J.: Stability and performance of switching Kalman filters. International Journal of Control 75(16/17), 1269–1281 (2002)
10. Chow/Studden/69 Chow, Y.S., Studden, W.J.: Monotonicity of the variance under truncation and variations of Jensen's inequality. The Annals of Mathematical Statistics 40(3), 1106–1108 (1969)
11. Ferrari-Trecate, G., Mignone, D., Morari, M.: Moving horizon estimation for hybrid systems. IEEE Trans. on Automatic Control 47(10), 1663–1676 (2002)
12. Fletcher, R.: Practical methods of optimization, 2nd edn. Wiley Interscience Publication, Chichester (1987)
13. Heemels, W.P.M.H., De Schutter, B., Bemporad, A.: Equivalence of hybrid dynamical models. Automatica 37(7), 1085–1091 (2001)
14. Kamen, E.W.: Study of linear time-varying discrete-time systems in terms of timecompressed models. In: Proc. 31th IEEE Conf. on Decision and Control,Tucson, Arizona, pp. 3070–3075 (1992)
15. Mazor, E., Averbuch, A., Bar-Shalom, Y., Dayan, J.: Interacting multiple model methods in target tracking: A survey. IEEE Trans. on Aerospace and Electronic Systems 34(1), 103–123 (1998)
16. Pina, L., Ayala Botto, M.: Simultaneous state and input estimation of hybrid systems with unknown inputs. Automatica 42(5), 755–762 (2006)
17. Pina, L., Ayala Botto, M.: On the characterization of discrete mode uncertainty in hybrid state estimation. In: Proceedings of the 15th IFAC World Congress on Automatic Control, Seoul, Korea (2008)
18. Pina, L.: Hybrid State Estimation. PhD thesis, Instituto Superior T'ecnico, Lisbon, Portugal (2007)
19. Sontag, E.D.: Nonlinear regulation: The piecewise linear approach. IEEE Trans. on Automatic Control 26(2), 346–358 (1981)
20. Torrisi, F.D., Bemporad, A.: Discrete-time hybrid modeling and verification. In: Proc. 40th IEEE Conf. on Decision and Control, Orlando, Florida, pp. 2899–2904 (2001)

Part I

Intelligent Control Systems and Optimization

Adaptive Fuzzy Controller for Output Power Maximization of Induction Generators

Durval de A. Souza[1], José A. Dominguez Navarro[2], and Jesús Sallán Arasanz[2]

[1] Federal Center for Technology Education of Bahia – CEFET-BA/ UE de Vitória da Conquista, 3150, Zabelê, 45030-220, Vitória da Conquista, Bahia, Brazil
durval@cefetba.br

[2] University of Zaragoza, Department of Electrical Engineering, Calle Maria de Luna 3 50018, Campus del Actur, Zaragoza, Spain
jadona@unizar.es, jsallan@unizar.es

Abstract. This note presents a new technique for efficiency optimization of induction generators working at variable speed and load. The technique combines two distinct control methods, namely, on-line search of the optimal operating point and a model based efficiency control. For a given operating condition, characterized by a given turbine speed (ω_T) and electric torque (T_e), the search control is implemented via the "Rosenbrock" method, which determines the flux level that results in maximum output power. Once the optimal flux level has been found, this information is used to update the rule base of a fuzzy controller, which plays the role of an implicit mathematical model of the system. Initially, for any load condition, the rule base yields the rated flux value. As the optimum points associated with the different operating conditions are identified, the rule base is progressively updated, so that the fuzzy controller learns to model the optimal operating conditions for the entire torque-speed plane. After every rule base update, the Rosenbrock controller output is reset, but it is kept active to track possible minor deviations of the optimum point.

1 Introduction

In the last decade, wind power is integrated into the electrical grid and accounts for a noticeable share of the total power generation [1]. In this context the inverter-fed induction generator has been identified as a possible source of energy to be used in modern micro and high power applications [2].

The mechanic power captured by a wind turbine is given by:

$$P_m = \frac{1}{2} \cdot C_p \cdot \rho \cdot A \cdot v_{wind}^3 \qquad (1)$$

where Cp = power coefficient; ρ = air density; A = swept area, v_{wind} = average wind speed.

The theoretical maximum value of Cp is 0,593, also known as Betz limit. However the power coefficient is a function of the tip speed ratio λ, which is defined by equation (2).

$$\lambda = \frac{\omega_r \cdot R}{v_{vento}} \qquad (2)$$

J.A. Cetto et al. (eds.): Informatics in Control, Automation and Robotics, LNEE 37, pp. 49–60.
springerlink.com

where ω_r= rotor speed and R = turbine radius. For one given wind speed, if the power coefficient is to be kept constant at its maximum value, the tip speed ratio must be $\lambda_{optimum}$ and consequently, $\omega_{optimum}$ can be determined. Thus, there will be an optimum mechanical power and a torque for any given wind speed, as defined by equations (3) and (4), respectively.

$$P_m = k_{otima} \cdot \omega^3 \tag{3}$$

$$T_m = k_{otimo} \cdot \omega^2 \tag{4}$$

Figure 1 shows the family of transferred power curves for the wind turbine. Such power is associated with the different wind and turbine speeds (P_m=f (v_{wind}, ω_r)), where for each wind speed there is an operation point where the transferred power is maximum. It is the reason because the variable speed in the wind generation is relevant. However, although the turbine works at optimum rotational speed, the system will be working with sub optimum efficiency. The optimal condition is obtained with all the system (turbine and generator) working with maximization of their efficiency.

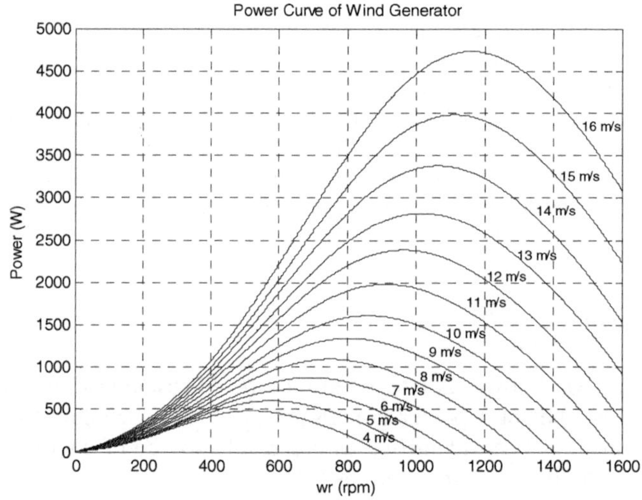

Fig. 1. Curves of transferred power of the wind turbine

The presence of a converter in a drive system enables an extra degree of freedom, namely, speed and flux adjustment. In fact, efficiency optimization in adjustable speed drives is usually obtained by machine flux control. This is due to the fact that in electric machines, maximum efficiency is achieved when the copper losses become equal to the core losses. Typically, under partial load operation, rated flux condition results in relatively large core losses, small copper losses, and poor efficiency. By decreasing the flux, core losses are reduced, whereas an increase in copper losses takes place. The total losses, however, are reduced, and the efficiency is improved [3]. Figure 2 shows the scheme of the Inverter-Fed Induction Generator used in this research.

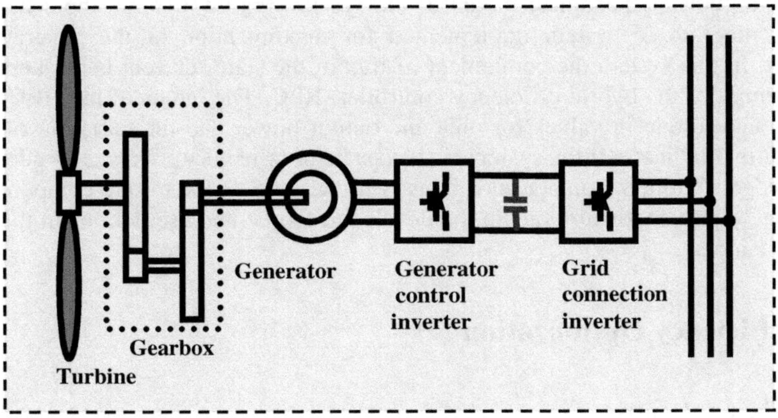

Fig. 2. Inverter-fed induction generator connected to the grid

In this text, a new efficiency optimization technique is introduced. It is applicable to any adjustable speed drive, but it is illustrated here for an induction generator under field-oriented control. The technique combines two distinct control strategies, namely, on-line search and model base control. For a given operating condition, characterized by a given turbine speed (ω_T) and electric torque (T_e), the search control is implemented via the Rosenbrock method, which determines the flux level that results in maximum output power. Once the optimal flux level has been found, this information is used to update the rule base of a fuzzy controller, which plays the role

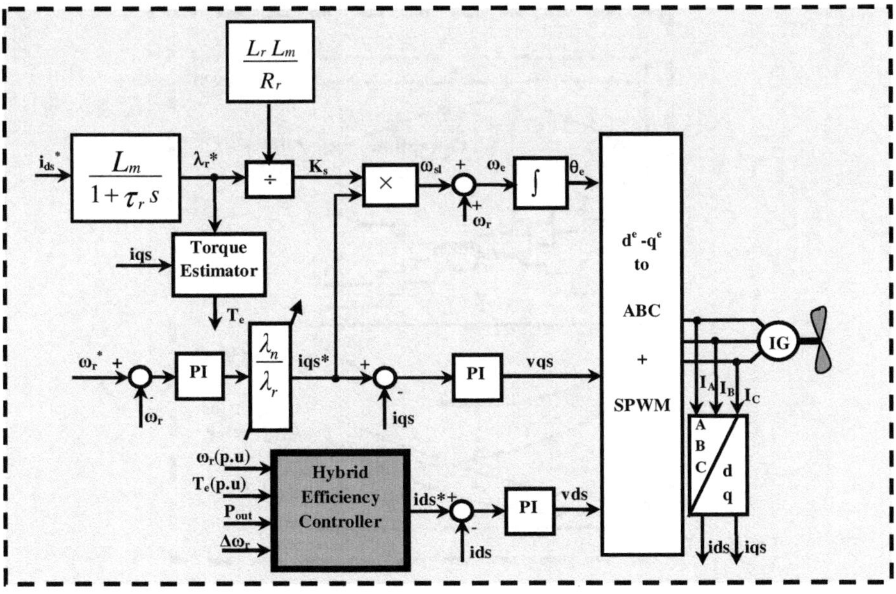

Fig. 3. Block diagram of the system of control of the induction generator

of an implicit mathematical model of the system. Figure 3 shows the scheme of the block diagram of system implemented for maximization of the generator output power. In this system the component of flux of the stator current is not kept constant and comes of the hybrid efficiency controller, HEC. The inputs of the HEC are: rotor speed and torque in values for unit; the output power and the variation of the rotor speed (that indicates if the system is stead state or transitory). Thus, a torque estimator is needed. A torque compensator is used in the loop of the torque component of the current to prevent disturbious in the developed torque and oscillations in the speed of the machine.

2 Efficiency Optimization

2.1 Search Control

In steady state with constant torque and speed, the flux component of the current is decreased whereas the torque component is increased. Initially, a reduction in total stator current occurs ($I_s = i_{qs} - j\ i_{ds}$), and consequently a reduction in stator copper losses, but rotor copper losses are increased. As the reduction in both stator and iron losses are higher than the increase in rotor losses, the total losses are reduced. If i_{ds} is continuously reduced, a reduction in the total losses will occur until the moment when the increase in copper losses becomes higher that the reduction in core losses, that is, the point minimum losses will be exceeded. This philosophy is illustrated in Figure 4. The determination of this point of minimum losses, which corresponds to optimum efficiency, can be performed using different procedures.

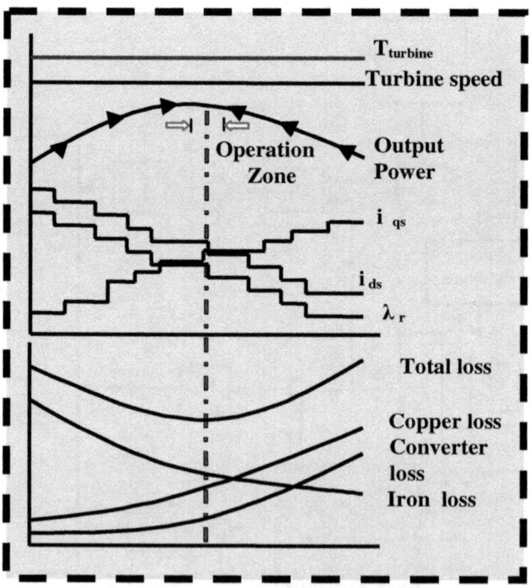

Fig. 4. Philosophy of the efficiency optimization search method

A turbine is submitted to a load proportional to the square of speed of the wind. Thus, in weak wind conditions, below 7 m/s, the generator works typically with light rotor load. In such conditions the intensity of the rotor flux of the induction generator (IG), commanded in vector control for i_{ds}, can be reduced to values below nominal flux, reducing reactive circulation, diminishing iron losses and consequently increasing global efficiency of both inverter and machine [3], something essential at low wind speed to improve power extraction capacity [4]. As mentioned before, rated flux results in excessive core losses and poor efficiency under light load conditions. Another aspect worth mentioning is the need to prevent machine torque disturbances during the efficiency optimization control. Under vector control, the developed torque can be expressed as:

$$T_e = k_t \lambda_r i_{qs} \tag{5}$$

where: λ_r is the rotor flux and i_{qs} is the torque component of the stator current, and k_t is a constant. If the flux is reduced to improve efficiency, i_{qs} must be increased accordingly, such that their product remains constant at any given time.

2.2 The Rosenbrock Method

This is a very simple method, and guaranteed to converge. The reference for the flux component of the stator current (i_{ds}^*) is modified in small steps in a given direction, while the system approaches the optimum efficiency point, i.e., the measured change in output power in the n-th step is positive ($\Delta P(n) > 0$). When the method recognizes that an "overshoot" has occurred ($\Delta P(n) < 0$), it reverses the search direction, with a reduced step size [9]. The search process can be mathematically expressed as shown in (6):

$$i_{ds}^*(n+1) = i_{ds}^*(n) + k\Delta i_{ds}^*(n); \qquad \begin{cases} k = 1; & \text{if } \Delta P(n) > 0 \\ k = -\dfrac{1}{2}; & \text{if } \Delta P(n) < 0 \end{cases} \tag{6}$$

where: $\Delta P(n) = P(n) - P(n\text{-}1)$ and $\Delta i_{ds}^*(n) = i_{ds}^*(n) - i_{ds}^*(n\text{-}1)$.

3 The Proposed System

The indirect method of vector control is applied to the IG. It derives the reference for the torque component of the stator current (i_{qs}^*) from the speed error, utilizing a conventional proportional-Integral (PI) controller. As the system operates with variable flux, a compensation block is introduced at the output of the speed PI controller. Essentially, this block multiplies the original PI controller output by the ratio rated flux / actual flux (estimate).

The reference for the flux component of the stator flux (i_{ds}^*) is not kept constant here, as in the majority of high performance IM drive systems. It is defined as the sum of two block outputs: $i_{ds}^*(k) = i_{ds}'^*(k) + \Sigma\Delta i_{ds}^{*'}$. The first term ($i_{ds}'^*$) is obtained from a fuzzy controller, that from two inputs (speed (ω_r) and estimate load torque (T_L), derives a preliminary reference ($i_{ds}'^*$) through fuzzy inference. The second one ($\Sigma\Delta i_{ds}'^*$) is the actual output of a search controller, based on the Rosenbrock method. Its value represents the accumulated control actions taken by the controller during the search process up to the current iteration (n), as can be seen in Figure 5.

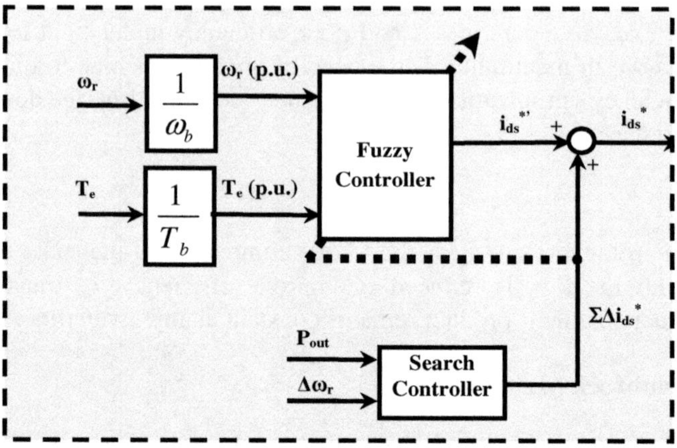

Fig. 5. Hybrid efficiency controller

When the system is turned on for the first time, the rule base of the fuzzy controller contains rated d-axis current reference ($i_{ds}'^*$) for all rules, i.e. for any speed and load torque point. When a steady state condition is detected, the search controller becomes active. After a few steps, it reaches the optimum efficiency point by imposing the $\Sigma\Delta i_{ds}'^*$ change to the original reference ($i_{ds}'^*$) from the fuzzy controller. Once the controller recognizes this optimum condition, the rule base can be updated to reflect the knowledge of the optimum flux level for this particular operating point (load torque and speed). At the same time, the search controller output must be reset, to prevent erroneous operation. When the optimum point is found, the rule base is updated, and the output of the search controller reset, such that, effectively, $i_{ds\ opt}^* = i_{ds}'^*$.

As the optimum efficiency points related to the several operating conditions are identified, the rule base is progressively updated, such that the fuzzy controller "learns" the optimum flux level for the entire torque-speed plane. Once the learning process is completed, the output of the fuzzy controller already reflects the optimum flux level, and the fuzzy controller is capable of driving the system at optimum efficiency without delays. To prevent sub-optimal operation, the search controller remains active to track possible deviations of the optimum point. Under transient conditions, the search process is cancelled, and the flux reference is solely derived from the fuzzy controller. It is worth noticing that no switching of strategies is required, since higher torques demands are normally met by imposing higher flux levels, i.e., the optimum level of flux for higher torques is close to the rated flux value. This methodology is summarized in Figure 6.

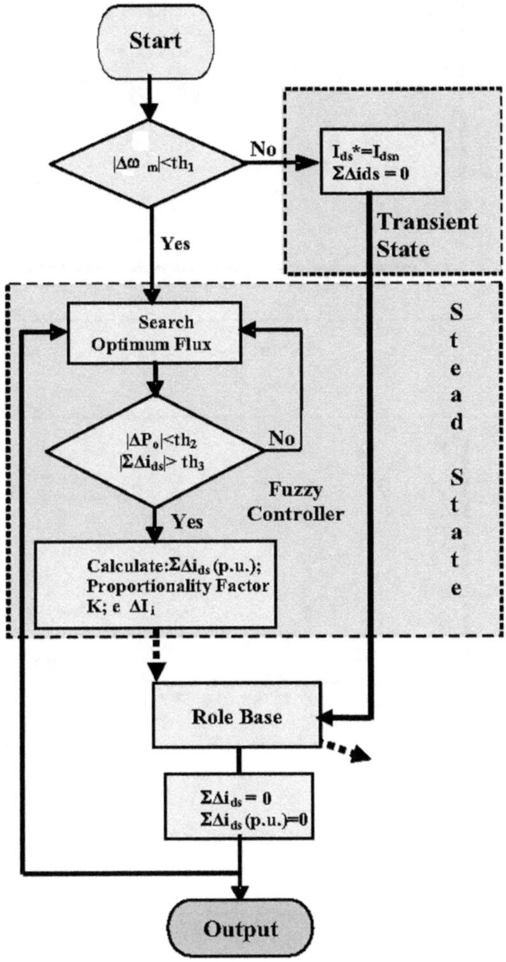

Fig. 6. State diagram for efficiency controller

3.1 The Fuzzy Efficiency Controller

The fuzzy sets for the input variables are shown in Figure 7. Both utilize normalized universes of discourses, to make the controller easier to port for different machine ratings. The output variable ($i_{ds}'^*$) is represented by singletons, and is shown in Figure 8. The rule base for the fuzzy controller is illustrated in Table 1. It is typically initialized with rated $i_{ds}'^*$ (1 p.u.), and it is progressively updated to incorporate the knowledge of the maximum efficiency points as they are found by the search controller, as previously described.

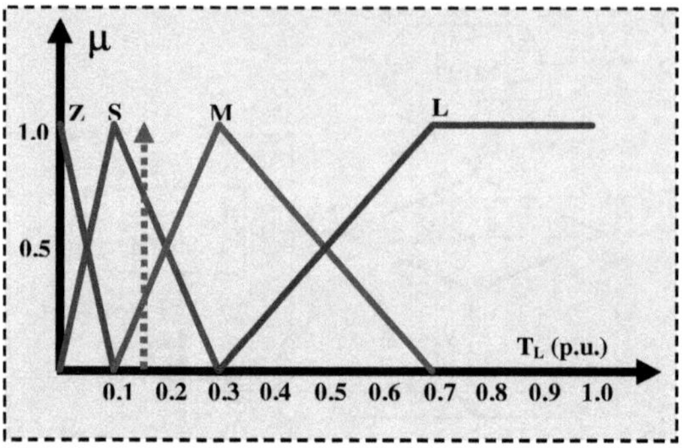

(a)

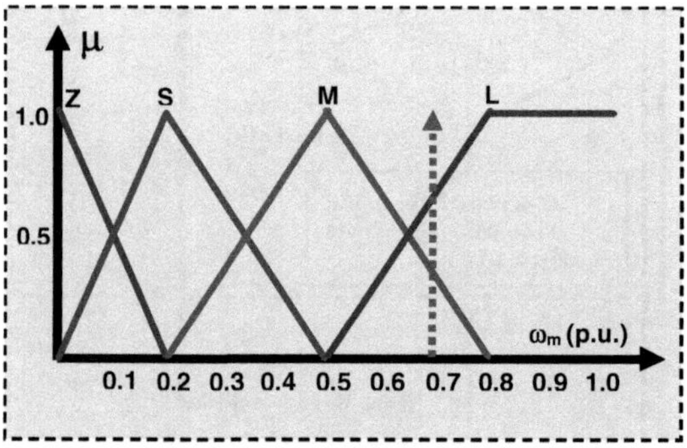

(b)

Fig. 7. Fuzzy sets for the input variables: (a) load torque and (b) speed

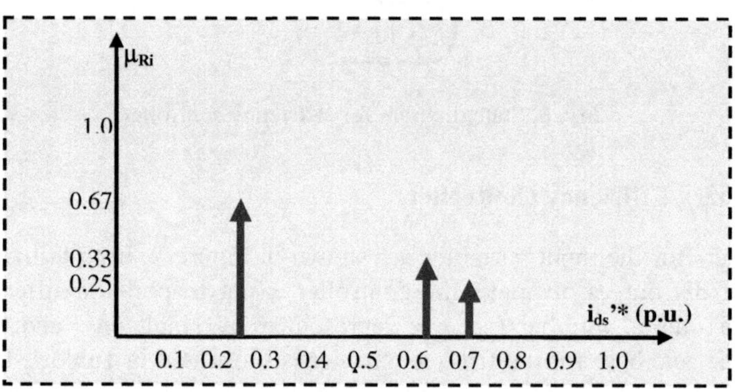

Fig. 8. The output variable

Table 1. Rule base for the fuzzy controller

T_L \ ω_m	Z	S	M	L
Z	1	1	1	1
S	1	1	$I_{A(n+1)}$	$I_{B(n+1)}$
M	1	1	$I_{C(n+1)}$	$I_{D(n+1)}$
L	1	1	1	1

The primary flux reference current $i_{ds}'^{*}$ is obtained by fuzzy sup-min inference, and the height method of defuzzification:

$$i_{ds}'^{*} = \frac{\sum_{i=A}^{D} I_{i(n+1)} \times \mu_{Ri}}{\sum_{i=A}^{D} \mu_{Ri}} \tag{7}$$

At steady state condition, whenever the search controller identifies an optimum flux level, the rule base must be updated. This process can be summarized as follows:

1) Identify the fired rules in the Rule Base (e.g., rules A,B,C,D in Table 1);

2) Compute the degree of truth for each rule, by applying the minimum (min) operator over the degree of membership for the input variables T_L and ω_r: $\mu_{Ri} = \min(\mu_{Tl}, \mu_{\omega r})$;

3) To evaluate the proportionality factor K, given by (8);

$$K = \frac{\sum_{i=A}^{D} \mu_{Ri} \times \sum \Delta i_{ds}(pu)}{\sum_{i=A}^{D} \mu_{Ri}^{2}} \tag{8}$$

4) Compute the correction term $\Delta I_i(n)=K x \mu_{Ri}$ for each fired rule as the product of its degree of truth and factor K;

5) Get the new value for each fired rule i (i=A,B,C,D) by (9).

$$I_i(n+1) = I_i(n) + \Delta I_i(n) \tag{9}$$

4 Simulation Results

In this simulation a 4 kW squirrel-cage induction machine was used (220/380V, p = 2). A reference speed step of 0.88 p.u, 0.40 p.u, 0.88 p.u and 0.40 p.u is applied at t = 1s, t = 30s, t = 45s and t = 55s, respectively, as shown in Figure 9 (a). After the initial transient, at t = 6s, the search begins. At t = 15s the controller identifies that an

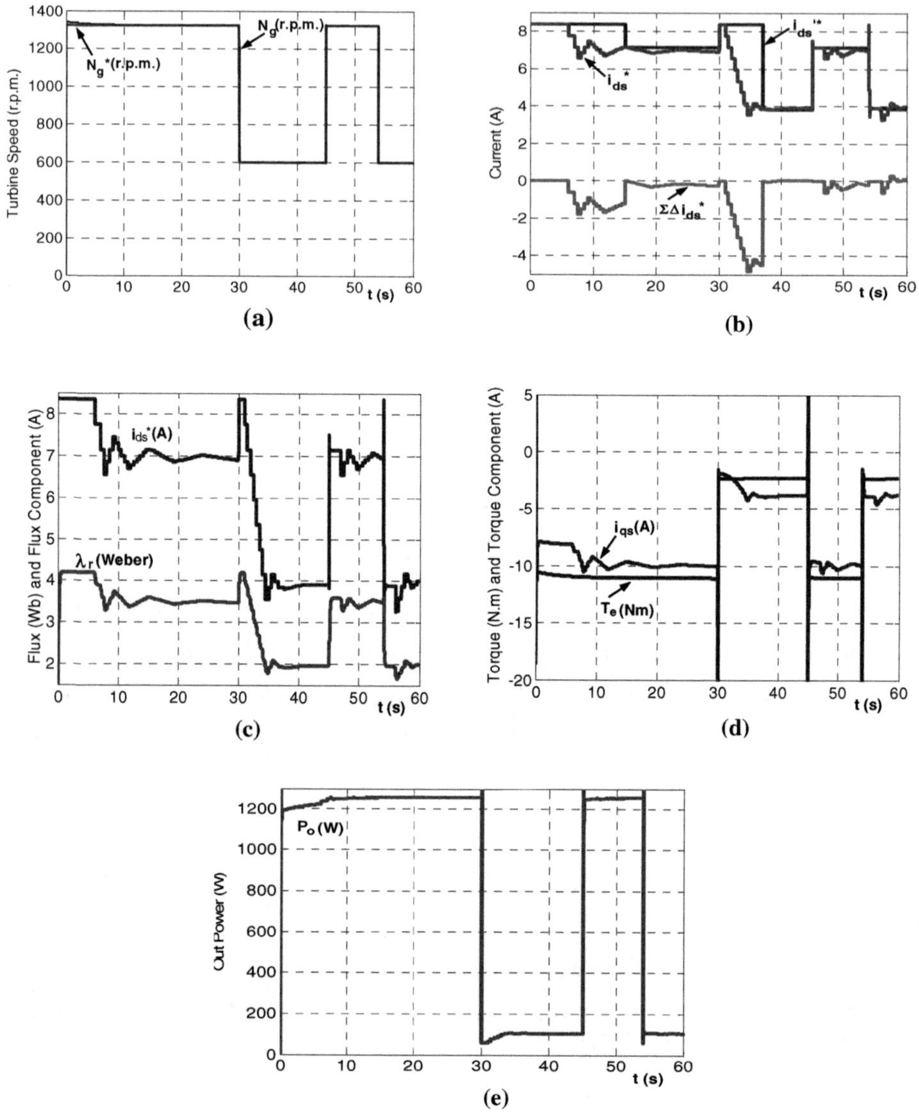

Fig. 9. Simulation results for the operation of the fuzzy controller: A) Reference and actual speed on the axis of turbine in r.p.m.; B) Components of *d-axis* current; C) Components of *d-axis* current and rotor flux; D) Components of *q-axis* current and torque estimate; E) Output power

optimum point has been found, and proceeds to update the rule base. Up to this point, the output of the fuzzy controller (i_{ds}'*) was the rated value for the magnetizing current, but from this time on, its output is made equal to the optimum value. Simultaneously, the output of the search controller is reset ($\sum\Delta i_{ds} = 0$), as can be seen in Figure 9 (b). The rotor flux response follows a first order filter profile of the reference current (i_{ds}*) as expected, and is shown here multiplied by a factor of 10, Figure 9 (c). The changes in flux level have a direct impact on the output power, Figure 9 (e), as well as in the torque component of stator current reference (i_{qs}^*), as expected, but the electromagnetic torque is unaffected, due to proper feed-forward compensation in i_{qs}^*, as shown in Figure 9 (d). At t = 45s and t = 55s when a previous reference speed step is applied, the rule base recognize and the system operates directly with optimum efficiency.

5 Conclusions

The proposed control strategy consists of a more effective way to implement the efficiency optimization via flux control in an induction machine. From the analysis of literature can be observed that in the comparative study of the diverse known search techniques, none of them results on a process as fast as the one achieved with the present method. This implies a great energy saving, because the system can be turned all the time operate at maximum efficiency. Another noteworthy point is that the transition from steady to transitory state occurs without abrupt changes in the system, or without any topological control change, since the displacement from a low torque point to one of higher torque or vice-versa is already programmed in the base of rules. The salient features of this technique are summarized next: i) It is applicable to any machine size, and does not require knowledge of machine parameters; ii) The rule base self tuning is progressive, and does not need any intervention from the operator; iii) Once tuned, the system is capable of operating all the time at optimum efficiency, without delay from one steady state condition to another, with significant energy saving; iv) During transients the rule base is kept active and, as a consequence, there is no switching from one control strategy (for steady state) to another (during transients), provided that the tuning has been completed; v) Proper disturbance compensation is included, such that no correction is needed to keep torque and speed constant during the optimization process; and vi) The system is capable of tracking slow parameter deviations, guaranteeing true optimum efficiency.

References

1. Kumar, V., Joshi, R.R.: Fuzzy Logic Based Light Load Efficiency Improvement of Matrix Converter Based Wind Generator System. Journal of Theoretical and Applied Information Technology, JATIT, 79–89 (2007)
2. Leidhold, R., Garcia, G., Valla, M.I.: Field-Oriented Controller Induction Generator With Loss Minimization. Transactions on Industrial Electronics 49(1), 147–156 (2002)
3. Sousa, G.C.D., Bose, B.K.: Fuzzy logic based on-line efficiency optimization control of an indirect vector controlled induction motor drive. Transactions on Industrial Electronics 42(2), 192–198 (1995)
4. Simões, M.G., Franceschetti, N.N., Bose, B.K.: Otimização de um Sistema de Geração de Energia Eólica Através de Controle Fuzzy. SBA Controle & Automação, Brasil. Jan., Fev., Mar., Abril de 1999. SBA 01, 48–58 (1999)
5. Rüncos, F., et al.: Geração de Energia Eólica - Tecnologias Atuais e Futuras. WEG Jaraguá do Sul. Brasil. WEG Máquinas S.A, http://www.weg.com.br

6. Sousa, G.C.D., Bose, B.K., Cleland, J.: Loss Modeling of Converter Induction Machine System for Variable Speed Drive. In: Proceedings of the IECON 1992, International Conference on Industrial Electronics, Control and Instrumentation, vol. 1, pp. 114–120 (1992)
7. Abrahamsen, F., Blaabjerg, F., Pedersen, J.K., Grabowski, P.Z., Thogersen, P.: On the Energy Optimized Control of Standard and High-Efficiency Induction Motors in CT and HVAC Applications. Transactions on Industry Applications 34(4), 822–831 (1998)
8. Abrahamsen, F., Blaabjerg, F., Pedersen, J.K.: State –of-the-Art of optimal Efficiency Control of Low Cost Induction Motor Drives. In: Power Electronics and Motion Control, PEMC 1996, vol. 2, pp. 163–170 (1996)
9. Moreno-Eguilaz, J., Cipolla, M., Peracula, J., da Costa Branco, P.J.: Induction Motor Optimum Flux Search Algorithms with Transient State Loss Minimization Using a Fuzzy Logic Based Supervisor. In: 28th Annual IEEE Power Electronics Specialists Conference, vol. 2, pp. 1302–1308 (1997)
10. de Almeida Souza, D., Aragão Filho, W.C.P., Sousa, G.C.D.: Adaptive Fuzzy Controller for Efficiency Optimization of Inductions Motors. Transactions on Industrial Electronics 54(4), 2157–2164 (2007)
11. Leidhold, R., Garcia, G., Valla, M.I.: Control para Máximo Rendimiento de Generadores Eólicos de Velocidad Variable, con Limitación de Velocidad y Potencia. XVI Congresso Brasileiro de Automática. Natal, RN. 2 a 5 de Setembro de, pp. 3121–3126. CBA (2002)
12. Deprez, W., et al.: Energy Efficiency of Small Induction Machines: Comparasion Between Motor and Generator Mode. European Copper Institute 3(1), 1–6 (2007)
13. Cassanellas, F.: Losses in PWM Inverter Using IGBTs. IEEE Proc. Electronics Power Applications 141(5), 235–239 (1994)
14. Bose, B.K., Patel, N.R., Rajashekara, K.: A Neuro-Fuzzy-Based On-Line Efficiency Optimization Control of a Stator Flux-Oriented Direct Vector Controlled Induction Motor Drive. IEEE Transactions on Industrial Electronics 44(2) (April 1997)
15. Kioskeridis, I., Margaris, N.: Loss Minimization in Induction Motor Adjustable-Speed Drives. IEEE Transactions on Industrial Eletronics 43(1) (February 1996)
16. Kusko, A., Galler, D.: Control Means for Minimization of Losses in AC and DC Motor Drives. IEEE Transactions on Industry Applications 19(4), 561–570 (1983)
17. Muljadi, E., Butterfield, C.P.: Wind Farm Power Model Development. NREL/CP 500-36199. USA (July 2004)
18. Vieira, J.P.A., et al.: Novas Estratégias de Controle Fuzzy Aplicadas ao Conversor do DFIG para Melhoria da Estabilidade Transitória em Esistemas Eólicos. IEEE Latin America Transactions 5(3) (June 2007)
19. Pierce, K.G., Migliori, P.G.: Maximizing Energy Capture of Fixed-pitch Variable-speed Wind Turbines. NREL, AIAA-200-0032. USA (July 2000)
20. Wang, J.B., Liaw, C.M.: Indirect field-oriented induction motor drive with fuzzy detuning correction and efficiency optimization controls. IEEE Power Electronics Applications 144(1) (January 1997)

Architectural Approach for the Implementation of a Position Control System for a Boat-Like Inspection Robot

Christoph Walter[1], Tino Krüger[2], Gunnar Strauß[1], and Norbert Elkmann[1]

[1] Fraunhofer Institute for Factory Operation and Automation
Sandtorstrasse 22, 39106 Magdeburg, Germany
{Christoph.Walter,Gunnar.Strauss,
Norbert.Elkmann}@iff.fraunhofer.de
[2] STILL GmbH, Berzeliusstrasse 10, 22113 Hamburg, Germany
Tino.Krueger@still.de

Abstract. Here we present the position control system of a swimming inspection robot for large underground concrete pipes that are partially filled with wastewater. The system's hardware consists of a laser-based measurement system for position determination and a mechanical rudder to move the robot laterally within the pipe. To automatically adapt the position control system to different environment conditions, a self-tuning controller is presented.

We show how the necessary software components were implemented as part of a non real-time CORBA-based [1] distributed system. The controller has hybrid requirements regarding latency and interarrival times of computed position values. We describe the supporting data-flow architecture and the fault tolerance mechanism. This control application integrates real-time and non real-time components.

Keywords: Distributed control system, Fault tolerance, Self-tuning controller.

1 Introduction

The Emschergenossenschaft based in Germany is currently planning the Emscher sewer system, arguably the largest residential water management project in Europe in years to come. The new Emscher sewer will have a total length of 51 km, with diameters between 1.4 and 2.8 m at depths up to 40 m under surface. The Emschergenossenschaft engaged the Fraunhofer Institute for Factory Operation and Automation (IFF) in Magdeburg, Germany, as the general contractor to develop automatic inspection and cleaning systems to meet the requirements imposed by legal guidelines.

Large underground concrete pipes that are partially filled with wastewater are a hazardous environment for humans. Nevertheless, such pipes must be inspected on a regular basis. Many of today's remote controlled inspection systems for underground sewer pipes consist of a single TV-camera and are designed for pipe diameters below one meter. The recent development of automatic inspection systems [2] equipped with advanced sensors makes it possible for an operator to perform this task from an outside position even for pipes with a diameter between 1.4 to 2.8 meters. The cable-guided damage surveying system (SEK) (see Fig. 1) was developed to be a versatile and easy-to-use tool to detect various kinds of damages above and below the water line with high accuracy.

J.A. Cetto et al. (eds.): Informatics in Control, Automation and Robotics, LNEE 37, pp. 61–73.
springerlink.com

Fig. 1. The cable-guided damage surveying system within a sewer pipe

To assist the inspection, it is our goal to keep the robot near the center of the pipe during the data acquisition phase. This must be done while operating under different environment conditions. These include varying water level, flow velocity and pipe diameter. A simple controller with a fixed parameter set was unable to maintain a stable position under different conditions. To solve this problem the position control system of the robot was fitted with a self-tuning controller that is able to identify the model based control parameters prior to data acquisition.

The main part of the controller was implemented using a software programmable logic controller (PLC). The input to the controller is the current lateral position relative to the pipe axis. An iterative algorithm using data from various sensors, mainly three banks of laser distance sensors, determines the position. The PC-based implementation of this algorithm has an unknown worst-case execution time (WCET). The actual execution time depends on the internal state of the software and the current input data. Further delays stem from the transmission of data over the Ethernet network between the PLC and the PC, as well as from the scheduling scheme of the position calculation task. While the actual controller is robust against a certain level of latency and jitter in the arrival of position data, we must consider cases when the position data does not arrive in time. This may happen when the position determination algorithm has an unusually long execution time or as a result of transient overload of the communication channel.

Another challenge is the second part of the self-tuning controller, which performs the identification of the system's behavior and which derives an adequate control parameter set. The algorithm used here is very sensitive to jitter in the interarrival times of position measurements and to missing samples.

An architectural approach for solving both problems is described in section 3. In section 4, the actual controller is presented in more detail. Section 5 discusses the specific dynamic quality of service approach used to avoid failure in cases of transient overload. In section 6, we present results from a real-world test of the system.

2 Related Work

A large number of architectural approaches, communication protocols, middleware, and development toolkits for robot applications have been conceived by researchers. These solutions often focus on the requirements of emerging robot applications such as mobile or service robotics. Examples of addressed challenges are task-planning and task-execution, localization, group communication, and system integration. Currently, several ongoing initiatives are trying to establish common standards for software architectures for academic and commercial robot applications. Examples are RT-Middleware [3], the RoSta initiative [4], and Microsoft Robotics Studio [5].

The design and execution of time critical applications such as control loops using distributed systems is a challenging task. On one hand, considerable work has been done to introduce time constraints or timely predictable behavior in distributed systems. On the other hand, a number of approaches have been published to relax time constraints of certain applications by introducing fault tolerance mechanisms.

In the field of communication middleware, the Object Management Group (OMG) has released extensions to the known CORBA-specification to extend it with real-time features [6]. The primary focus here is on the introduction of priorities when executing method calls, which requires appropriate system software support. Software libraries implementing this standard were also made available [7].

To relax time constraints, approaches that try to balance quality of service and required computation time were considered. This includes the classical "anytime" [8] or "flexible" [9] algorithms, where quality of service can be dynamically traded against shorter computation times. This scheme can be used to ensure the timeliness of computations. The authors of [10] have also dealt with meeting task deadlines. They use a dynamic task scheduling scheme with detection and handling of timing faults. Faults are handled via a task-pair strategy with invocation of an alternative exception task in case of a timing fault.

3 Data Acquisition and Processing

The main goal of the inspection system is to collect data about the condition of the sewer line. The data should be processed, saved to hard disk, and presented to the user as fast as possible (best effort, non real-time). The data collected here must be completely processed and saved in order to avoid overlooking potential damage. To achieve its main goal, the system must also be controlled. Data required to control the system does not have to be saved or analyzed completely. Instead, it is important that certain deadlines be met when reacting on external inputs or events (real-time). In the case of the SEK, position data is also shared between both real-time and best effort components.

3.1 Data Processing and Data Flow

We developed a service oriented CORBA-based software architecture focusing on data flow and distributed processing. In our system, an application is constructed by combining software modules that provide functionality to other modules via a generic interface. The modules are loosely coupled and may be arbitrarily distributed over a number of computers connected by a network. The interface of the services is data centric. It allows services to offer data via named buffers to other services or clients. Each buffer represents a time series of specific data objects (see Fig. 2). A service can represent a data source (e.g. a sensor), a processing module or a data consumer (e.g. an actuator, a database, or a GUI). Services that represent sensors or sensor systems may provide data that represents a single measurement, a time series of measurements, a vector, an image, or any other type of data that is generated by the sensor system.

The data flow of the application is modeled by connecting the services via the publish/subscribe paradigm. In this communication model, the services play different roles. The module providing data is called the service provider. The data consumer is called the service client. Please note that in our system a client does not subscribe to specific data elements but to notifications on when such elements become available.

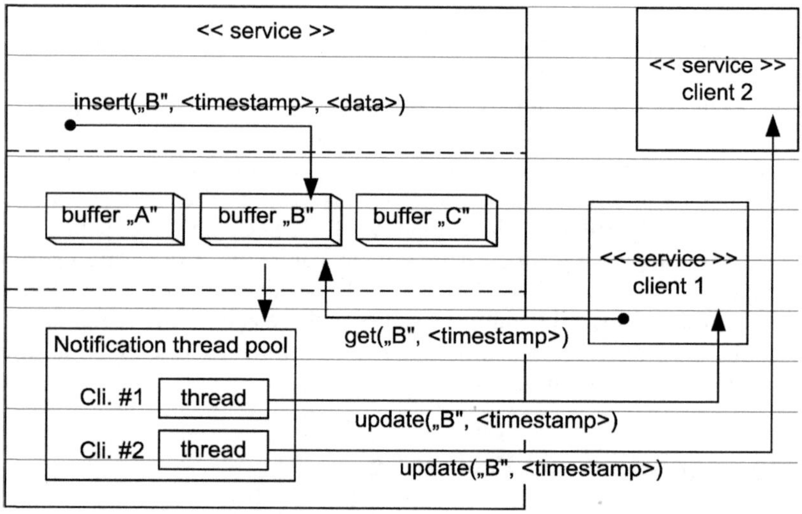

Fig. 2. A service provider offers data to clients via named buffers

The data objects that are being exchanged between services satisfy a given scheme. They consist of binary data and metadata in XML syntax. The most important element of the metadata is the timestamp, which indicates the recording time of the original data. The timestamp acts as a key for the data object within the given data buffer. The binary data can contain information in any format. A client can request individual data elements or arbitrary blocks of data elements from a buffer. This is achieved by sending structured queries with a simple syntax to the desired service. The client can then process these data items and/or eventually offer the results to further clients. The foundation of the used timestamps is the Newtonian time model. Timestamps are generated based on synchronized local clocks. Clock synchronization is carried out via the network by a dedicated service. The accuracy of synchronization between time-server and any client is better than 100 μs. This is well below the shortest interarrival times of data elements in our application. It was achieved using common PC-hardware, Gigabit-Ethernet, and a Windows NT-based operating system.

3.2 Implementation of Position Determination

Determining the position of the robot is of great importance. Knowledge of the position is not only required for position control, but also to process and interpret other sensors' data. This includes the registration of images from a high-resolution photographic camera system, the analysis of data from an ultrasonic underwater scanner, the determination of the magnitude of corrosion and incrustations, or the precise measurement of possible deviations in the position of individual pipe segments. These tasks cannot be performed without knowledge of the position and orientation of each sensor during data acquisition.

The heart of the position determination subsystem is an algorithm that was specifically developed for this purpose. It is implemented as a service (the position determination service; PDS). It is based on the measurements of a number of laser ranging sensors plus an inclination sensor. A discussion of the exact procedure can be found in [11].

Measurements from the laser sensors are acquired by the onboard software-PLC. The PLC is implemented as a real-time task on an embedded PC and is coupled to an interface service (IS) running as a non real-time task. The IS has access to the PLC's memory and is configured to acquire time-stamped measurements at a rate of 100 Hz. This rate is consistent with sensor capabilities and sufficient for our application.

The PDS is notified of the arrival of new data elements in the IS's buffer. Its goal is to process all data elements that are produced by the IS. To do this the PDS uses the following strategy of data access: Upon notification of a new data element, it asks the IS for all elements that succeed the last element the PDS has already processed. This is done via a single "get"-operation. By using this technique, it is possible for the PDS to catch up in case of a previous transient overload. For each received data element, the system's position is calculated and eventually made available for further processing by other clients. The calculation time may vary depending on the current input data, the state of the PDS, and the state of the underlying operating system.

4 Self-tuning Controller

This section describes the design of the self-tuning controller. We also discuss the implementation of its essential parts as software modules and their interactions.

4.1 Controller Design

One challenge in the context of stable data acquisition between real time and non-real time systems (hybrid systems) is the implementation of non-linear controllers. This chapter describes the motivation of using a non-linear controller in this area. In addition, it gives a short introduction about calculating the parameters for such non-linear controllers.

When moving through a partially filled sewer, the position of the SEK is controlled by a rudder arm mounted on top of the robot. The rudder can be turned relative to the longitudinal axis of the pipe, thus forcing the SEK to change its position. Due to disturbances in the water flow, a controller has to be integrated for closed-loop navigation. Self-tuning controllers belong to the denomination of adaptive controllers. They are used to fit the plant on a variable environment. In our case, the flow velocity of the waste-water, the water level, and the pipe diameter may change depending on the actual sewer and on current conditions like the time of day or the weather situation. These high levels of variability lead to the adoption of an adaptive controller for the robot. Adaptive controllers, in contrast to fixed structure designs can be used in a broad field of applications. Here we will focus on a self-tuning controller as the specific type of indirect adaptive controller. Intermediate steps are required for the calculation of the controller parameters (see Fig. 3).

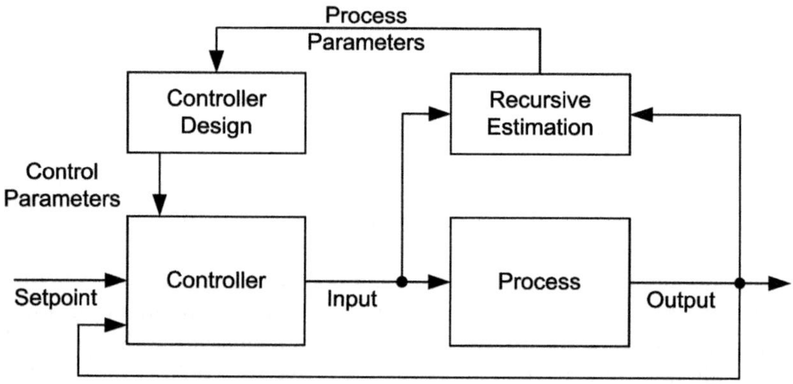

Fig. 3. Schematic diagram of a self-tuning controller

The calculation of a self-tuning controller contains the following steps:

1. Recursive estimation of process parameters (system identification)
2. Suitable determination of a control law with adjustable parameters
3. Calculation of controller parameters based on the desired closed-loop system
4. Parameter update using the IS

The advantage of this method lies in the possibility of completing each step of the calculation before the controller parameters are used. In addition, the closed loop can be pushed to a desired behavior while the parameters of the plant are changing.

The plant is depicted by a discrete autoregressive moving average with an extra input model (ARMAX-Model) of fourth-order. The advantage of ARMAX-Models compared to simpler models is the adequate freedom in describing the properties of the disturbance term. That gives the model:

$$\begin{aligned} &y(t) + a_1 y(t-1) + \ldots + a_3 y(t-3) = \\ &b_1 u(t-1) + \ldots b_3 u(t-3) + c_1 e(t) + \ldots + c_3 e(t-3) \end{aligned} \tag{1}$$

$$\theta = [a_1 \ldots a_3 \quad b_1 \ldots b_3 \quad c_1 \ldots c_3] \tag{2}$$

Where y(t) is the plant output, u(t) is the control input and e(t) the prediction error. The parameters used to estimate are captured by θ. In this case, the least–squares criterion is used to identify the transmission behavior of the process. The parameters are time varying in a small range during operation. Hence, it is desirable to base the identification on the most recent data rather than on the older data. This can be achieved by exponential discounting (λ) of old data. Then, it can be shown [12] that the extended recursive least square scheme (ERLS) becomes:

$$\begin{aligned} \theta(t) &= \theta(t-1) + L(t)[y(t) - \varphi^T(t)\,\theta(t-1)] \\ L(t) &= \frac{P(t-1)\varphi(t)}{(\lambda + \varphi^T(t)P(t-1)\varphi(t))} \\ P(t) &= \frac{1}{\lambda}\left(P(t-1) - \frac{P(t-1)\varphi^T(t)P(t-1)}{\lambda + \varphi^T(t)P(t-1)\varphi(t)}\right) \end{aligned} \tag{3}$$

The estimated process parameters from separately recorded test series can be verified using the indirect method. The controller parameters are then calculated after a successful verification.

When using ERLS it is advisable to continuously excite the system in order for the parameters to converge. Fig. 4 depicts exemplary the change of positions predicted by a model calculated from measured data. The rudder angle as input to the ERLS was oscillating between minus ten and ten degrees. The calculated model is afterwards the starting point for the controller design. The controller is designed so as to compensate the plant with the inverse system and the desired transmission behavior.

Assuming that the desired close loop behavior as shown in equation 4 where z-1 is backward shift operator.

$$M(z^{-1}) = \frac{\hat{B}(z^{-1})}{\hat{A}(z^{-1})} \tag{4}$$

The actual closed loop is given in equation 5 with the controller $K(z^{-1})$ and the plant behavior $S(z^{-1})$. The plant behavior is the solution of the extended recursive least square scheme. The suitable controller law is then given by equation 8 without solving a Diophantine equation.

$$G(z^{-1}) = \frac{K(z^{-1})S(z^{-1})}{1 + K(z^{-1})S(z^{-1})} \tag{5}$$

$$K(z^{-1}) = \frac{D(z^{-1})}{C(z^{-1})} \tag{6}$$

$$S(z^{-1}) = \frac{B(z^{-1})}{A(z^{-1})} \tag{7}$$

$$\begin{aligned} D(z^{-1}) &= A(z^{-1})\hat{B}(z^{-1}) \\ C(z^{-1}) &= B(z^{-1})(\hat{A}(z^{-1}) - \hat{B}(z^{-1})) \end{aligned} \tag{8}$$

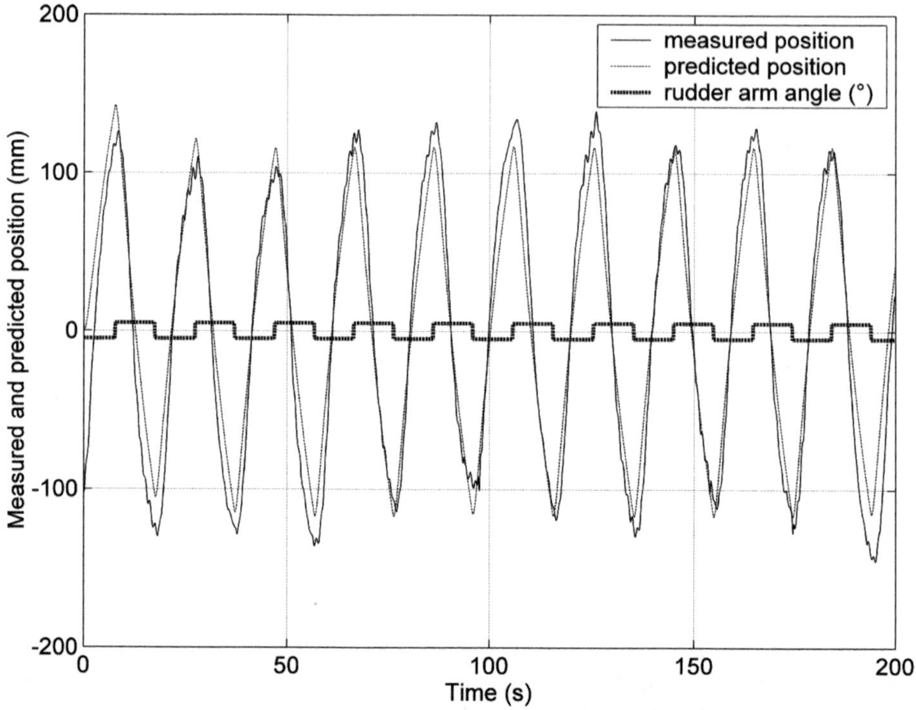

Fig. 4. Measured and predicted positions of the SEK

In terms of the causality of the controller law, we get the following conditions for the implementation:

1. The transmission behavior of the plant has to be stable, respect to the inverse implementation in the controller law.
2. The polynomial of the decoupled model shown in equation 9 should only have zeros within the unit circle.
3. The estimated plant model should not include a minimum phase system or must be corrected because of the stability of the open loop.
4. The controller law has to be causal.

$$\hat{A}(z^{-1}) - \hat{B}(z^{-1}) \tag{9}$$

The calculated controller is therefore less sensitive to noise and jitter compared to other designs like the pole placement variant, which is described by Aström and Wittenmark [13]. The critical part of the self-tuning controller is the calculation of the plant behavior $S(z^{-1})$ as the solution from ERLS. There is a small convergence radius of the scheme, which depends directly on the deterministic data acquisitions. In case there is a bad quality of the estimation, the system identification can be run again or a simple robust controller can be used.

4.2 Implementation

The self-tuning controller was implemented in the form of two main modules. The first module is the actual controller. It runs as a real-time task within the onboard PLC. Position-data is fed into it by the IS as described in section 3.2. The IS acts also as a client similar to the PDS but with a different strategy for data access: The controller doesn't want to process all data elements that are produced by the PDS, but needs the most recent one. For this reason the IS upon notification of new data requests the most recent data element from the PDS and feeds it into the PLC.

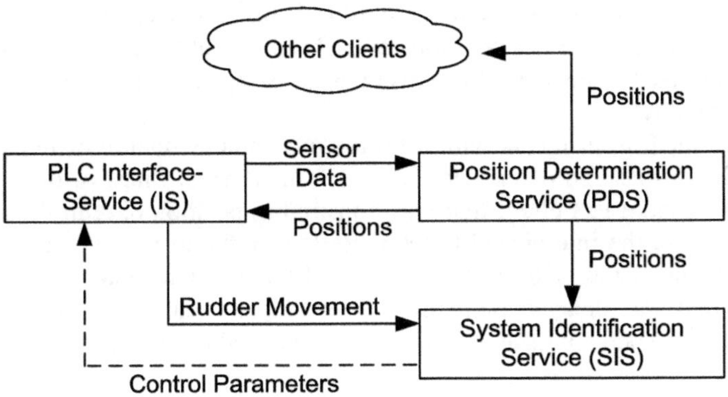

Fig. 5. Structure and data flow of controller implementation

The second module is a service, which computes the control parameter set (system identification service; SIS). It needs two kinds of data elements as input: (1) all sampled rudder arm angles during the system identification process and (2) the corresponding system position. The arrival of these data is not time critical but it is necessary that all samples are processed. It therefore uses a similar data access strategy as the PDS to gather all data elements that describe the rudder movement from the IS. Furthermore, it requests for each such element the two closest system positions from the PDS and interpolates between them to get the corresponding system position. Requesting closest elements given a certain timestamp is another feature of the data-access mechanism of our architecture. We assume that the time series of data elements buffered at the data sources is large enough to answer even requests that are delayed by several seconds. After the SIS has collected enough data, it can compute the control parameters and send them to the main control module via the IS. The described data flow is shown in Fig. 5.

5 Dynamic Quality of Service

Even though we have successfully implemented the system identification part of the controller as a non real-time service while guaranteeing the required data quality, the actual controller may still suffer from excessive latencies in the arrival of position data from the PDS. This latency is the result of the iterative algorithm used in the PDS and may also occur because of transient overloading of the communication channel.

To avoid system failure in such cases we integrated a dynamic quality of service approach that uses the timestamps of the position data available to the controller implemented in the PLC to determine the age of that data. If a certain age is exceeded, the controller uses an alternative position that it computes by a simple prediction mechanism based on the last known positions. This fallback mechanism is triggered if the position data available to the PLC is older than 90 ms. While this is a crude method and could easily be improved by using a more advanced approach, it is sufficient to deal with short disruptions to the communication. If however the position data becomes older than 900 ms, the controller is put into a safe state. In this state, the rudder is moved to its center position, whereby the likelihood of a collision with the wall is minimized. The controller resumes as soon as more recent position data becomes available.

We performed experiments with a prototypic implementation of the system under operating conditions. To determine the actual interarrival times of data produced by the PDS, we used a test-client that was consuming position data in the same way as the IS. In Fig. 6, the interarrival times recorded by the test-service are shown. Well under 5 percent of the position values arrived slightly later than usual. Less than 1 percent of that was delayed more than 90 ms. In the case of a delay over 90 ms, approximately 50 percent of those values were delayed more than 900 ms.

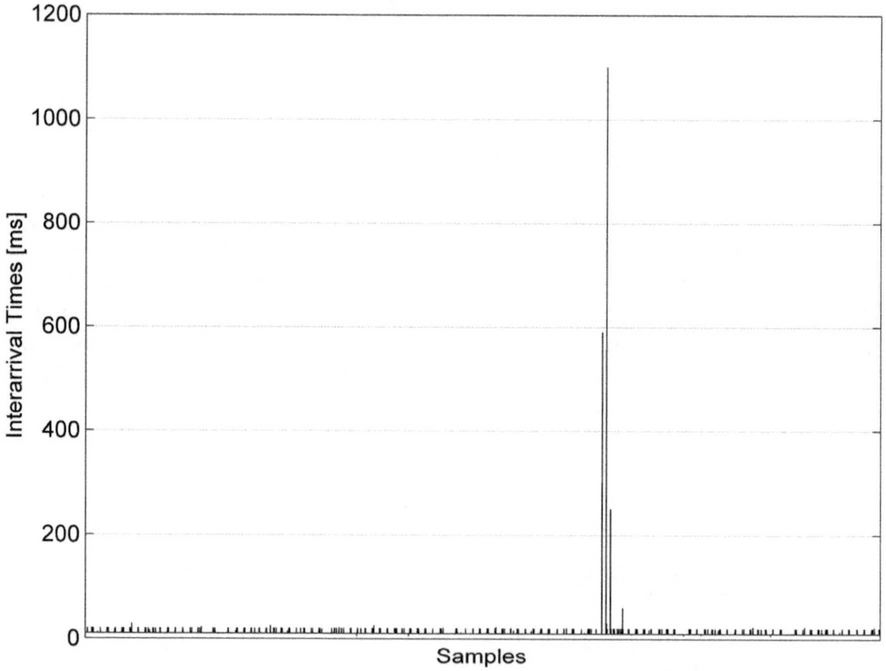

Fig. 6. Interarrival times of position data at Test-Service

When observing the behavior of the robot during the experiment, no significant deviation from the ideal position could be noticed. The system maintained a position near the center of the pipe, even when missing the 90 ms deadline. This was also the case when the 900 ms deadline was briefly missed. This can be attributed to the fact that the water flow had relatively small turbulences that required only slight corrections by the position control system. However, we expect that missing the 900 ms deadline may have a significant effect when more severe turbulence is present in the water flow.

6 System Behavior under Real-World Conditions

Due to the high costs of tests with the robot, first tests were made with a physical model of the SEK in a flow channel. The positive results of the system identification and the controller design with the model motivated us to test the system under real-world conditions.

Since the position of the SEK during an inspection should always be close to the middle of the pipe, the system identification, which requires a constant excitation, takes place in a separate phase before the inspection. This has the added advantage that during an identification phase the controller doesn't disturb the system and there are no direct dependencies between the excitation signal and the system output.

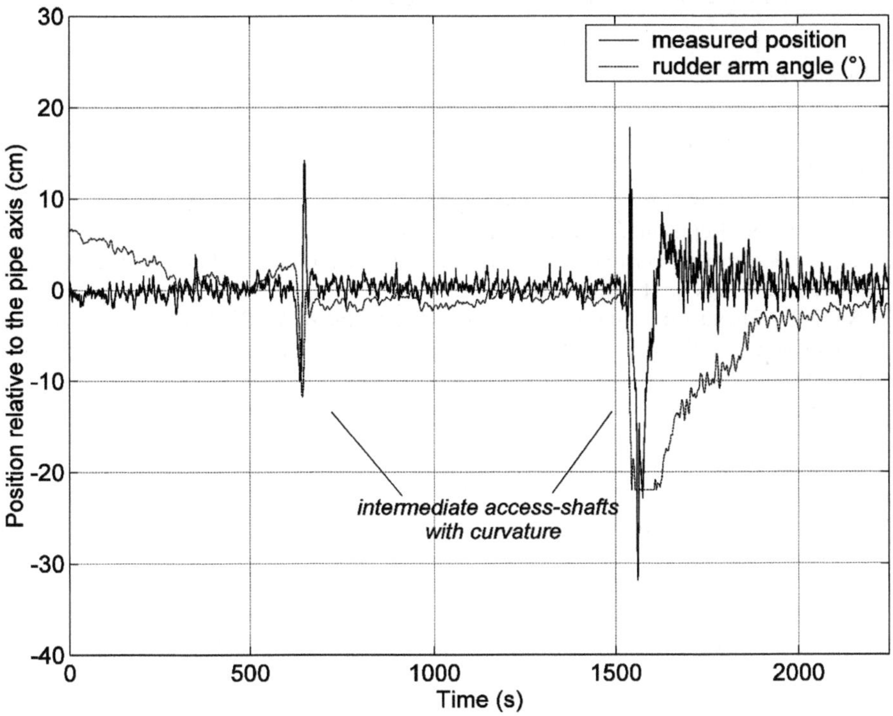

Fig. 7. Measured position of the SEK during an inspection

Fig. 7 depicts the measured lateral position of the robot relative to the pipe axis during testing. The system was in a sewer pipe with a diameter of 1.8 m and a length of approximately 200 m. In large parts, the deviation from the middle position is less than 3 cm and the requirements upon the accuracy of the positioning were fulfilled. A larger deviation could be observed near intermediate access-shafts with considerable curvature. This is acceptable since we do not expect these kinds of access shafts when performing regular inspections in the new Emscher sewer.

7 Conclusions

Here we presented an application of a distributed control system for the position control of a boat-like inspection robot for partially filled sewer pipes. We discussed the hybrid requirements of a self-tuning controller regarding timeliness and interarrival times of its input data. We also discussed how software components implementing the controller were integrated into a non real-time service-oriented system. This was achieved by implementing a scheme for detecting and handling excessive time delays to prevent system failure in case of transient overload. We conclude that, in the case of our application, it is feasible to use soft real-time or non real-time components in the control loop. Prerequisites are (1) a deterministic data acquisition with time stamping and (2) a hard real-time controller core that implements the mechanism that handles timing faults.

References

1. Object Management Group (eds.): Common object request broker architecture (CORBA/IIOP), `http://www.omg.org/technology/documents/corba_spec_catalog.htm`
2. Elkmann, N., Reimann, B., Schulenburg, E., Althoff, H.: Automated inspection system for large under-ground concrete pipes under operating conditions. In: International Conference on Field and Service Robotics (2005)
3. Ando, N., Suehiro, T., Kitagaki, K., Kotoku, T., Yoon, W.K.: RT-Middleware: Distributed Component Middleware for RT (Robot Technology). In: International Conference on Intelligent Robots and Systems, pp. 3555–3560 (2005)
4. Robot Standards and Reference Architectures, `http://www.robot-standards.org`
5. Microsoft Robotics Studio, `http://msdn.microsoft.com/robotics`
6. Wolfe, V.F., DiPippo, L.C., Cooper, G., Johnston, R., Kortmann, P., Thuraisingham, B.M.: Real-Time CORBA. In: Third IEEE Real-Time Technology and Applications Symposium (1997)
7. Schmidt, D.C., Levine, D.L., Mungee, S.: The Design of the TAO Real-Time Object Request Broker. Computer Communications 21(4), 294–324 (1998)
8. Dean, T.L., Boddy, M.: An analysis of time-dependent planning. In: Seventh National Conference on Artificial Intelligence, pp. 49–54 (1988)
9. Hendler, J.A.: Real-time planning. In: Working Notes of the AAAI Spring Symposium on Planning and Search (1989)

10. Nett, E., Gergeleit, M., Mock, M.: An adaptive approach to object-oriented real-time computing. In: 1st IEEE International Symposium on Object-oriented Real-time Distributed Computing, pp. 342–349 (1998)
11. Elkmann, N., Kutzner, S., Saenz, J., Reimann, B., Schultke, F., Althoff, H.: Fully automatic inspection systems for large underground concrete pipes partially filled with wastewater. In: International Conference Intelligent Robots and Systems (2006)
12. Ljung, L.: System Identification. PTR Prentice Hall, Englewood Cliffs (1999)
13. Astrom, K., Wittenmark, B.: Adaptive Control. Pearson Education, London (1992)

Fast and Compact Encoding of Numerical Controllers Using OBDDs

Giuseppe Della Penna[1], Benedetto Intrigila[2], Nadia Lauri[1], and Daniele Magazzeni[1]

[1] Department of Computer Science, University of L'Aquila, Italy
{dellapenna,lauri,magazzeni}@di.univaq.it
[2] Department of Mathematics, University of Rome "Tor Vergata", Italy
intrigil@mat.uniroma2.it

Abstract. In the last years, the use of control systems has become very common, especially in the *embedded systems* contained in a growing number of everyday products. Therefore, the problem of the *automatic synthesis* of control systems is extremely important. However, most of the current techniques for the automatic generation of controllers, such as *cell-to-cell mapping*, *dynamic programming*, *set oriented approach* or *model checking*, typically generate *numerical controllers* that cannot be embedded in limited hardware devices due to their size.

A possible solution to this problem is to *compress* the controller. However, most of the common lossless compression algorithms, such as LZ77, would decrease the controller performances due to their decompression overhead.

In this paper we propose a new, completely automatic OBDD-based compression technique that is capable of reducing the size of any numerical controller up to a *space savings of 90%* without any noticeable decrease in the controller performances.

Moreover, the output of this compression process can be easily embedded in software/hardware devices by automatically translating it in C or VHDL code.

1 Introduction

Control systems (or, shortly, *controllers*) are small hardware/software components that control the behavior of larger systems, the *plants*. A controller continuously looks at the plant *state variables* and possibly adjusts some of its *control variables* to keep the system in the *setpoint*, which usually represents its *correct* behavior.

In the last years, the use of controllers has become very common in robotics, critical systems and, in general, in the *embedded systems* contained in a growing number of everyday products. Therefore, the problem of the *automatic synthesis* of control systems is extremely important.

To this aim, several techniques have been developed, based on a more or less systematic exploration of the plant state space. One can mention, among others, *cell-to-cell mapping* techniques [1], *dynamic programming* [2] and *set oriented approach* [3]. Recently, *model checking* techniques have also been applied [4, 5] in the field of automatic controller generation.

The controllers generated using all these techniques are typically *numerical controllers*, i.e. tables indexed by the plant states, whose entries are commands for the

J.A. Cetto et al. (eds.): Informatics in Control, Automation and Robotics, LNEE 37, pp. 75–87.
springerlink.com

plant. These commands are used to set the control variables in order to reach the set-point from the corresponding states. Namely, when the controller reads a state from the plant, it looks up the action described in the associated table entry and sends it to the plant.

However, a main problem of this kind of controllers is the *size* of the table, which for complex systems may contain millions of entries, since it should be embedded in the control system hardware that is usually very limited.

A possible solution to these problems is to derive, from the huge numerical information contained in the table, a small *fuzzy control system*. This solution is natural since fuzzy rules are very flexible and can be adapted to cope with any kind of system. Moreover, there are a number of well-established techniques to guide the choice of fuzzy rules by statistical considerations, such as in Kosko space clustering method [6], by abstracting them from a neural network [7], by clustering the trajectories obtained from the cell mapping dynamics [1] and finally by using genetic algorithms [8].

However, two considerations are to be made with respect to this approach. First, the detection of the fuzzy rules requires (at least) a *tuning phase* which is not completely automatic but involves some human intervention. Second, the fuzzy rules appear not to be suitable when a very high degree of precision is needed. This is the case, e.g., of the *truck and trailer parking problem* when *obstacles* are to be avoided in the parking lot [4]. In this case, when the truck is near to an obstacle, only a very precise manoeuvre can park it, without hitting the obstacle. So, to achieve the required precision, an exceeding number of fuzzy rules would be necessary.

Therefore, it seems reasonable to pursue another possible approach, that is *to directly compress the control table*. By this we mean *data compression* in the usual sense of *reduction of the number of bits* needed to represent the controller table in the computer RAM memory [9, 10].

Observe, however, that in this case the compression algorithm should be constrained as follows:

1. the logical content of the table, that is the relation between the states and the corresponding control actions, must be preserved without any loss of information (i.e., we need some kind of *lossless* compression algorithm);
2. the access time to the table must be comparable with the one obtained with a direct representation of the table in the computer memory (i.e., the *decompression overhead* must be minimal).

Consider as an example the straightforward idea of directly compressing the table by mean of (some variant of) the well known Lempel-Ziv algorithm [11]. Since this is a lossless compression, the first requirement above is fulfilled, but not the second one, since the access time to the table would be linear in the *uncompressed* size of the table itself [10].

Since all the common compression algorithms have similar problems when applied to controllers, we decided to develop a new *ad-hoc* compression algorithm based on the well known *Ordered Binary Decision Diagrams (OBDDs)* [12].

Decision diagrams as well as *decision trees* are of wide use in data manipulation as well as in data mining [13]. Indeed, our choice of OBDDs was motivated by their capability of compressing large state spaces, that is at the hearth of *symbolic model checking* and of its great achievements [14, 15]. Our hypothesis has been that this capability is transferrable to the compression of the controller table which is, in a sense, an augmented state space representation.

Our experimental results show that this is indeed the case. For very large tables we obtain a quite good compression ratio (around 10:1). Moreover, not only the access time is good, but it is often *even better* than the one obtained by a direct representation of the table in memory. This phenomenon is due to the fact that, unless we accept a very sparse table, with huge waste of space, we need to represent the controller using some kind of *open addressing* table accessed through a *hash function*, with a consequent worsening of the access time [16].

A final but very important point to be stressed is that the implemented compression algorithm, which transforms the controller table in an OBDD, operates in a completely automatic way. The only parameter the user should set is the BDD variable reordering method (see Section 4), which can effect the compression ratio. After that, the compression operates in a "zip"-like fashion. However, the best setting of the above parameter can also be discovered by the algorithm by trying all the possible reorderings, thus making the compression process completely automatic. Finally, a further automatic step can also be applied to transform the OBDD into a C library that can be incorporated in any software applications or into VHDL code, to directly synthesize an integrated circuit.

The paper is organized as follows: in Section 2 we give a description of OBDDs and in Section 3 we show how we use them to encode numerical controllers. In Section 4 we present our experimental results and Section 5 concludes the paper.

2 Ordered Binary Decision Diagrams

A Binary Decision Diagram (*BDD*) is a data structure used to represent a boolean function [12]. Indeed, any boolean function f can be represented as a binary tree having two kind of leaf values: F (boolean false) or T (boolean true). Each node of the tree (*decision node*) is labeled by a variable of the formula f. The two edges outgoing a

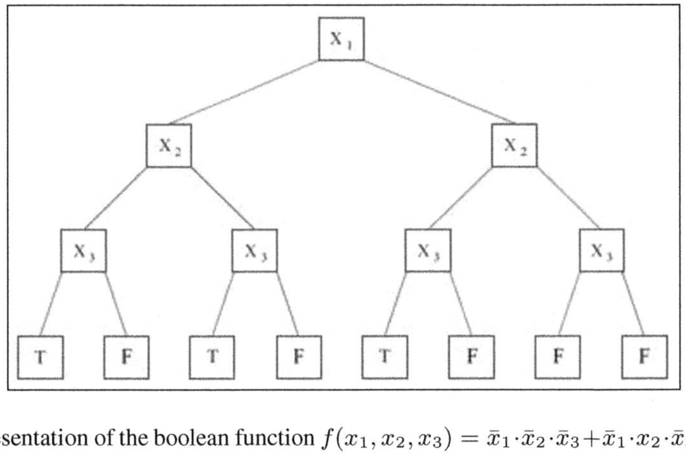

Fig. 1. Representation of the boolean function $f(x_1, x_2, x_3) = \bar{x}_1 \cdot \bar{x}_2 \cdot \bar{x}_3 + \bar{x}_1 \cdot x_2 \cdot \bar{x}_3 + x_1 \cdot \bar{x}_2 \cdot \bar{x}_3$

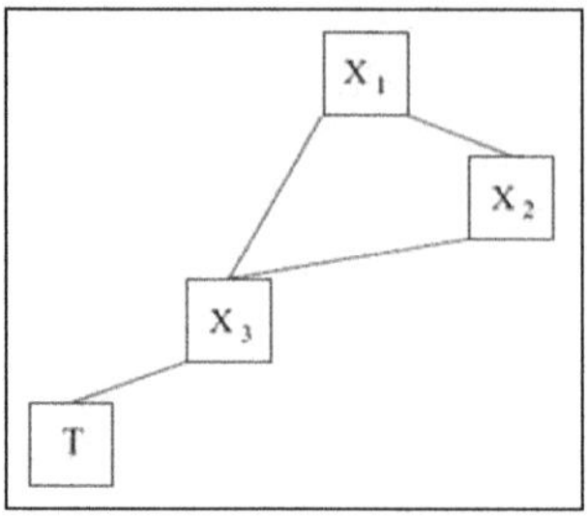

Fig. 2. The OBDD in Figure 1 after removal of duplicate nodes and redundant tests

decision node represent an assignment of the corresponding variable to false or true, respectively. Therefore, a path from the BDD root to one of its leaves represents a (possibly partial) variable assignment for f , and the corresponding leaf value is the value of f for the given assignment.

An advantage of BDDs is that many logical operations, like conjunction, disjunction, negation or abstraction can be implemented by polynomial-time graph manipulation algorithms. Indeed, BDDs are extensively used in the software tools, e.g., to synthesize circuits or to perform formal verifications.

If variables appear in the same order on all paths (or, in other words, if all the nodes on the same tree level refer to the same variable) the BDD is called *ordered* (OBDD).

As an example, Figure 1 shows the OBDD for the boolean function $f(x_1, x_2, x_3) = \bar{x}_1 \cdot \bar{x}_2 \cdot \bar{x}_3 + \bar{x}_1 \cdot x_2 \cdot \bar{x}_3 + x_1 \cdot \bar{x}_2 \cdot \bar{x}_3$.

Usually, OBDDs are also *reduced* (ROBBD) by merging isomorphic subgraphs and eliminating nodes having two isomorphic children. In this way, any boolean formula f can be represented by an *unique* and very compact rooted, acyclic and directed graph.

The size of a reduced BDD is determined both by the function being represented and by the chosen *ordering of the variables*. Thus, a correct variable ordering is of crucial importance to gain the best "reduction factor". The problem of finding the best variable ordering is NP-hard, but there exist efficient heuristics to tackle the problem.

Figure 2 shows the reduced version of the OBDD in Figure 1. Note that all the paths leading to F have been actually removed from the graph, since they are simply the complement of those leading to T.

A useful way to see BDDs, that will be used in this paper, is that they *encode the compressed representation of a relation*. However, unlike other compression techniques, the actual operations on BDDs are performed *directly on that compressed representation*, i.e. without decompression. Details on this issue are given in Section 3.

3 Boolean Encoding of Numerical Controllers

OBDDs can be used to create some kind of compressed representation of any data that can be encoded as a *logical expression*. In this Section, we describe how a numerical controller can be seen as a boolean formula, and give details on the actual

OBDD-based compression algorithm that has been implemented. Moreover, we give some information on how the OBDD-compressed controller can be efficiently queried and transformed in a hardware-embeddable form.

3.1 Boolean Encoding of the Controller Table

As already said, an OBDD can be actually seen as a compact representation of a relation. On the other hand, a controller table containing a set of (state,action) pairs represents a relation $R = \{(s,a)|a \text{ is the action associated to } s \text{ in the controller table}\}$ between states and actions. Since BDDs encode formulas, it may be useful to represent R through its characteristic function C_R defined as follows:

$$C_R(s,a) = \begin{cases} T & \text{if } (s,a) \in R \\ F & \text{otherwise} \end{cases}$$

Now, to write a definition of C_R as a boolean formula, we first have to represent its arguments, i.e., states and actions, in terms of logic variables. To this aim, we expand them to their binary memory representation.

Let suppose that states are n-bit values and actions are m-bit values. We write $s[i]$ and $a[i]$ to denote the ith bit of state s and action a,respectively.

Let x_i, $i = 1 \dots n$ and y_j, $j = 1 \dots m$ be $n + m$ boolean variables. A state s is then be represented by the formula

$$f_s(x_1, \dots, x_n) = \bigwedge_{i=1 \dots n} l_i \text{ where } l_i = \begin{cases} x_i & \text{if } s[i] = 1 \\ \bar{x}_i & \text{if } s[i] = 0 \end{cases}$$

Each f_s is a boolean formula in n variables that is true if and only if its variables are assigned with the bits of s (denoting, as usual, the boolean true with 1 and the boolean false with 0). In the same way, an action a corresponds to the formula

$$f_a(y_1, \dots, y_m) = \bigwedge_{i=1 \dots m} l_i \text{ where } l_i = \begin{cases} y_i & \text{if } a[i] = 1 \\ \bar{y}_i & \text{if } a[i] = 0 \end{cases}$$

Therefore, the controller characteristic function C_R can be encoded by the boolean formula

$$f_R(x_1, \dots, x_n, y_1, \dots, y_m) = \bigvee_{(s,a) \in R} (f_s \wedge f_a)$$

f_R is a boolean formula in $n + m$ variables that is true if and only if the variable assignment corresponds to the encoding of a (s, a) pair for which $R(s, a)$ holds.

For example, assume that the controller table contains the following 2-bit states $s = 00, s' = 01, s'' = 10$, with the following associated 1-bit actions $u = 0, u' = 0, u'' = 1$. Then the formula for the characteristic relation would be $f_R = \bar{x}_1 \cdot \bar{x}_2 \cdot \bar{y}_1 + \bar{x}_1 \cdot x_2 \cdot \bar{y}_1 + x_1 \cdot \bar{x}_2 \cdot y_1$.

```
BDD BDD_encoding(controller_table CTRL) {
 read number N of entries in CTRL;
 read number n of bits in each state of CTRL;
 read number m of bits in each action of CTRL;
 foreach j = 1...n create boolean variable x_j;
 foreach j = 1...m create boolean variable y_j;
 BDD f_R;
 foreach i = 1...N {
  BDD E, f_s, f_a;
  foreach j = 1...n //encode state bits
   if (bit(CTRL[i].state,j) == 1) f_s = f_s ∧ x_j;
   else f_s = f_s ∧ x̄_j;
  foreach j = 1...m //encode action bits
   if (bit(CTRL[i].action,j) == 1) f_a = f_a ∧ y_j;
   else f_a = f_a ∧ ȳ_j;
  E = f_s ∧ f_a;
  f_R = f_R ∨ E; //disjunction of entries
 }
 return f_R;}
```

Fig. 3. Algorithm for the logic encoding of numerical controllers

3.2 Algorithm for the Logic Encoding of Numerical Controllers

The encoding algorithm, whose pseudocode is shown in Figure 3, implements the technique described in the previous Section.

After reading the number of bits in the controller states and actions, the `BDD_encoding` procedure creates the corresponding set of boolean variables x_i and y_i, respectively.

Then, for each entry of the controller table, a new BDD E is created as the appropriate conjunction of the state and action variables. In particular, the code checks every bit in the state and adds to the BDD f_s a conjunction with the corresponding variable, in its positive or negated form, based on the value of the bit. The process is then repeated for the action bits in the BDD f_a, and finally E is obtained as $f_s \wedge f_a$.

The BDD E is then added to the final BDD f_R using a disjunction. When all the controller entries have been processed, `BDD_encoding` returns f_R.

As we can see, the algorithm is actually very simple, since all the BDD manipulation is done by the external BDD package. In particular, our implementation uses the CUDD [17] BDD manipulation package. Such package provides a large set of operations on BDDs and many variable dynamic reordering methods, which are crucial to gain the highest possible compression factor (Section 2).

Note that, to ensure the correctness of our approach, we also developed a parallel-query algorithm that tests for correctness and completeness the BDD-encoded controller f_R with respect to the original numerical controller $CTRL$. This procedure simply compares the results obtained by querying the uncompressed and the compressed controller with all the states in the controller table.

3.3 Querying the Encoded Controller

Once we have compressed the controller table in a BDD f_R, we must show how this information can be accessed by the controller itself to perform its task. There are several ways to do this.

A first method relies on the fact that the generated BDD actually encodes a function from states to actions. In other words, given a state, there is only one action associated to such state in the controller table. At the BDD level, this means that if we *restrict* the BDD by assigning to the variables $x_1 \ldots x_n$ the bits of a particular state s, then the resulting BDD has only one *satisfying assignment*, i.e., the one assigning to $y_1 \ldots y_m$ the bits of the corresponding action a.

Thus the action associated to a state can be found by applying two operations (restriction and then deduction of satisfying assignments) on the BDD of f_R. However, since controllers must work in the quickest and simplest possible way, we may consider an alternative query method that requires less runtime overhead.

From the BDD of f_R, we extract m different BDDs $f_R^1 \ldots f_R^m$, one for each bit of the action, in the following way.

$$f_R^i(x_1, \ldots, x_n) = \begin{array}{l} \exists\,(y_1 \ldots y_{i-1}, y_{i+1}, \ldots y_m) \\ f_R(x_1, \ldots, x_n, y_1 \ldots y_{i-1}, 1, y_{i+1}, \ldots y_m) \end{array}$$

In particular, in each f_R^i we existentially abstract all the action bit variables $y_j, j = 1 \ldots m$ except for y_i, which is assigned to the constant 1. The resulting formula f_R^i has n free variables $x_1 \ldots x_n$, which correspond to the state bits.

If we assign $x_1 \ldots x_n$ to the binary representation of a state s, then f_R^i is true if and only if there exists an action associated to s whose ith bit is 1. Since the action associated to s, if it exists, is unique, f_R^i actually returns the ith bit of the action associated to s (assuming, as usual, that the logical true and false correspond to the values 1 and 0, respectively).

Having these functions at hand, we can rebuild the binary representation of the action a associated to the state s by simply applying each f_R^i to the encoding of s, without any further runtime BDD manipulation. The execution overhead is minimal in this case, since the "computation" of the BDD value for a particular variable assignment requires only a visit of the associated graph.

3.4 BDD into C Code Translation

Embedding and querying the BDD-compressed numerical controller within a small hardware or software device is also an issue that can be addressed in various ways.

Obviously, we cannot require CUDD or another BDD-manipulation package to be present in the controller. However, a BDD (and in particular the set of BDDs obtained using the technique described in Section 3.3) can be simply translated in a C code fragment composed by nested if-then-else statements.

Indeed, let n be a node of the OBDD associated with the logic variable $V(n)$, and let $T(n)$ and $E(n)$ be the two children of n for $V(n) = \mathit{true}$ and $V(n) = \mathit{false}$, respectively. Then, we can define the C-translation of n as follows:

$$CT(n) = \begin{array}{l} \texttt{IF } (V(n)) \\ \texttt{THEN } CT(T(n)) \\ \texttt{ELSE } CT(E(n)) \end{array}$$

This translation process is very straightforward, requires only a visit of the OBDD graph and generates a block of code whose size is linear with respect to the OBDD size.

However, to perform this translation we need to process the OBDD structure directly in memory, without relying on the high level API of the CUDD package. Indeed, this API exposes a virtual view of the OBDD that hides most of the diagram optimizations which are the core of our compression methodology.

Moreover, the generated code may be still too complex to be handled by most of the common C compilers, in particular the `gcc` compiler [18] used in our experiments. Therefore we need to split the if-then-else expression in several sub-expressions and enclose them in separate auxiliary functions spread over different files. Then, a main function calls in turn each auxiliary function and aggregates their return values in the final result.

To obtain the best splitting strategy, we performed several experiments combining different splitting criteria and compiler optimization algorithms (i.e. the `-Ox` switch of `gcc`). The adopted heuristics combines the number of OBDD levels encoded by the if-then-else expression with the number of incoming edges of the corresponding sub-diagrams and splits the expression when the resulting value reaches a particular threshold.

The output of this process is a standard C library which exposes a function for each action bit. In particular these functions take as input an array of bits, corresponding to a state representation, and return the value of the corresponding action bit.

Moreover, this translation process can be adapted to encode the OBDD in different formats. For instance, we developed a variant of the algorithm above which is able to output a fragment of VHDL code [19] that defines the integrated circuits computing each action bit.

Indeed, the C library is suitable to embed the controller in a software application, whereas the VHDL code can be used to easily embed it in a (small) hardware device resulting in good time performances.

4 Experimental Results

To setup our experiments, we first have to fix some BDD encoding parameters, namely the variable ordering in the boolean formulas and the dynamic reordering method used by the BDD package.

Indeed, the BDD structure and therefore the compression ratio can be influenced by the original ordering of the variables in the boolean formulas presented to the BDD package. In particular, we recall that the variables in our BDDs are the state bit variables, namely $x_i, i = 1 \ldots n$, and the action bit variables, $y_i, i = 1 \ldots m$. Thus, we may consider the variable orderings arising from all the possible combinations of the following conditions:

- the state bit variables and the action bit variables can be ordered with different endianness, that is from the most significant bit to the least or vice-versa;
- the state bit variables can be placed before the action bit variables, after them or interleaved.

Namely, we can write the function f_R of Section 3 with any of the ten variable orderings $O1 \ldots O10$ shown in Table 1. Note that in $O9$ and $O10$ we assume $n > m$.

Table 1. Possible initial variable orderings

O1	$x_1 \cdots x_n y_1 \cdots y_m$	O6	$y_1 \cdots y_m x_n \cdots x_1$
O2	$x_1 \cdots x_n y_m \cdots y_1$	O7	$y_m \cdots y_1 x_1 \cdots x_n$
O3	$x_n \cdots x_1 y_1 \cdots y_m$	O8	$y_m \cdots y_1 x_n \cdots x_1$
O4	$x_n \cdots x_1 y_m \cdots y_1$	O9	$x_1 y_1 x_2 y_2 \cdots x_m y_m x_{m+1} \cdots x_n$
O5	$y_1 \cdots y_m x_1 \cdots x_n$	O10	$x_n, y_m x_{n-1} y_{m-1} \cdots x_{m-n} y_1 x_{m-n-1} \cdots x_1$

Moreover, variables can be dynamically reordered by the BDD package during the construction of the final BDD. In our experiments, we used the fourteen dynamic reordering methods $R1 \ldots R14$ offered by the CUDD package and shown in Table 2.

Table 2. CUDD variable reordering methods

R1	Random reordering	R8	Converging group sifting
R2	Random pivot reordering	R9	Window permutation (size 2)
R3	Sift	R10	Window permutation (size 3)
R4	Converging sifting	R11	Window permutation (size 4)
R5	Symmetric sifting	R12	Converging window permutation (size 2)
R6	Converging symmetric sifting	R13	Converging window permutation (size 3)
R7	Group sifting	R14	Converging window permutation (size 4)

In the following experiments we tried all the possible combinations between initial variable orderings and dynamic reordering methods, choosing each time the one that produces the better compression ratio.

All the experiments were performed on a 2.66GHz Pentium 4 with 1GB RAM.

4.1 Inverted Pendulum Controller

The first case study is the numerical controller for the inverted pendulum problem [2], where the controller has to bring the pendulum to equilibrium by applying a torque in the shaft.

The optimal numerical controller was generated using the dynamic programming method described in [2]. Table 3 shows the details of such controller. In the table, row *Entries* indicates the number of controller entries (i.e., controlled states) and row *Size*-column *Normal* indicates the controller table size in Kilobytes. Moreover, to compare the effectiveness of the BDD compression with respect to the common compression

Table 3. Numerical controller for the inverted pendulum

	Normal	LZ
Entries	311618	
Size	3043	1295 (42.6%)
Time	8 ms	657 ms

Table 4. Number of nodes in the BDD for the inverted pendulum controller with respect to different variable orderings

	O1	O2	O3	O4	O5	O6	O7	O8	O9	O10
R1	92490	108474	65382	64012	156872	83808	149943	86957	92013	145877
R2	142012	145654	66711	78029	130315	79575	144583	89528	148037	142432
R3	65842	61588	65842	61584	138377	75360	145181	143181	65854	127241
R4	65842	71314	70691	61584	130355	71313	75427	135580	65842	65843
R5	65842	65848	70774	70770	144310	70773	144422	70776	65865	142677
R6	65842	65860	70777	70770	135575	70773	135573	70773	65860	130348
R7	66759	66555	92830	61798	144268	61586	136870	61594	69198	65865
R8	65842	61594	75420	61584	145172	61586	135575	61586	65868	65863
R9	92295	92279	65441	61873	163149	121083	255309	121307	101164	178482
R10	76724	77231	**61584**	61588	134758	122539	143860	78044	79227	142072
R11	72724	72781	61584	61584	134023	122157	134254	106597	77295	134723
R12	81329	98992	61630	61641	134162	128731	138481	79723	86284	136761
R13	77034	65868	61584	61584	134023	123547	134275	111460	66122	134257
R14	65842	65868	61584	61584	134275	77960	134275	116568	68455	134079

Table 5. BDD-compressed numerical controller for the inverted pendulum

	BDD
Entries	311618
Size	489 (16.1%)
Time	1 ms

techniques, we also show the size of the controller table compressed by the LZ77-based [11] algorithm of gzip [20] (row *Size*-column *LZ*, where the relative size of the compressed file is also shown as a percentage) and the average controller access time in milliseconds for both the uncompressed and the compressed representations (row *Time*-column *normal* and row *Time*-column *LZ*, respectively). Note that, as we expect, access times for the LZ-compressed controller are very high since, in the worst case, the controller must be completely decompressed to find the required table entry.

When BDD compression is applied to the controller, we obtain BDDs whose size (in number of graph nodes) is shown in Table 4 for each combination of variable initial ordering ($O1 \ldots O10$) and dynamic reordering ($R1 \ldots R14$).

Here, the smallest BDD (indicated by the bold number in Table 4) has 61584 nodes. The actual size in Kilobytes and the average access time for the BDD-compressed controller are shown in Table 5. As we can see, BDD compression reduces the controller 26.5% more than LZ, and has also better access times than the LZ-compressed version, since it does not require any decompression to read the table entries.

4.2 Truck and Trailer Obstacles Avoiding Controller

In the second case study, we consider the controller for the *truck and trailer obstacles avoiding* problem. Namely, the controller has to back a truck with a trailer up to a

Table 6. Results for the truck and trailer obstacles avoiding controller

	Normal	LZ	BDD
Entries	3256855		
Size	71650	22644 (31.6%)	7038 (9.8%)
Time	89 ms	3173 ms	108 ms

specified parking place starting from any initial position in the parking lot. Moreover, the parking lot contains some obstacles, which have to be avoided by the truck while maneuvering to reach the parking place. Corrective maneuvers are not allowed, that is the truck cannot move forward to *backtrack* from an erroneous move.

The numerical controller was generated with the CGMurφ tool [21, 5]. Results are in Table 6. As we can see, the controller has a very big size. However, the best BDD compression scheme ($O5$,$R5$) is able to reduce the size of the controller up to a 90.2% space savings, that is 21.8% more than using LZ77 compression. Moreover, the BDD compression continues to win also with respect to the access times.

4.3 Inverted Pendulum on a Cart Controller

The last case study is the numerical controller for the inverted pendulum on a cart problem [22]. The system consists of a planar inverted pendulum on a cart that moves under an applied horizontal force, constituting the control. The position of the pendulum is measured relative to the position of the cart as the offset angle from the vertical upright position. The controller goal is to set such angle to zero.

Table 7. Results for the inverted pendulum on a cart controller

	Normal	LZ	BDD
Entries	151394		
Size	1478	90 (6.1%)	215 (14.6%)
Time	3 ms	206 ms	1 ms

The numerical controller was generated with the CGMurφ tool. Results are in Table 7. The number of controller entries is very small with respect to the previous two case studies. We see that on a small controller the BDD compression has a lower compression ratio than LZ, but always better access times (1ms vs. 206ms), since it does not require any decompression to read the table entries.

5 Conclusions

In this paper, we presented an OBDD based compression algorithm for numerical controllers. The compression algorithm is completely automatic and can be applied to the (state,action) tables generated by any numerical controller generation technique.

Our experiments show that this new algorithm has a very high compression ratio (up to 10:1), that is often more than the ratio obtained on the same data by the most common lossless compression techniques, such as LZ77. Indeed, OBDD compression is not actually lossless, but rather "relation-invariant". That is, the OBDD compression leaves intact the behavior of the state-action relation stored in the table. However, by working on the logic encoding of the relation, the OBDD is capable of *optimizing the representation of the relation*, so reducing its size.

Moreover, accessing the entries of an OBDD-compressed numerical controller does not require any data decompression (as it would happen, e.g., with LZ77), so the controller performances are very good. Also in this case, the optimized representation generated by the OBDD sometimes allows to achieve access times that are even better that those of an hash function used on the uncompressed table.

Therefore, the BDD compression is a technique that can be actually exploited to reduce the size of numerical controllers, generating a compact structure that is easy to store and query. This would allow, e.g., to use *high precision controllers* even in *limited devices*.

Indeed, the OBDD-to-C and OBDD-to-VHDL translations sketched in Section 3.4 complete our methodology offering a completely automatic process to generate, from any numerical controller, a small executable definition ready to be embedded in software applications or hardware devices.

As a further step of our research, we intend to investigate the relationship between the compression obtained with the use of the OBDDs and the reduction of fuzzy control systems by means of a *hierarchical* approach (see e.g. [23]). Indeed, there is an evident correspondence between, on one hand, the search of the best ordering of variables needed for the OBDD compression and, on the other, the hierarchical decomposition into subsystems of a fuzzy system.

References

1. Leu, M.C., Kim, T.Q.: Cell mapping based fuzzy control of car parking. In: ICRA, pp. 2494–2499 (1998)
2. Kreisselmeier, G., Birkholzer, T.: Numerical nonlinear regulator design. IEEE Transactions on Automatic Control 39(1), 33–46 (1994)
3. Grüne, L., Junge, O.: A set oriented approach to optimal feedback stabilization. Systems Control Lett. 54(2), 169–180 (2005)
4. Della Penna, G., Intrigila, B., Magazzeni, D., Melatti, I., Tofani, A., Tronci, E.: Automatic generation of optimal controllers through model checking techniques. In: Proceedings of 3rd International Conference on Informatics in Control, Automation and Robotics, ICINCO2006 (2006) (to be published in Informatics in Control, Automation and Robotics III), `http://www.di.univaq.it/magazzeni/cgmurphi.php`
5. Della Penna, G., Magazzeni, D., Tofani, A., Intrigila, B., Melatti, I., Tronci, E.: Automatic synthesis of robust numerical controllers. In: ICAS 2007, p. 4. IEEE Computer Society, Los Alamitos (2007)
6. Kosko, B.: Neural Networks and Fuzzy Systems. Prentice-Hall, Englewood Cliffs (1992)
7. Sekine, S., Imasaki, N., Tsunekazu, E.: Application of fuzzy neural network control to automatic train operation and tuning of its control rules. In: Proc. IEEE Int. Conf. on Fuzzy Systems 1993, Yokohama, pp. 1741–1746 (1995)

8. Della Penna, G., Fallucchi, F., Intrigila, B., Magazzeni, D.: A genetic approach to the automatic generation of fuzzy control systems from numerical controllers. In: Basili, R., Pazienza, M.T. (eds.) AI*IA 2007. LNCS (LNAI), vol. 4733, pp. 230–241. Springer, Heidelberg (2007)
9. Cover, T.M., Thomas, J.A.: Elements of Information Theory. Wiley, Chichester (2006)
10. Nelson, M., Gailly, J.-l.: The Data Compression Book. MT Books (1995)
11. Ziv, J., Lempel, A.: A universal algorithm for sequential data compression. IEEE Transactions on Information Theory 23(3), 337–343 (1977)
12. Bryant, R.: Graph-based algorithms for boolean function manipulation. IEEE Trans. on Computers C-35(8), 677–691 (1986)
13. Hand, D.J., Mannila, H., Smyth, P.: Principles of Data Mining. MIT Press, Cambridge (2001)
14. Burch, J.R., Clarke, E.M., McMillan, K.L., Dill, D.L., Hwang, L.J.: Symbolic model checking: 10^{20} states and beyond. Inf. Comput. 98(2), 142–170 (1992)
15. Clarke, E.M., Grumberg, O., Peled, D.A.: Model Checking. MIT Press, Cambridge (1999)
16. Cormen, T.H., Leiserson, C.D., Rivest, R.L., Stein, C.: Introduction to Algorithms. MIT Press, Cambridge (2001)
17. CUDD Web Page (2007), `http://vlsi.colorado.edu/~fabio/`
18. GCC Compiler, `http://gcc.gnu.org/`
19. Pedroni, V.A.: Circuit Design with VHDL. MIT Press, Cambridge (2004)
20. GZip Web Page (2007), `http://www.gzip.org/`
21. CGMurphi Web Page (2006), `http://www.di.univaq.it/magazzeni/cgmurphi.php`
22. Junge, O., Osinga, H.: A set oriented approach to global optimal control. ESAIM Control Optim. Calc. Var. 10(2), 259–270 (2004)
23. Stufflebeam, J., Prasad, N.R.: Hierarchical fuzzy control. In: Proceedings of IEEE International Fuzzy Systems Conference, pp. 498–503 (1999)

Part II

Robotics and Automation

Dynamic-Based Simulation for Humanoid Robot Walking Using Walking Support System

Aiman Musa M. Omer[1], Yu Ogura[1], Hideki Kondo[1], Hun-ok Lim[2], and Atsuo Takanishi[3]

[1] Graduate School of Science and Engineering, Waseda University, Tokyo, Japan
aiman@fuji.waseda.jp
[2] Department of Mechanical Engineering, Kanagawa University, Yokohama, Japan
holim@ieee.org
[3] Department of Mechanical Engineering / Humanoid Robotics Institute
Waseda University, Tokyo, Japan
takanisi@waseda.jp

Abstract. A new humanoid bipedal robot WABIAN-2R was developed to simulate human motion. WABAIN-2R is able to perform similar human-like walking motion. Moreover, the robot is able to perform walking motions with a passive walk-assist machine. However, walking with an active walk-assist machine is expected to be unstable. Conducting this experiment is highly risky and costly. Therefore, we had developed a dynamic simulator in order to test walking robot with walk-assist machine before conducting it in real simulation.

Keywords: Dynamic simulation, Humanoid robot, Biped robot, Walking assist machine.

1 Introduction

With the rapid aging of society in recent times, the number of people with limb disabilities is increasing. According to the research by the Health, Labour and Welfare Ministry, Japan, there are around 1,749,000 people with limb disabilities; this accounts for more than half of the total number of disabled people (3,245,000 handicapped people)[1]. The majority of these people suffer from lower-limb disabilities. Therefore, the demands for establishing a human walking model that can be adapted to clinical medical treatment are increasing. Moreover, this model is required for facilitating the development of rehabilitation and medical welfare instruments such as walking machines for assistance or training (Fig. 1(a)). However, experiments that are carried out to estimate the effectiveness of such machines by the elderly or handicapped could result in serious bodily injury.

Many research groups have been studying biped humanoid robots in order to realize the robots that can coexist with humans and perform a variety of tasks. For examples, a research group of HONDA has developed the humanoid robots—P2, P3, and ASIMO[2]. The Japanese National Institute of Advanced Industrial Science and Technology (AIST) and Kawada Industries, Inc. have developed HRP-2P. The University of Tokyo developed H6 and H7, and the Technical University of Munich developed Johnnie. Waseda University developed the WABIAN series that realized various walking motions

J.A. Cetto et al. (eds.): Informatics in Control, Automation and Robotics, LNEE 37, pp. 91–100.
springerlink.com

by using moment compensation. Korea Advanced Institute of Science and Technology (KAIST) also developed a 41-DOF humanoid robot— KHR-2 [3].

The above mentioned human-size biped robots achieved dynamic walking. If these humanoid robots can use rehabilitation or welfare instruments (as shown in Fig. 1(b)), they will be able to help in testing such instruments quantitatively. The main advantages of the human simulator can be considered to be as follows: (1) The measurement of the angle and the torque required at each joint can be measured easily and quantitatively as compared to the corresponding values in the case of a human measurement. (2) Experiments using such robots can help identify leg defects of a human from an engineering point of view. (3) A robot can replace humans as experimental subjects in various dangerous situations: experiments involving the possibility of falling, tests with incomplete prototype instruments, simulations of paralytic walks with temporarily locked joints.

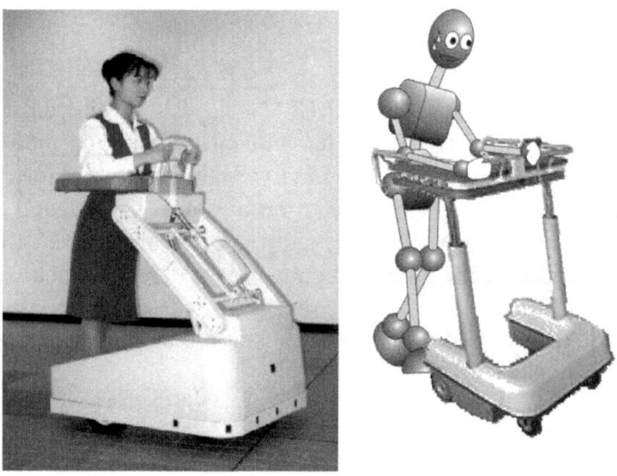

Fig. 1. Walk-assist machine used by Human and Robot

Such experiments require a humanoid robot that enables it to closely replicate a human. However, humans have more redundant DOFs than conventional biped humanoid robots; this feature enables them to achieve various motions. Therefore, a DOF configuration that is necessary to reproduce such motions is one of the very important issues in the development of a humanoid robot [4].

The Waseda Bipedal Humanoid Robot WABIAN-2R has been developed to simulate human motion. WABIAN-2R performed human-like walking motions (Fig. 2). Moreover, WABIAN-2R achieved to perform walking motion using walk-assist machine. However, the walk-assist machine was freely rolling without activating its wheels motors. In this case, the robot faced the minimum resistance or disturbance case by the walk-assist machine. On the other hand, activating the walk-assist machine may create a large disturbance for robot due to separate control for each of them. Conducting this experiment may be highly risky.

As we develop humanoid robot to coexist in the human environment, we need to conduct many experiments such as robot walking on uneven surface, climbing the stairs, and robot interact with other machine and instruments. Doing any new type of experiment using WABIAN-2 might be risky. Therefore, we need find a safer method for initial experimental testing. Using a dynamic simulation is useful method due to some reasons such as: (1) It is safer in terms of cost and risk. (2) It is easy to monitor and view motion outputs. (3) It can show the variation cased by any external disturbances. In this paper, a dynamic simulator is described, which is able to easily simulate any new type of walking. Using the dynamic simulator, we can monitor the motion performance and output all needed data that is useful for further development. This paper is aimed to simulate the walking motions of WABIAN-2 using walk-assist machine.

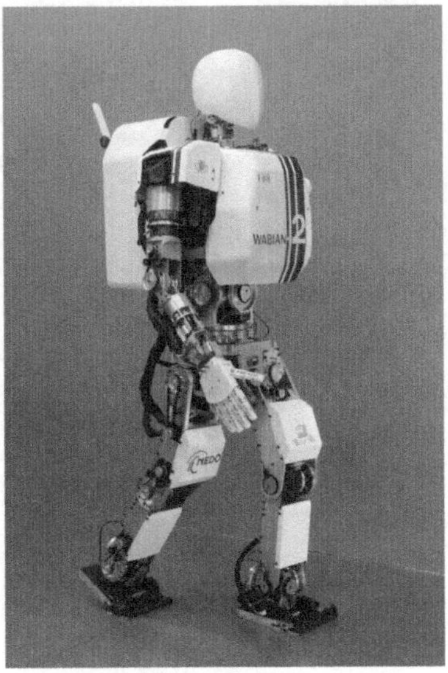

Fig. 2. WABIAN-2R

2 Dynamic Simulation

Dynamic simulation could be used to simulate the dynamic motion of a mechanical structured model. It can analyze the effects of the surrounding environment on the mechanisms and objects. In robotics researches, simulation software are used for robotic simulation. There are many software used for robotics simulation in different applications. Most of those software are for industrial robot applications. However, there are some software used for mobile robot simulation. For examples, RoboWorks, SD/FAST, OpenHRP, and Yobotics are used for mobile and legged robot simulation. Webots is high and advanced simulation software used in Robotics simulation. It is use for prototyping and simulation of mobile robots. It has many advanced functions and techniques. Webots is very easy to use and implement. Therefore, we choose it as simulation software [5].

2.1 Modeling

In order to develop a dynamic simulation, we need to go through several steps. First is modeling where we set up the simulation environment and initial parameters. We set up a full structure of WABIAN-2, based on the specifications (size, shape, mass distribution, friction, .etc) of components of WABIAN-2 (Fig. 3).

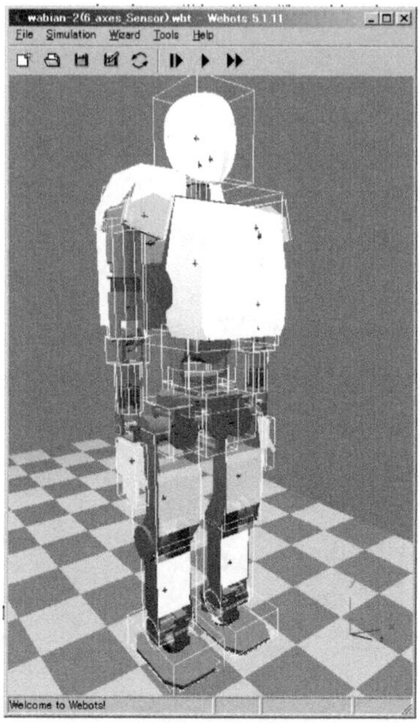

Fig. 3. Modeled WABIAN-2R in the simulation world

2.2 Controlling

Second is controlling, which identifies simulation objects and controls the simulation procedures. The controller is some how similar to the WABIAN-2R control. It gets the input data from the CSV pattern file, and sets the position angle of each joint through inverse kinematics techniques. Moreover, the controller sets the simulate time step and the measurement of data.

2.3 Running

Lastly is the running of the simulation and checking the dynamic motion. We can view the simulation from different view sides which gives us a clear idea about the simulation performance. Moreover, most of the needed data could be measured through several functions.

3 Walking with Walking Assist Machine

WABIAN-2 performed some walking experiments using walking assist machine. The performance was conducted by leaning its arms on the walking assist machine holder. The walking assist machine moves passively without generating its own motion. The robot was able to walk and push the walking assist machine forward. The experiments were conducted with different walking styles and different heights of arm rest (Fig. 4).

The walking performance of WABIAN-2 using an active walking assist machine, expected to be unstable. The walk-assist machine has its own control system, not connected to WABIAN-2 control system. The walking assist machine moves with constant velocity in a forward direction, while the robot moves by setting its position. The robot arms may displace from its position on the arm rest of the machine which will case external forces on WABIAN-2. In order to stabilize the walking, the external force has to be minimized.

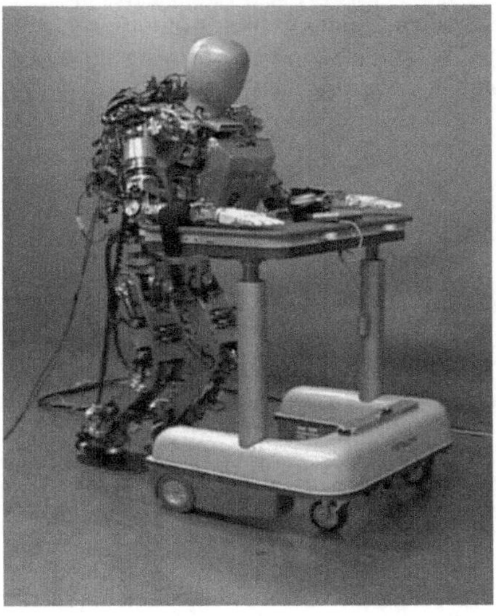

Fig. 4. WABIAN-2 with Walking Assist Machine

3.1 Force Sensor

The real walking assist machine is developed to sense the force applied by the load on the arm rest. A force sensor is attached on the top of the arm rest consisting of four displacement sensors. The displacement sensor is simply a spring mechanism. It senses forward and vertical forces and turns toque by determining relative displacements between the upper frame and the lower one (Fig. 5). We can develop the system that can adjust the velocity of the walking assist machine in order to minimize the displacement.

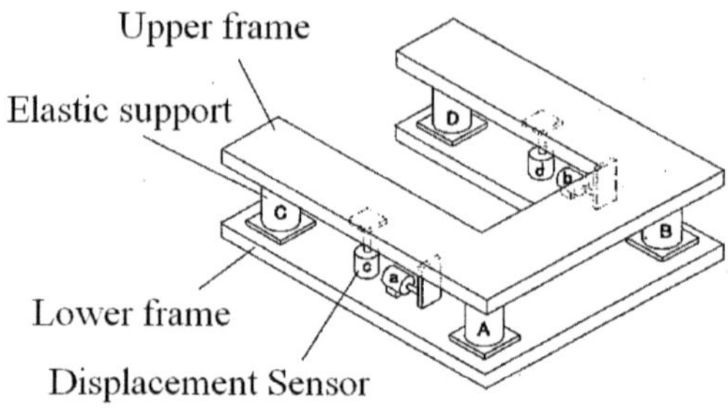

Fig. 5. Force Sensor

3.2 Velocity Control

There were some developments made on the walking assist machine control system to adjust its speed according to the force applied on the arm rest [7]. The arm rest is designed to measure the force and torques applied by the user of the machine (Fig. 6). The controller uses those measure data as an input data to set the velocity of each motor of the machine (Fig. 7). The force f_y and the turning moment m which applied by the arm of the user is calculated in the sensor by the following equations:

$$m_z = m + s_x f_y \tag{1}$$

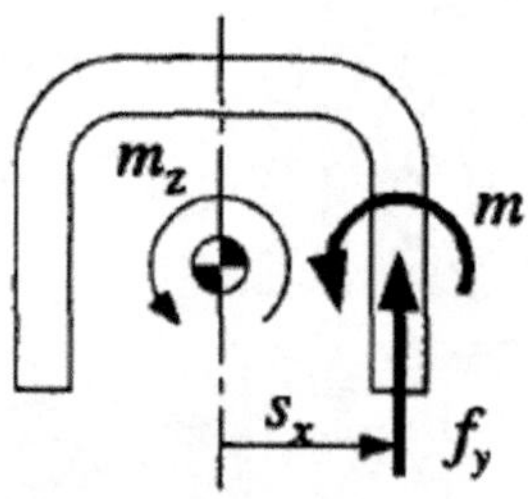

Fig 6. Force and Moment Applied to Arm rest

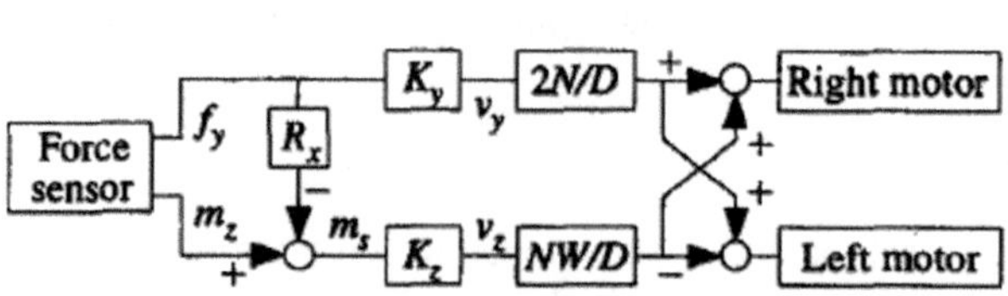

Fig. 7. Block Diagram for Control System

where m_z is the moment measured by the sensor, s_x the distance shifted from the arm position to the sensor position. The values for m_z and f_y are the input data for the controller that set the velocity of each wheel motor [7].

In this study, we have introduced a new control system model that controls the velocity of the walking assist machine. The system adjusts the velocity according to the force measured by the force sensor. The new adjusted velocity is based on current velocity and the displacement with WABIAN-2.

Developing the equations of the modeled system, we can have the following equation:

$$F_y = ma \tag{2}$$

where m is the total mass of the walking assist machine, a is the acceleration, and F_y is the force measured by the spring. The force is the result of displacement of the spring mechanism, which can be expressed as

$$F_y = C\mathrm{x} \tag{3}$$

where C is the spring constant and x is the amount of displacement. Substitute equation (3) in (2), we will have

$$a = (C/m)\ \mathrm{x} \tag{4}$$

the acceleration is the derivative of velocity. Approximately, it is equal to the difference in velocity over step, which could be express as

$$\mathrm{a(t) = (v\ (t + \Delta\ t) - v\ (t))\ /\ \Delta t} \tag{5}$$

since we are dealing with discrete time, we can rearrange equation (5) to

$$\mathrm{a(k) = (v(k+1) - v(k))\ /\ T} \tag{6}$$

where v(k) is the current velocity, v(k+1) is the next velocity, and T is the step time. Substitute equation (4) in (6), we will have

$$\mathrm{v\ (k+1)} = (C\ \mathrm{T}\ /m)\ \mathrm{x(k) + v(k)} \tag{7}$$

where x(k) refer to the displacement measured by the spring of the sensor. This equation represents the velocity control process in the system.

4 Simulation Result

We test several types of motions performed by WABIAN-2. The simulator simulates the walking performance of conventional walking and stretch walking (Fig. 8). Moreover, it simulates some other motions as the input pattern. The dynamic simulation has given us a simulation motion just like the real simulation. We monitor the simulation from different viewpoints. Moreover, we could measure some output data.

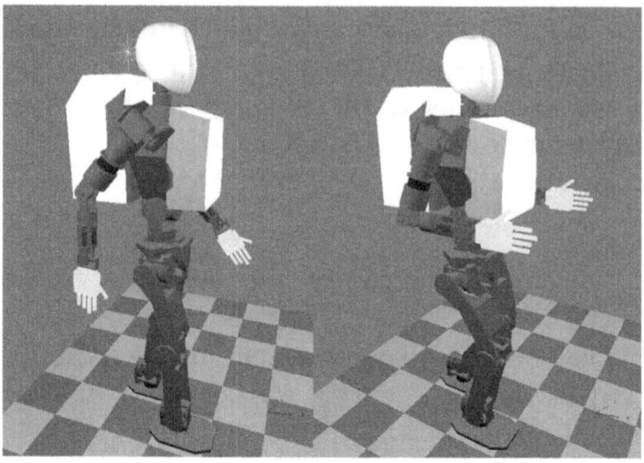

Fig. 8. Simulation of different type of walking

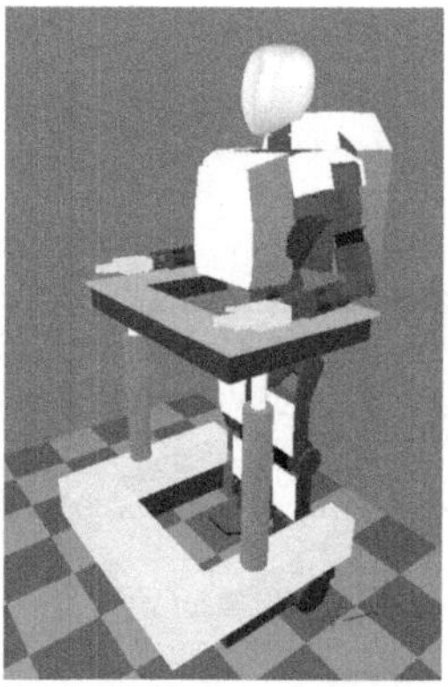

Fig. 9. Simulation of walking with walk-assist machine

We conducted some simulation experiments of walking using the walking assist machine. The robot is able to walk stably with a passive walking assist machine just like the real experiment (Fig. 9). But it was not possible to achieve the same result when we conducted the experiment using an active walking assist. As expected, the robot was affected by the external force produced by its contact with the walking assist machine. The robot became unstable during its walking, and in some experiments it fell down (Fig. 10).

By adding the force sensor to the simulated walking assist machine, we were able to measure the amount of external force acting on the robot. Using these measurements with the velocity control we had developed, the robot could walk with the active walking assist machine. The amount of holding torque we set to the walking assist machine wheels could increase from 0.5 N.m to 0.75 N.m by using this new velocity control in the control system of the dynamic simulator we had developed (Fig. 11).

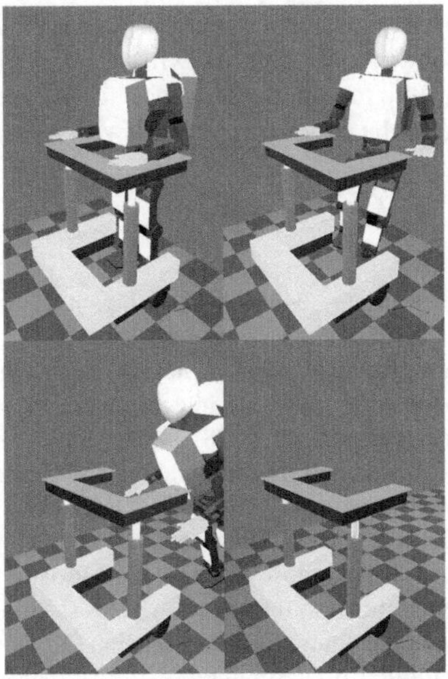

Fig. 10. Simulation of walking with active walking machine

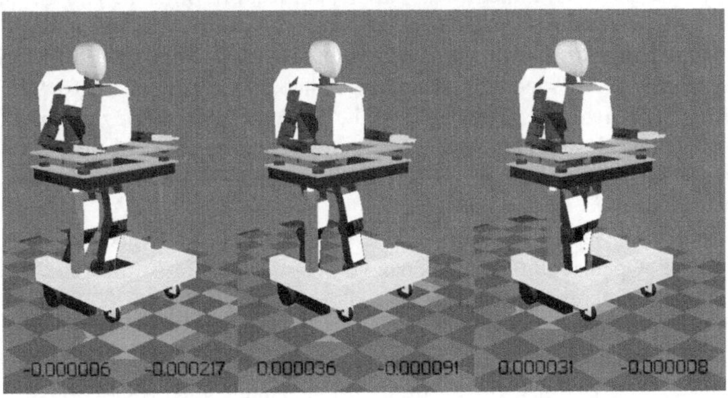

Fig. 11. Simulation of walking with the walking machine using velocity control

5 Conclusions and Future Work

This paper describes the simulation of walking by WABIAN-2R with the walking assist machine. The dynamic simulation is very important to check the motion of any new pattern generated. Using the dynamic simulation we can see the effect of the walking assist on WABIAN-2R. As expected, the walking was unstable due to the effect of external forces created from the arm rest. By using the velocity control in the control system of the simulation, the robot is able to walk stably with the walking assist machine.

In the near future, it is important to develop WABIAN-2R system to be stabilized during walking. The stabilization control will be based on Zero Moment Point. Moreover, it is necessary to develop the robot to interact with other objects and equipments. This will make the robot can interact with its surrounding environment.

References

1. Health, Labour and Welfare Ministry of Japan, `http://www.mhlw.go.jp/english/wp/wphw/vol1/p2c4s2.h-tml`
2. Sakagami, Y., Watanabe, R., Aoyama, C., Matsunaga, S., Higaki, N., Fujimura, K.: The intelligent ASIMO: System overview and integration. In: Proc. IEEE/RSJ Int. Conference on Intelligent Robots and Systems, pp. 2478–2483 (2002)
3. Omer, A.M.M., Ogura, Y., Kondo, H., Morishima, A., Carbone, G., Ceccarelli, M., Lim, H.-o., Takanishi, A.: Development of A Humanoid Robot Having 2-DOF Waist and 2-DOF Trunk. In: Humanoid 2005, Tsukuba (December 2005)
4. Ogura, Y., Aikawa, H., Shimomura, K., Kondo, H., Morishima, A., Lim, H.-o., Takanishi, A.: Development of a New Humanoid Robot WABIAN-2. In: Proceedings of the 2006 IEEE International Conference on Robotics and Automation Orlando, Florida (May 2006)
5. Webots, Commercial Mobile Robot Simulation Software, http://www.cyberbotics.com
6. Mojon, S.: Realization of a Physic Simulation for a Biped Robot. Semester Project at BIRG laboratory Swiss Federal Institute of Technology (Summer 2003)
7. Egawa, S., Nemoto, Y., Fujie, M.G., Koseki, A., Hattori, S., Ishii, T., Egawa, S., Nemoto, Y., Fujie, M.G.: Power-Assisted Walking Support System With Imbalance Compensation Control For Hemiplegics. In: Proceedings of The Rrst Joint BMES/EMBS Conference Serving Humanity, Advancing Technology, Athn$, GA, USA, o& 1&16, 99
8. Egawa, S., Takeuchi, I., Koseki, A., Ishi, T.: Force-sensing Device for Power-assisted Walking Support System. In: System Integration Conference (December 2002)
9. Klopsteg, P.E., Wilson, P.D., et al.: Human Limbs and Their Substitutes, New York Hafner (1963)
10. Kanehiro, F., Fujiwara, K., Kajita, S., Yokoi, K., Kaneko, K., Hirukawa, H., Nakamura, Y., Yamane, K.: Open architecture humanoid robotics platform. In: Proceedings of IEEE International Conference on Robotics and Automation, ICRA 2002, May 11-15, vol. 1 (2002)
11. Sardain, P., Bessonnet, G.: Force Acting on a Biped Robot Center of Pressure-Zero Moment Point. IEEE Transactions On Systrems, Man, And Cybernetics-Part A: Systems And Huamns 34(5) (September 2004)

People Tracking and Identification Using Laser Features and Colour Distributions

Andreas Kräußling, Bernd Brüggemann, and Dirk Schulz

Research Establishment of Applied Sciences (FGAN)
Department of Communication, Information Processing and Ergonomics (FKIE)
Bad Neuenahrer Straße 20, 53343 Wachtberg, Germany
{a.kraeussling,brueggemann,schulz}@fgan.de
http://www.fgan.de

Abstract. Tracking multiple crossing people is a great challenge, since common algorithms tend to loose some of the persons or to interchange their identities when they get close to each other and split up again. In several consecutive papers it was possible to develop an algorithm using data from laser range scanners which is able to track an arbitrary number of crossing people without any loss of track. In this paper we address the problem of rediscovering the identities of the persons after a crossing. Therefore, a system of two cameras is used. An infrared camera detects the people in the observation area and then a charge–coupled device (CCD) camera is used to extract the colour information about those people. The colour information is represented by colour histograms, which are computed within the HSV colour space. Before the crossing the system learns the parameters of a Dirichlet distribution for each person. After the crossing the system relocates the identities by comparing the actually measured colour distributions with the distributions, which have been learnt before the crossing. The most probably assignment of the identities is then found using Munkres' Hungarian algorithm. It is demonstrated using data from real world experiments that our approach can reliably reassign the identities of the tracked persons after a crossing.

Keywords. People tracking, Crossing targets, Laser range scanners, Vision, HSV colour space, Dirichlet distribution.

1 Introduction

Multirobot systems and service robots need cooperative capabilities to some extend. This holds for robots that are supposed to move in a certain formation or the interaction with people or robots. Apparently, there has to be knowledge about the objects we wish to interact with. More precisely, we want to know where these objects are. Thus, we need to track their positions.

Target tracking deals with the state estimation of one or more objects. This is a well studied problem in the field of aerial surveillance with radar devices [1] as well as in mobile robotics [2, 3, 4, 5, 6, 7]. Because of their high accuracy and relatively low price, laser range scanners are standard in our days for mobile robotic applications such as collision avoidance, navigation, localisation and map building [8, 9].

J.A. Cetto et al. (eds.): Informatics in Control, Automation and Robotics, LNEE 37, pp. 101–112.
springerlink.com

Due to the high resolution of these sensors, one target is usually the source of multiple returns of a laser scan. This conflicts with the assumption of punctiform targets like in the field of radar tracking, where each target is the origin of exactly one measurement. Therefore, one needs to be able to assign the obtained measurements to extended targets.

Tracking extended targets has similarities to tracking punctiform objects in clutter, which denotes false alarms nearby a target that do not originate from a target. We observe multiple readings which can be associated with the same target for punctiform objects in clutter as well as for extended targets.

A very important characteristic of tracking in the field of mobile robotics is the occurrence of crossing or interacting people. This means that two or more persons get very close to each other, so that they cannot be separated by common tracking algorithms [10, 11, 12, 13, 14]. This situation is the central topic of this work and can appear e.g. when some humans meet, talk to each other and split up again. It is a well known problem in mobile robotics [2, 15]. There are several methods for tracking punctiform crossing targets in clutter:

1. the MHT (Multi Hypothesis Tracker) introduced by Reid in 1979 [16].
2. the JPDAF (Joint Probabilistic Data Association Filter) introduced by Fortmann, Bar–Shalom and Scheffe in 1983 [10].

Of course, an extension of these algorithms for tracking extended crossing objects is straightforward. But there are several reasons, why such approaches might fail:

- in most cases there are several measurements from the same target.
- the crossing can last for a longer time period.
- one of the objects might be occluded by the other objects for some time.
- the objects can accomplish very abrupt manoeuvres during the crossing or especially at the end of the crossing.

These difficulties are well known in the mobile robotics community:

- Tracking moving objects whose trajectories cross each other is a very general problem ... Problems of this type cannot be eliminated even by more sophisticated methods ... [2].
- Tracks are lost when people walk too closely together ... [15].

Due to these circumstances our research group has tried successfully to develop methods which can track crossing people without loss of track. In [17] we proposed a first method which can track two interacting persons without loss of track. In [11, 13] an improved method for two persons has been introduced which is faster and more accurate than the method proposed in [17]. Finally, in [18] we were able to present a method which can track an arbitrary number of people without loss of track. All those methods use the Viterbi algorithm [19, 20] and some geometrical properties, that characterise the problem.

Since these results we have concentrated our efforts on the problem of the reassignment of the identities of the persons after a crossing. Above all there are four approaches to deal with this problem:

1. different colours and surface–textures of the pairs of trousers people wear might result in different intensities of the reflected laser beams.
2. people could wear badges sending out for instance infrared or ultrasound signals which can be received by specific sensors to identify the tracked people [21].
3. usually people wear clothing with different colours. This information could be exploited for the identification of the people using a camera network.
4. people do often differ in the size of their bodies. These differences could be used for the identification as it has been proposed in [22].

Since methods 2 and 4 already have been investigated we concentrated our work on method 1 and 3. Thereby, it has emerged that both methods are feasible. We will investigate method 3 in this work and show how this approach can be used to identify the persons correctly after a crossing. Furthermore we have combined this approach and the tracking algorithm proposed in [18] to develop an algorithm that can track an arbitrary number of interacting humans without any loss of track which is able to identify the subjects correctly after an interaction.

The remainder of this paper is organised as follows. In section 2 the combined method for tracking multiple interacting persons is described. It consists of a tracking algorithm based on the Viterbi algorithm and an additional clustering routine and an identity reassignment method which exploits colour information and uses the Hungarian algorithm. In section 3 the experiments are given. They verify, that our method can track interacting humans without loss of track and without mixing the identities of the persons after a crossing. In section 4 we conclude.

2 The Method

There are two problems that have to be solved when tracking multiple interacting persons: they have to be tracked without any loss of track and they have to be identified when they split up again after a crossing. For the solution of the first problem laser range scanners are used. They measure the distance to the next reflecting object. These data are then processed with a so called Cluster Sorting algorithm. Thus, it is possible to estimate the positions of the persons without any loss of track. A possible solution of the second problem using cameras and colour information will be described later.

2.1 The Cluster Sorting Algorithm

The Cluster Sorting algorithm uses a hidden Gauß–Markov chain. Thus, the Kalman filter can be applied for calculating the position estimates. Thereby, a major problem is the discrimination of these measurements which are from the actually considered person from those which are from other persons or obstacles like walls. For this purpose a validation gate is calculated [1]. In doing so the distinction from other persons is only possible as long as there is no crossing. In this work we apply the following two ways to make use of the measurements in the validation gate, which we call selected measurements. The first way is to calculate an unweighted mean of these measurements and to use this mean in the update equation of the Kalman filter. The corresponding algorithm is called Kalman filter algorithm (KFA). As has been shown in [12, 13, 14] it is very fast and gives good information about the position of the target, but cannot reproduce multimodal probability distributions. Thus it is not able to handle multiple interacting people.

The second approach is based on the Viterbi algorithm. It calculates for every old selected measurement a separate position estimate and validation gate. The new selected measurements are then these measurements, which lie in at least one of those validation gates. The corresponding algorithm is called Viterbi based algorithm (VBA) and has been introduced in [23]. By this procedure it is ensured, that the associated tracking algorithm can deal with multimodal probability distributions to some extend, which is a major improvement when dealing with multiple interacting targets. This is due to the following behaviour of the VBA. When the crossing people split up again, the VBA starts to calculate for each person a cloud of position estimates, so that no information about the persons gets lost. In contrast, common tracking algorithms make a hard decision which person they track. When they are deterministic, they associate all tracks with the path of the same person, so that the information about all other humans gets lost. When they have a random component like the SJPDAF [3], it occasionally happens, that the algorithm chooses for each track the path of a different person, so that no persons get lost [24]. But this behaviour is not stable so that such approaches are not sufficient for tracking multiple interacting humans. Because the VBA provides the whole information about all the persons after a crossing, one just has to reassign the clusters of position estimates to the persons. This problem is solved by the Cluster Sorting Algorithm, which is proposed in the following. The disadvantages of the VBA are a great computational burden and a poor information about the position of the tracked objects [12, 13, 14], so that it should not be used for tracking single humans.

The Cluster Sorting algorithm uses two classes of objects:

- single targets.
- clusters, which represent at least two interacting persons, i.e. humans that are moving very close to each other.

Single targets are tracked with the KFA, since there is no need for representing multimodal probability distributions. Furthermore, this algorithm is very fast and gives very accurate information about the position of the tracked object.

Clusters are tracked with the VBA, since multimodal distributions have to be represented. This approach guarantees that none of the objects that are associated with the cluster is lost. This fact is important especially when the objects split and start to move separately again.

Three different events have to be regarded when tracking multiple interacting people:

1. The merging of two single persons. This means that two single targets get very close to each other. This is the case, if at least one measurement is located in the validation gates of both targets. Then the algorithm stops to track the two single targets with the KFA and starts tracking a cluster, which contains both targets, using the VBA. Therefore, it uses the measurements located in the validation gates of at least one target.
2. The merging of a single human and a cluster. This means that a single person and a cluster get very close to each other. This happens, if at least one measurement is located in the validation gates of the person and the cluster. In this case the

algorithm stops to track the single human and the cluster separately. Instead it starts tracking a combined cluster. Therefore, it uses the measurements located in the validation gates of either the single target or the previously considered cluster or both.

3. The merging of two clusters. This means that two clusters get very close to each other. This is the case, if at least one measurement is located in the validation gates of both clusters. If this is true, the algorithm stops to track the two clusters and starts tracking a combined cluster. Therefore, it uses the measurements located in the validation gates of at least one of the previously considered clusters.

Note, that whenever a merging takes place, the algorithm remembers the humans which correspond to the new combined cluster.

For each tracked cluster, we have to detect if it splits into its single persons. For this, we define three conditions, that are checked by the algorithm:

1. The position estimates corresponding to the measurements in the validation gates are separated into subclusters. For this purpose, we select the first estimate, which then is associated with the first subcluster. For all other estimates associated with the cluster, the Euclidean distance to the first estimate is calculated. If this distance is below a certain threshold, the estimate is associated with the first subcluster. In our experiments, we set the threshold to $150\,cm$, which corresponds to the maximum distance between the legs of a walking person. We then have to consider the estimates, for which the Euclidean distance to the first subcluster exceeds this manually chosen threshold. Using the same procedure we applied for building the first subcluster, we now construct subclusters until all estimates are associated with one of these smaller clusters. If the number of subclusters equals the number of humans which were merged into this cluster, the first condition for the dispersion of the cluster is fulfilled. Then, we proceed with step 2.
2. We now check the distance between the subclusters pairwise. If the distance is above a manually chosen bound, we regard these clusters as separated. We chose the value of that bound to be $300\,cm$. The second condition is fulfilled, if the number of pairs of separated subclusters equals $\frac{n(n-1)}{2}$. Thereby, n is the number of single persons associated with the cluster. Hence, we are checking if all subclusters are pairwise separated.
3. Above this, we can separate single subclusters from the cluster and visualise them in the graphics. This follows the same logic as in step 1. Note, the algorithm is not able to determine, how many targets are represented by a single subcluster.

If conditions 1 and 2 are met, the n subclusters are associated with the n single targets, that are therefrom tracked with the KFA. When separating targets from clusters, we cannot guarantee that the target association is the same as before merging the targets into the cluster. In the following subsection, we will describe, how the target association can be retained after the clusters have been split again, based on colour statistics for each track, which we learn for single targets while they are tracked using the KFA.

Fig. 1. The robot equipped with the laser scanners and the cameras

2.2 Colour-Based Identity Assignment

To overcome this problem that track associations are lost after the cluster tracking phase, we additionally take information about the colour of the clothes the persons wear into account. We use a CCD camera in combination with an infrared camera to obtain this colour information during tracking. The two cameras are mounted next to each other on the robot, the setup is shown in Figure 1. People can be detected in an infrared image based on their higher temperature compared to their surrounding. The small displacement of the two cameras now easily allows to correlate the pixels of the two images and to extract the pixels originating from a person from the CCD images as well. Our system searches the image for persons in a left-to-right fashion. Features in the laser data are detected from left to right as well. As long as there is no occlusion and all persons are visible in the camera images and the laser data, this ordering gives a one-to-one correspondence between regions in the image originating from humans and features in the laser data.

In order to distinguish the tracks after clusters have been split-up, we learn a simple colour model for each Kalman filter, while the individual persons are separated. The model consists of a distribution over relative colour frequencies in the images of each person. Therefore, we compute a colour frequency vector $\mathbf{v}_j = \langle v_1, \ldots, v_L \rangle_j$, for each person j in each image, where $0 \leq v_l \leq 1.0$ and the v_l sum up to 1. For this purpose, the CCD pixels originating from a person are converted to the HSV colour space [25] and we discretised the hue (H) values into L bins. The current implementation uses $L = 6$ bins each one representing a primary colour as indicated in table 1.

The underlying assumption of our model is, that the pixels of a person in an image are drawn from a multinomial distribution, which describes the persons appearance in that image. As the shape of a person changes between images and as the colour distribution is discretised, the multinomial will vary across images. We assume that these variations are themselves characteristic for the individual tracks and are drawn from a distribution over multinomials. In Bayesian statistics, the Dirichlet distribution is the conjugate prior of the multinomial distribution [26]. Therefore, we assume that

Table 1. The values of the hue in the HSV colour space

Hue	red	yellow	green	cyan	blue	magenta
Degrees	0	60	120	180	240	300

the colour frequency vectors $\mathbf{v}_{j,k_1:k_2}$ of a track j memorised between the time steps k_1 and k_2 are approximately Dirichlet distributed with parameters α_j, i.e.

$$\mathbf{v}_{j,k} \sim \mathrm{Dir}(\alpha_k), \tag{1}$$

where

$$\mathrm{Dir}(\mathbf{v}, \alpha_k) = \frac{\Gamma(\sum_{i=1}^{L} \alpha_{k,i})}{\prod_{i=1}^{L} \Gamma(\alpha_{k,i})} \prod_{i=1}^{L} v_i^{\alpha_{k,i}-1} \tag{2}$$

is the probability density function of the Dirichlet distribution. As soon as a crossing is detected, we compute the maximum-likelihood estimate $\hat{\alpha}_j$ of the parameters of the Dirichlet distribution[1] of the stored relative frequency vectors for each track j.

The tracks j, which are calculated by the tracking algorithm, are normally interchanged during a crossing. After the clusters have split, we try to reassign the correct person ids to the different tracks based on the colour statistics we have learnt. For this purpose, we assume that the colour frequencies of a track j in subsequent images are independently Dirichlet distributed given the assignment to a person (or track before the crossing) i. So if we observe colour frequencies $\mathbf{v}_{j,k_1}, \ldots, \mathbf{v}_{j,k_2}$ for a track j after the crossing, we can compute assignment likelihoods

$$p_{i,j,k_1:k_2} = \prod_{k=k_1}^{k_2} \mathrm{Dir}(\mathbf{v}_{j,k}, \hat{\alpha}_i). \tag{3}$$

The goal now is to compute the joint assignment σ of all the tracks created from a fragmented cluster to the persons associated with this cluster that maximises the joint assignment likelihood

$$\Pi_\sigma = \prod_{l=1}^{n} p_{l,\sigma(l),k_1:k_2}. \tag{4}$$

A joint assignment σ is a permutation that maps tracks to persons. To determine the optimal joint assignment, we make use of Munkres' Hungarian algorithm [27], which computes optimal assignments based on an assignment cost matrix. To compute the cost matrix, we change from likelihoods $p_{i,j,k_1:k_2}$ to negative log-likelihoods $-\log(p_{i,j,k_1:k_2})$; those can be interpreted as individual assignment costs and we arrange them into a $n \times n$ assignment cost matrix.

[1] We use Thomas Minka's fastfit MATLAB toolbox for this purpose. http://research.microsoft.com/~minka/software/fastfit/

3 Experiments

We conducted experiments with three persons in a real world scenario in our laboratory. The first person was wearing a blue cardigan and a blue pair of trousers. The second person was wearing a yellow cardigan and blue pair of trousers. The third person was wearing a red shirt and a red pair of trousers.

Throughout the experiments we used colour histograms with six bins and Figure 2 shows the corresponding colour distributions for the three persons. The first subject has his maximum in the blue domain, the second has it in the yellow domain and the third in the red one. Thus, the measured colour distributions show a good coincidence with the real colours.

It is difficult to visualise six dimensional Dirichlet distributions, we therefore decided to show the marginal distributions for the individual $v_{i,l}$, instead of the joint Dirichlet distribution over $\mathbf{v}_i$ for each person i. These marginals are Beta distributions which give a good indication of how the proportions of each of the six colours varied across the training data for a person. Figures 3, 4 and 5 present the Beta distributions of the six colours for each person. From Figure 3 it can be concluded, that the first person is most likely wearing clothes from cyan and blue. Thereby it should be noted, that cyan is very similar to blue. Also, the yellow colour is represented a little bit. From figure 4 it can be seen, that the most likely colour of the clothes of person two is yellow, whereas the by far most likely colour of the third person is red (figure 5). Thus, the learnt distributions coincide very well with the true colours of the clothes, which the subjects wear.

The experiments were accomplished with a B21 robot platform shown in figure 1. On the top the camera system is mounted. The left camera is the infrared camera and the right camera is the CCD camera. There are two laser range scanners armed back to back at the robot, so that there is a 360 degree field of view.

The number of possible permutations of three objects is $3! = 6$. Thus, we conducted six experiments, for each permutation one experiment. We defined position 1 as the

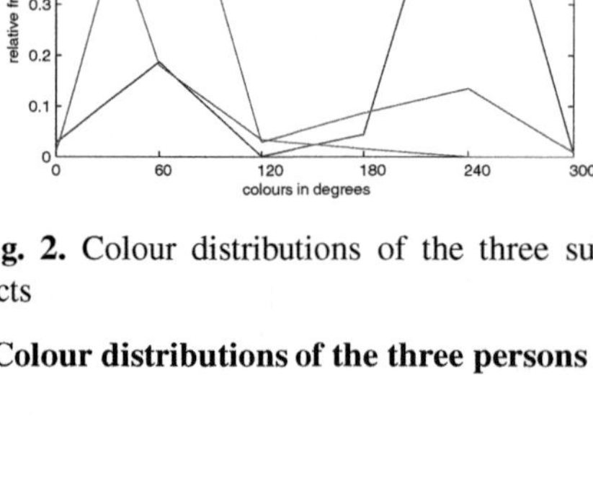

Fig. 2. Colour distributions of the three subjects

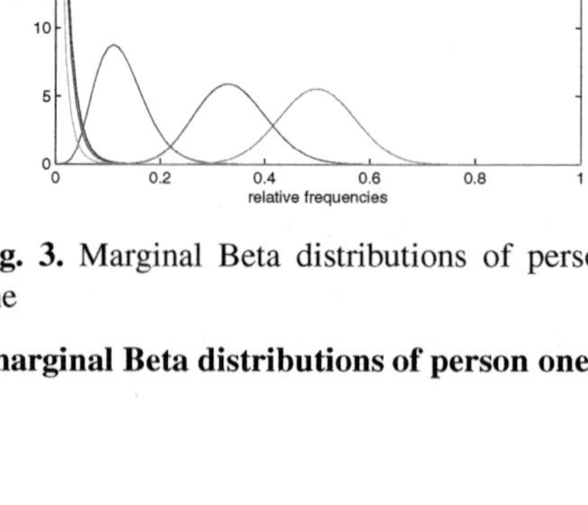

Fig. 3. Marginal Beta distributions of person one

Colour distributions of the three persons and marginal Beta distributions of person one.

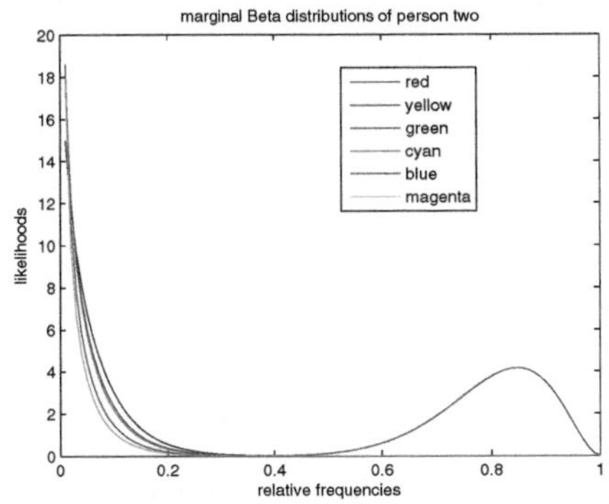

Fig. 4. Marginal Beta distributions of person two

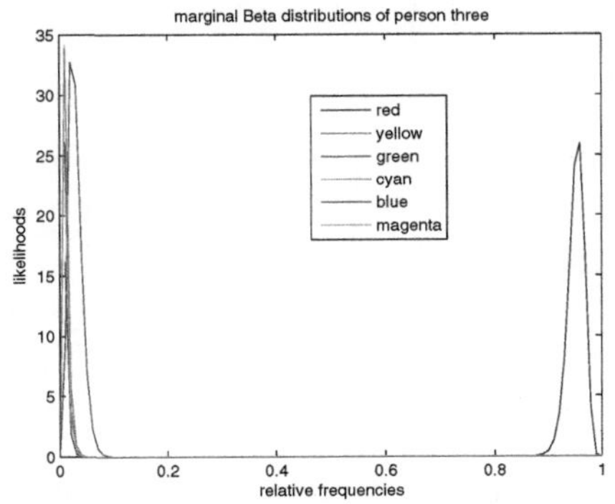

Fig. 5. Marginal Beta distributions of person three

Marginal Beta distributions of persons two and three.

right upper part of the surveillance area, position 2 as the left upper part of the surveillance area and position 3 as the lower middle part. The persons are indexed in the order they appear in the surveillance area. In the six experiments the person indexed 1 occupied position 1 before the crossing, the person indexed 2 occupied position 2 and the person indexed 3 occupied position 3. After the crossing they occupied different positions corresponding to the six possible permutations. The experiment for the permutation $123 \mapsto 231$ is described in detail. At first, the three persons are occupying their start positions and are indexed in the order of their appearance in the surveillance area (figure 6). Thereby, as long as the identities are known, in the graphics the corresponding number is drawn at the place, where the algorithm locates the subject. Next, persons one and two merge to a cluster. This cluster is reproduced by a blue ellipse in the corresponding figure 7.

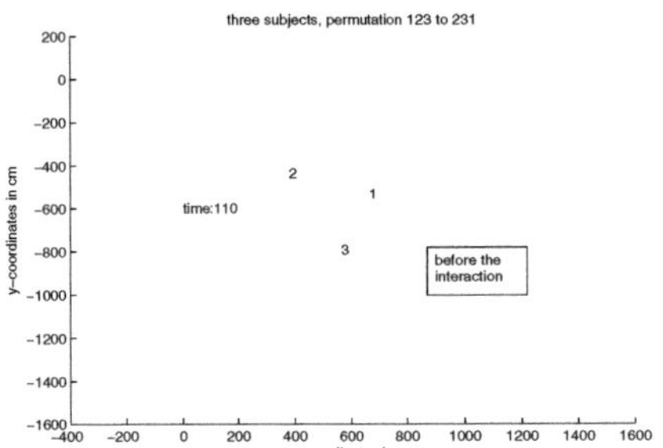

Fig. 6. The three persons before the interaction

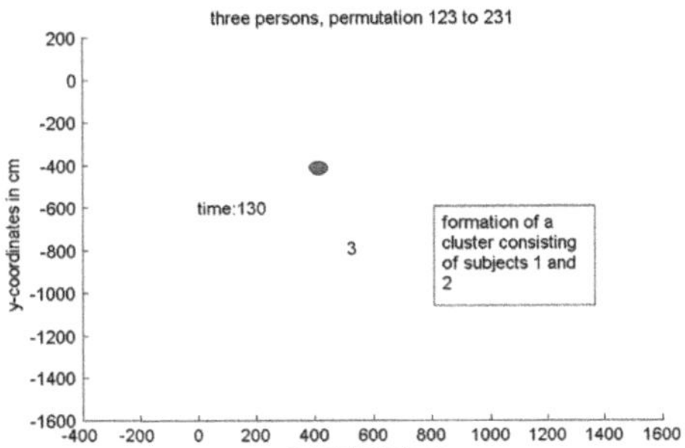

Fig. 7. The persons one and two have merged into a cluster

A real experiment with three persons.

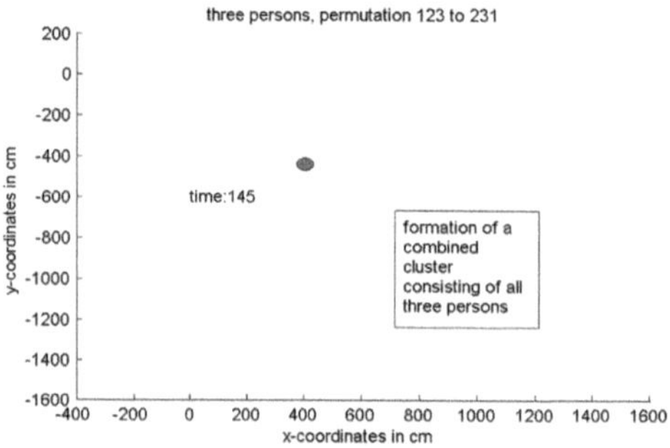

Fig. 8. The third person has joint the other two

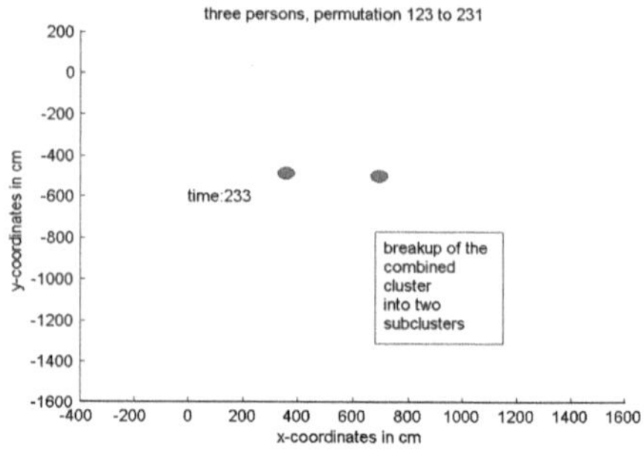

Fig. 9. The combined cluster has split up into two subclusters

A real experiment with three persons.

Then, person three joins the other two people, so that the merging of a single person and a cluster to a combined cluster, which is again reproduced by a blue ellipse (figure 8), occurs.

Next, the cluster splits up into two subclusters (figure 9), each consisting of an unknown number of persons.

Then, the cluster splits up into three subclusters, each of them representing one person (figure 10). At this stage of the experiment condition two of the breakup of a cluster is not fulfilled yet.

Finally, the combined cluster breaks up into its three single persons, because this condition is fulfilled. Thus, the algorithm reassigns the identities, which he has learnt before the crossing, to the single persons. These identities are indicated in figure 11. Obviously, the algorithm tracks the three persons without loss of track and reassigns the identities correctly after the crossing. The same holds for the experiments for the other five permutations.

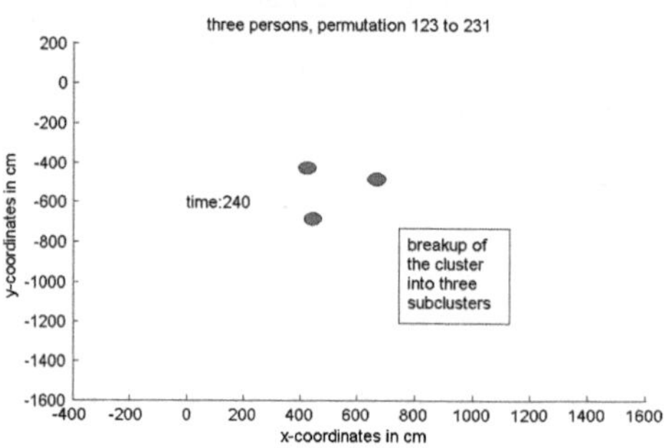

Fig. 10. There are three distinct subclusters

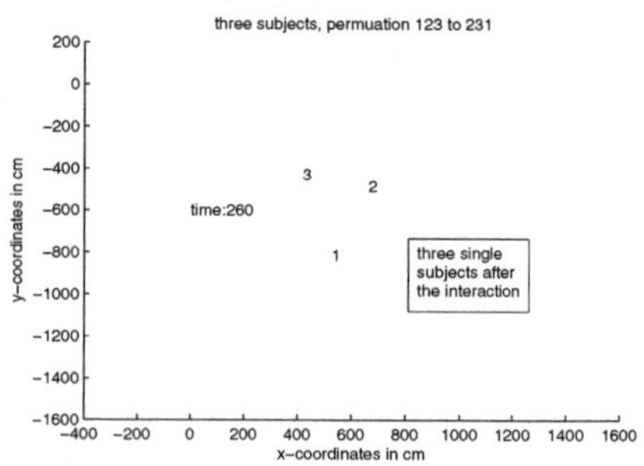

Fig. 11. There are three independent single persons

A real experiment with three persons.

4 Conclusions

In this paper the problem of tracking multiple interacting humans has been addressed. This challenge has been recommended as a difficult task by several authors. There are two jobs which have to be solved:

1. there must not be any loss of track when the people split again.
2. the identities of the persons have to be reassigned correctly after a crossing.

A solution in form of a complex algorithm has been proposed. The first problem is solved by a tracking algorithm, which uses data from two laser scanners and a simple Kalman filter for tracking single persons and an extension of the Viterbi algorithm for tracking clusters when people interact. A clustering routine is then used to distribute the clusters to the humans at the end of a crossing. The second problem is tackled using data from an infrared and a CCD camera. The camera system provides the robot with colour information about the tracked persons. The most likely assignment of the identities is then found by usage of the Hungarian algorithm. Several experiments show that both problems are solved sufficiently by the proposed algorithm.

Of course, there are limitations, when the colours of the clothes people wear are too similar. This situation might especially occur in those situations, where are lot of persons have to be tracked. Moreover, if a human moves away in the shadow of the other persons in the cluster during a crossing, the tracking algorithm will loose this person and the end of the crossing would never be detected. This drawback can be eliminated by the usage of additional robots, so that a full coverage of the scene is provided.

Right now we are doing experiments, where the robot follows a certain person, which is interacting with other persons. Thereby, the robot platform searches out the right person by analysing the reflectance properties of the pairs of trousers people wear.

References

1. Bar-Shalom, Y., Fortmann, T.: Tracking and Data Association. Academic Press Professional, Inc., San Diego (1988)
2. Prassler, E., Scholz, J., Elfes, E.: Tracking people in a railway station during rush-hour. In: Christensen, H.I. (ed.) ICVS 1999. LNCS, vol. 1542, pp. 162–179. Springer, Heidelberg (1998)
3. Schulz, D., Burgard, W., Fox, D., Cremers, A.B.: Tracking multiple moving objects with a mobile robot. In: Proceedings of the 2001 IEEE Computer Society Conference on Computer Vision and Pattern Recognition (CVPR 2001), Kauai, Hawaii, USA, vol. 1, pp. 371–377 (2001)
4. Fod, A., Howard, A., Mataric, M.J.: Laser–based people tracking. In: Proceedings of the, IEEE International Conference on Robotics and Automation (ICRA 2002), Washington, DC, USA, pp. 3024–3029 (2002)
5. Romera, M.M., Vazquez, M.A.S., Garcia, J.C.G.: Tracking multiple and dynamic objects with an extended particle filter and an adapted k–means clustering algorithm. In: Proceedings of the 5th IFAC/EURON Symposium on Intelligent Autonomous Vehicles (IAV 2004), Lisbon, Portugal (2004) CD–ROM
6. Zhao, H., Shibasaki, R.: A novel system for tracking pedestrians using multiple single–row laser–range scanners. IEEE Transactions on Systems, Man and Cybernetics — Part A: Systems and Humans 35(2), 283–291 (2005)

7. Bellotto, N., Hu, H.: People tracking with a mobile robot: A comparison of Kalman and particle filters. In: Proceedings of the 13th IASTED International Conference on Robotics and Applications, Wurzburg, Germany, pp. 388–393. International Association of Science and Technology for Development (2007)
8. Thrun, S.: Learning metric–topological maps for indoor mobile robot navigation. Artificial Intelligence 99(1), 21–71 (1998)
9. Thrun, S., Fox, D., Burgard, W.: Markov localization for mobile robots in dynamic environments. Journal of Artificial Intelligence Research 11, 391–427 (1999)
10. Fortmann, T.E., Bar-Shalom, Y., Scheffe, M.: Sonar tracking of multiple targets using joint probabilistic data association. IEEE Journal of Oceanic Engineering 8(3), 173–184 (1983)
11. Kräußling, A., Schneider, F.E., Wildermuth, D.: A switching algorithm for tracking extended targets. In: Proceedings of the 2nd International Conference on Informatics in Control, Automation and Robotics (ICINCO 2005), Barcelona, Spain, pp. 126–133 (2005)
12. Kräußling, A.: Tracking extended moving objects with a mobile robot. In: Proceedings of the 3rd IEEE Conference on Intelligent Systems (IS 2006), London, UK (2006) CD–ROM
13. Kräußling, A., Schneider, F.E., Wildermuth, D., Sehestedt, S.: A switching algorithm for tracking extended targets. In: Informatics in Control, Automation and Robotics, vol. II, pp. 117–128. Springer, Heidelberg (2007)
14. Kräußling, A.: Tracking Extended Moving Objects with a Mobile Robot. In: Intelligent Techniques and Tools for Novel System Architectures. Studies in Computational Intelligence Edition, vol. 109, pp. 513–530. Springer, Heidelberg (2008)
15. Schumitch, B., Thrun, S., Bradski, G., Olukotun, K.: The information–form data association filter. In: Proceedings of the 2005 Conference on Neural Information Processing Systems (NIPS 2005), Vancouver, British Columbia, Canada. MIT Press, Cambridge (2006)
16. Reid, D.B.: An algorithm for tracking multiple targets. IEEE Trans. Automatic Control 24, 843–854 (1979)
17. Kräußling, A., Schneider, F.E., Wildermuth, D.: Tracking of extended crossing objects using the Viterbi algorithm. In: Proceedings of the 1st International Conference on Informatics in Control, Automation and Robotics (ICINCO 2004), Setubal, Portugal, pp. 142–149 (2004b)
18. Kräußling, A., Schneider, F.E., Sehestedt, S.: Tracking multiple objects using the Viterbi algorithm. In: Proceedings of the 3rd International Conference on Informatics in Control, Automation and Robotics (ICINCO 2006), Setubal, Portugal, pp. 18–25 (2006); also to be published in the Springer book of best papers of ICINCO 2006
19. Viterbi, A.J.: Error bounds for convolutional codes and an asymptotically optimum decoding algorithm. IEEE Transactions on Information Theory 13(2), 260–269 (1967)
20. Forney Jr., G.D.: The Viterbi algorithm. Proceedings of the IEEE 61(3), 268–278 (1973)
21. Schulz, D., Fox, D., Hightower, J.: People tracking with anonymous and ID–sensors using Rao–Blackwellised particle filters. In: Proceedings of the 18th International Joint Conference on Artificial Intelligence (IJCAI 2003), Acapulco, Mexico, pp. 921–928 (2003)
22. Schulz, D.: A probabilistic exemplar approach to combine laser and vision for person tracking. In: Proceedings of Robotics: Science and Systems II, Philadelphia, Pennsylvania, USA (2006), `http://www.roboticsproceedings.org/rss02`
23. Kräußling, A., Schneider, F.E., Wildermuth, D.: Tracking expanded objects using the Viterbi algorithm. In: Proceedings of the 2nd IEEE Conference on Intelligent Systems (IS 2004), Varna, Bulgaria (2004a) CD–ROM
24. Kräußling, A., Schulz, D.: Tracking extended targets — a switching algorithm versus the SJPDAF. In: Proceedings of the 9th IEEE International Conference on Information Fusion (FUSION 2006), Florence, Italy (2006) CD–ROM
25. Gonzalez, R.C., Woods, R.E.: Digitale Image Processing. Addison–Wesley Publishing Company, Inc., Munich (1992)
26. Gelman, A., Carlin, J.B., Stern, H.S.: Bayesian Data Analysis. Chapman & Hall, Boca Raton (2003)
27. Munkres, J.: Algorithms for assignement and transportation problems. Journal of the Society for Industrial and Applied Mathematics 5(1), 32–38 (1957)

Robotic Wheelchair Trajectory Control Considering User Comfort

Razvan Solea[1,2] and Urbano Nunes[1]

[1] ISR - Institute of Systems and Robotics
Department of Electrical and Computer Engineering
University of Coimbra, 3030-290, Coimbra, Portugal
{razvan,urbano}@isr.uc.pt

[2] Department of Automation and Industrial Informatics
"Dunarea de Jos" University of Galati, 800-008, Galati, Romania
razvan.solea@ugal.ro

Abstract. This paper analyzes the comfort of wheelchair users when a sliding-mode trajectory-tracking controller is used. The transmission of the horizontal (fore-and-aft) vibration to the head-neck complex (HNC) in the seated human body may cause unacceptable discomfort and motion sickness. A double-inverted pendulum model with two degrees of freedom is considered as a model for the HNC. The user comfort is examined not only in the time domain (using the fourth power vibration dose value), but also in the frequency domain (using the cross-spectral density method). For measuring the acceleration of the wheelchair, along the trajectory, an inertial measurement unit was used.

1 Introduction

Robotic wheelchairs, such as RobChair [16],[12], are human-centred robots used to assist humans with reduced mobility. Inappropriate wheelchair dynamic accelerations may be a source of unsafe motions and user discomfort. Few studies have been performed addressing how dynamic acceleration affects wheelchairs users.

Most current literature focus on the vibration exposure of a seated occupant. To this end, standards have been developed by the International Organization for Standardization (ISO) to quantify how much exposure is allowable for various frequencies of exposure. To standardize the methods of data collection for whole-body vibration, the ISO introduced the ISO-2631 [1]. The boundaries in ISO-2631 are based on cumulative root-mean-square (rms) amplitude over a single day, specified for frequencies between 1 and 80 Hz.

The human body is a complex dynamic system, the properties of which vary from moment to moment and from one individual to another. From the results of large amount of experimental data, various biomechanical models have been developed to describe the human motion. These models can be grouped as lumped or distributed parameter models. The lumped parameter models consider the human body as several rigid bodies, springs and dampers [2], [3], [4]. Some distributed models treat the spine as a layered structure of rigid elements representing the vertebral bodies, and deformable elements representing the intervertebral disc by finite elements [5].

J.A. Cetto et al. (eds.): Informatics in Control, Automation and Robotics, LNEE 37, pp. 113–125.
springerlink.com

The dynamic response of seated subjects exposed to vibration has been widely assessed in terms of the driving point impedance, apparent mass and transmissibility (transmission of motion through the human body). The transmission of the acceleration to the head-neck complex (HNC) in the seated human body may be the cause of discomfort and motion sickness in wheelchairs. The seat back, by limiting the horizontal and rotational motion of the trunk, increases the transmission of the trunk horizontal acceleration to the HNC. This may has considerable influence on discomfort.

The present study focuses specifically on the influence of sliding-mode tra-jectory-tracking (SM-TT) controller action on user comfort. The user comfort is examined not only in the time domain (using the transmissibility parameter), but also in the frequency domain. For measuring accelerations of the wheelchair, a three-dimensional inertial sensor was used. The analysis of user comfort is made in three different situations: i) SM-TT control under odometry navigation; ii) when the odometric data is fused with absolute position data from magnetic markers (using an EKF-based fusion in the on-line pose estimation); iii) SM-TT control with purposely-incorrectly-tuned parameters.

2 Control of Wheeled Mobile Robots

The application of sliding mode control strategies in nonlinear systems has received considerable attention in recent years [6], [7], [8], [9]. A well-studied example of a non-holonomic system is a wheeled mobile robot (WMR) that is subject to the *rolling without slipping* constraint.

In trajectory-tracking, is an objective to control the non-holonomic WMR to follow a desired path, with a given orientation relatively to the path tangent, even when disturbances exist. In the case of trajectory-tracking the path is to be followed under time constraints. The path has an associated velocity profile, with each point of the trajectory embedding spatiotemporal information that is to be satisfied by the WMR along the path. By other words, trajectory tracking is formulated as having the WMR following a virtual target WMR which is assumed to move exactly along the path with specified velocity profile.

2.1 Kinematic Model of a Unicycle-Type Mobile Robot

Let the pose of the mobile robot be defined by the vector $q_r = [x_r, y_r, \theta_r]^T$, where $[x_r, y_r]^T$ denotes the robot position on the plane and θ_r the heading angle with respect to the x-axis. In addition, v_r denotes the translational velocity of the robot, and ω_r the angular velocity around the vertical axis. For a unicycle WMR rolling on a horizontal plane without slipping, the kinematic model can be expressed by

$$\begin{bmatrix} \dot{x}_r \\ \dot{y}_r \\ \dot{\theta}_r \end{bmatrix} = \begin{bmatrix} cos\theta_r & 0 \\ sin\theta_r & 0 \\ 0 & 1 \end{bmatrix} \cdot \begin{bmatrix} v_r \\ \omega_r \end{bmatrix} \tag{1}$$

which represents a non-linear system.

Controllability of the system (1) is easily checked using the Lie algebra rank condition for nonlinear systems. However, the Taylor linearization of the system about the origin is not controllable, thus excluding the application of classical linear design approaches.

2.2 Trajectory Tracking Model

Without loss of generality, it can be assumed that the desired trajectory $q_d(t) = [x_d(t), y_d(t), \theta_d(t)]^T$ is generated by a virtual unicycle mobile robot. The kinematic relationship between the virtual configuration q_d and the corresponding reference velocity inputs $[v_d, \omega_d]^T$ is similar to (1). From the error vector [10],

$$\begin{bmatrix} x_e \\ y_e \\ \theta_e \end{bmatrix} = \begin{bmatrix} cos\theta_d & sin\theta_d & 0 \\ -sin\theta_d & cos\theta_d & 0 \\ 0 & 0 & 1 \end{bmatrix} \cdot \begin{bmatrix} x_r - x_d \\ y_r - y_d \\ \theta_r - \theta_d \end{bmatrix} \tag{2}$$

we get the error dynamics:

$$\begin{cases} \dot{x}_e = -v_d + v_r \cdot cos\theta_e + \omega_d \cdot y_e \\ \dot{y}_e = v_r \cdot sin\theta_e - \omega_d \cdot x_e \\ \dot{\theta}_e = \omega_r - \omega_d \end{cases} \tag{3}$$

2.3 Trajectory Planner

A trajectory planner for human-transport robots must generate smooth velocity profiles (linear and angular) with low associated accelerations. The trajectory planning process can be divided into two separate parts. First, a continuous collision-free path is generated. In a second step, called trajectory generation, a velocity profile along the path is determined. A method to generate a velocity profile, respecting human body comfort, for any two-dimensional path in static environments was proposed in [10].

Figures 1 - 2 show an example of a planned trajectory using the method described in [10] were the goal was to obtain an overall rms acceleration in the range of "not uncomfortable" (see Table 1). The overall rms acceleration is defined as:

$$a_w = \sqrt{k_x^2 \cdot a_{wx}^2 + k_y^2 \cdot a_{wy}^2 + k_z^2 \cdot a_{wz}^2} \tag{4}$$

where a_{wx}, a_{wy}, a_{wz}, are the rms accelerations on x, y, z axes respectively, and k_x, k_y, k_z, are multiplying factors. For a seated person $k_x = k_y = 1.4$, $k_z = 1$. For motion on the x-y plane, $a_{wz} = 0$. The local coordinate system is chosen so that the x-axis and y-axis are aligned with the longitudinal and lateral directions, respectively.

Table 1. ISO 2631-1 STANDARD

Overall Acceleration	Consequence
$a_w < 0.315m/s^2$	not uncomfortable
$0.315 < a_w < 0.63m/s^2$	a little uncomfortable
$0.5 < a_w < 1m/s^2$	fairly uncomfortable
$0.8 < a_w < 1.6m/s^2$	uncomfortable
$1.25 < a_w < 2.5m/s^2$	very uncomfortable
$a_w > 2.5m/s^2$	extremely uncomfortable

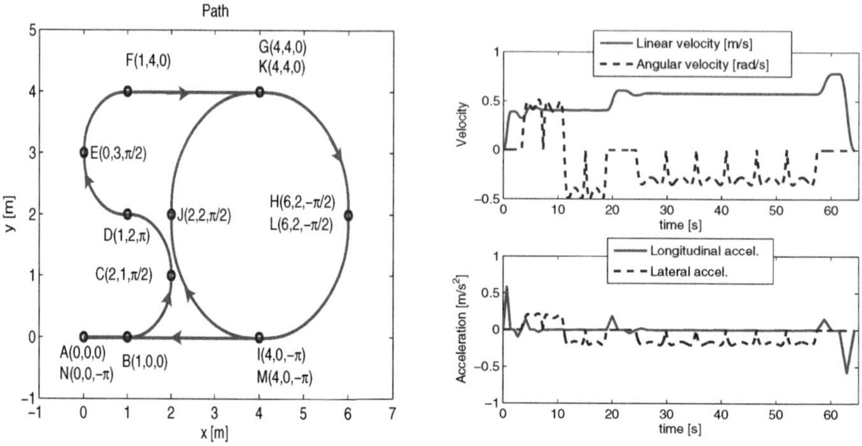

Fig. 1. A) Path example composed by thirteen path segments calculated by the trajectory planner, from the fourteen waypoints A to N. B) Velocity and acceleration profiles.

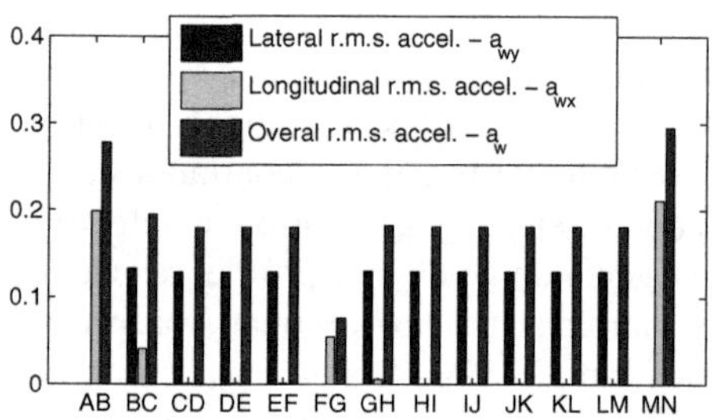

Fig. 2. Acceleration rms values for each path segment (AB, BC, ..., MN) of path depicted in Fig. 1

The ISO 2631-1 standard [1] (Table 1) relates comfort with the overall rms acceleration acting on the human body.

Figure 1.A shows the generated path, where the larger circles represent the used fourteen waypoints (A, B, ..., N). Each waypoint is defined by a position, in meters, and an orientation, in radians. The generated velocity and acceleration profiles are shown in Fig. 1.B. As can be observed in Fig.2, the rms overall accelerations, in each path segment, are below the imposed acceleration constraint of $a_{w(i,i+1)} < 0.31m/s^2$.

2.4 Sliding-Mode Controller

The objective of SMC is the same as for classical controllers, i.e., to force the output states to follow the desired input states. However, the SMC is a model-based control strategy in which the controller structure and gains are designed based on the system model.

Uncertainties which exist in real mobile robot applications degrade the control performance significantly, and accordingly, need to be compensated. In [9] a SM-TT controller for WMRs is proposed, where trajectory-tracking is achieved even in the presence of large initial pose errors and disturbances.

Let us define the sliding surface $s = [s_1\ s_2]^T$ as

$$\begin{aligned} s_1 &= \dot{x}_e + k_1 \cdot x_e, \\ s_2 &= \dot{y}_e + k_2 \cdot y_e + k_0 \cdot sgn(y_e) \cdot \theta_e. \end{aligned} \tag{5}$$

where k_0, k_1, k_2 are positive constant parameters, x_e, y_e and θ_e are the tra-jectory-tracking errors defined in (3). If s_1 converges to zero, trivially x_e converges to zero. If s_2 converges to zero, in steady-state it becomes $\dot{y}_e = -k_2 \cdot y_e - k_0 \cdot sgn(y_e) \cdot \theta_e$. For $y_e < 0$, $\dot{y}_e > 0$ if only if $k_0 < k_2 \cdot |y_e| / |\theta_e|$. For $y_e > 0$, $\dot{y}_e < 0$ if only if $k_0 < k_2 \cdot |y_e| / |\theta_e|$. Finally, it can be known from s_2 that convergence of y_e and $\dot{y}_e$ leads to convergence of θ_e to zero. Using the reaching law defined in [11]

$$\dot{s} = -Q \cdot s - P \cdot sgn(s) \tag{6}$$

$$\begin{aligned} &Q = diag\,[q_1, q_2]\,,\ \ q_i > 0, P = diag\,[p_1, p_2]\,,\ \ p_i > 0,\ i = 1, 2 \\ &sgn(s) = [sgn(s_1), sgn(s_2)]^T \end{aligned}$$

together with (5), and after some mathematical manipulation, we get the commands for trajectory-tracking,

$$\dot{v}_c = \frac{1}{cos\theta_e}(-q_1 s_1 - p_1 sgn(s_1) - k_1 \dot{x}_e - y_e \dot{\omega}_d - \dot{y}_e \omega_d + v_r \dot{\theta}_e sin\theta_e + \dot{v}_d) \tag{7}$$

$$\begin{aligned} \omega_c = &\frac{1}{v_r cos\theta_e + k_0 sgn(y_e)}(-q_2 s_2 - p_2 sgn(s_2) - k_2 \dot{y}_e - \dot{v}_r sin\theta_e + \\ &+ x_e \dot{\omega}_d + \dot{x}_e \omega_d) + \omega_d. \end{aligned} \tag{8}$$

The SM-TT architecture with a on-line robot's pose estimator, fusing odometry with absolute position data as described in [12], is depicted in Fig. 3.

RobChair has a two-level control architecture. High-level control algorithms (including reference motion generation) are written in C and run with a sampling time of $T_s = 50$ ms on a embedded PC, which also provides a user interface with real-time visualization and a simulation environment. The PC communicates through a CAN bus with several devices. Wheel velocity commands,

$$\omega_R = \frac{v_c + \frac{L}{2} \cdot \omega_c}{R}, \quad \omega_L = \frac{v_c - \frac{L}{2} \cdot \omega_c}{R} \tag{9}$$

where R is radius of the drive wheels and L the distance between drive wheels, are sent to the PI controllers, and encoder measures N_R and N_L are inputs of the robot's pose estimator.

The low-level control layer is in charge of the execution of the wheels velocity control. For each wheel, a microcontroller implements a digital PI with a cycle time of $T_c = 5$ ms. Two power amplifiers drive the motors with PWM voltage.

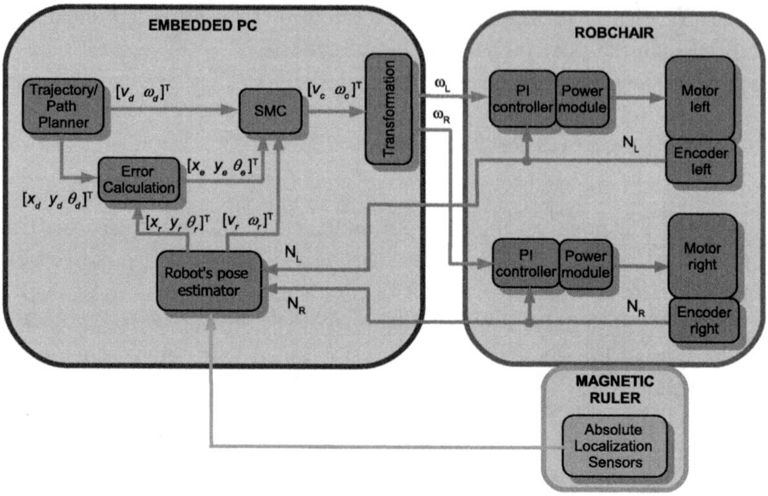

Fig. 3. SM-TT control architecture with a robot's pose estimator, fusing odometry and absolute position data

3 Human Head-Neck Complex Model and Evaluation of Comfort

In general, comfort while riding depends not only on the amplitude, but also on the frequency of wheelchair vibrations and accelerations. Oscillations have influence on users comfort and may affect users health. Moreover, natural frequency of the wheelchair and human's organs is strongly related with the user's discomfort while riding.

3.1 Model of Head-Neck Complex

A double-inverted pendulum model with two degrees-of-freedom is considered for the HNC model (Fig. 4.A). One of the centers of rotation of the model was assumed to be at C7-T1 (O_1 in Fig. 4.A), and the other at C0-C1 (O_2 in Fig. 4.A) of the cervical spine. Two lumped masses, indicating the mass of the neck and the mass of the head, were considered in the model. The center of mass of the neck was assumed to be exactly at the mid-point of the two centers of rotation. Moreover, the center of mass of the head was assumed to be exactly over the center of mass of the neck and the center of rotation (Fig. 4.A). The equation of motion using generalized coordinates can be expressed as:

$$M(q)\ddot{q} + C(q,\dot{q})\dot{q} + Kq + G(q) = Q_q \tag{10}$$

where: $q = \begin{bmatrix} \theta_1 \\ \theta_2 \end{bmatrix}$; $Q_q = 0$; $M(q) = \begin{bmatrix} M_{11} & M_{12} \\ M_{21} & M_{22} \end{bmatrix}$;
$M_{11} = m_1 l_1^2 + m_2 L_1^2 + J_1$; $M_{12} = m_2 L_1 l_2 cos(\theta_1 - \theta_2)$
$M_{21} = m_2 L_1 l_2 cos(\theta_1 - \theta_2)$; $M_{22} = m_2 l_2^2 + J_2$
$C(q,\dot{q}) = \begin{bmatrix} c_1 + c_2 & C_{12} \\ C_{21} & c_2 \end{bmatrix}$

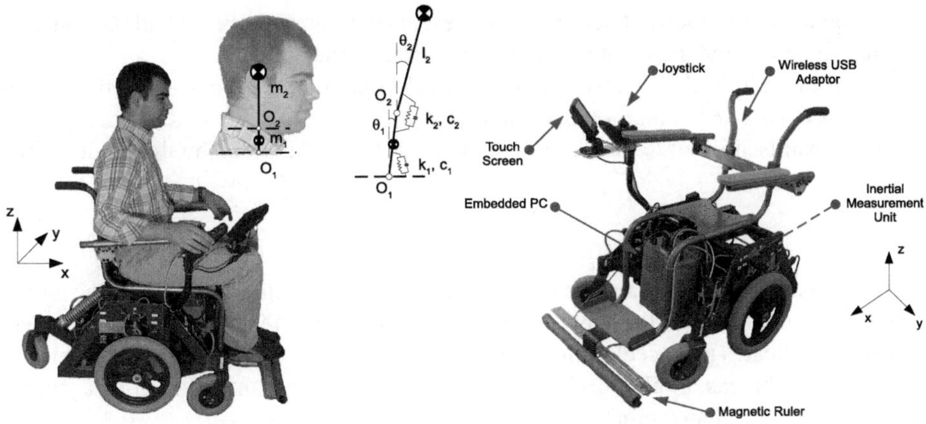

Fig. 4. A) Human head-neck model. B) RobChair platform.

Table 2. HNC model values used in our study

Element	Neck	Head
Length of segment $L_i[m]$	0.080	0.138
Center of gravity $l_i[m]$	0.040	0.069
Mass $m_i[kg]$	1.01	4.22
Moment of Inertia $J_i[kgm^2]$	0.0011	0.210
Spring constant $k_i[Nm/rad]$	14.04	10.29
Damping constant $c_i[Nms/rad]$	0.347	0.230

$$C_{12} = m_2 L_1 l_2 sin(\theta_1 - \theta_2)\dot{\theta}_2 - c_2; C_{21} = -m_2 L_1 l_2 sin(\theta_1 - \theta_2)\dot{\theta}_1 - c_2$$
$$K = \begin{bmatrix} k_1 + k_2 & -k_2 \\ -k_2 & k_2 \end{bmatrix}; G(q) = \begin{bmatrix} -(m_2 L_1 + m_1 l_1)\ddot{x}cos(\theta_1) \\ -m_2 l_2 \ddot{x} cos(\theta_2) \end{bmatrix}$$

In our study, the user characteristic elements, shown in Table 2 (from [2]), were used to model the dynamic behavior of the human HNC.

3.2 Time-Domain Calculations

The fourth power vibration dose value (VDV) is primarily a measurement procedure used to report the relative severity of complex oscillation exposures, being preferred to other measures due to its use of the duration and variability of the motion:

$$VDV = \left\{ \int_0^T [a_w(t)]^4 \cdot dt \right\}^{\frac{1}{4}} \tag{11}$$

in units of $ms^{-1.75}$, where the frequency-weighted acceleration is defined by $a_w(t)$, and T is the duration of an experiment. VDV was selected over the difference between peak accelerations (peak-to-peak value) and acceleration rms parameter. The problem with the peak-to-peak value is that it only represents one instance in time, rather than the entire signal, and it does not accurately represent oscillatory motion. And, as concerns the rms value it is independent of the duration time of the signal.

According to the EU directive on mechanical vibration [13] the average limit value of VDV is $9.1ms^{-1.75}$ and the upper limit is $21ms^{-1.75}$.

The transmissibility (T_r) is defined as the output VDV divided by the input VDV,

$$T_r = \frac{VDV_{out}}{VDV_{in}} \tag{12}$$

The transmissibility defines the performance of the wheelchair in terms of the amplification or attenuation of the vibration that is transmitted to the occupant. A value less than unity indicates that accelerations were attenuated by the combination of wheelchair and human, whereas a value higher than unity indicates an amplification of accelerations by the wheelchair-human system.

3.3 Frequency-Domain Calculations

Given the input, wheelchair acceleration, and the output, acceleration obtained from the HNC model, the transfer function is usually calculated using the cross-spectral density (CSD) method defined as:

$$H_{CSD}(f) = \frac{CSD_{in-out}(f)}{PSD_{in}(f)} \tag{13}$$

where $CSD_{in-out}(f)$ is the CSD of the input and output, and $PSD_{in}(f)$ is the power spectral density (PSD) of the input. The advantage of using the CSD method is that the function generates the phase of the response and it also only includes data at the input and output that are correlated, thus reducing the effects of noise in the measurement system.

4 Experimental Results

In order to validate the applicability of the sliding-mode controller for trajectory-tracking, real experiments have been performed using RobChair (intelligent platform developed in ISR-UC [16], [12]). RobChair, shown in Fig. 4.B, has two differentially driven rear wheels and two passive castor front wheels. There is also a fifth rear wheel connected to the back of the wheelchair with a damper used for stability. It is powered by two 12-V batteries (60 Ah) and reaches a maximum speed of 7 Km/h. It has been equipped, in ISR-UC, with several devices such as: two power driver modules, which provide an independent control of each motor, optical encoders, laser range finders, an inertial sensor and a magnetic sensing ruler, developed at ISR-UC, that detects robustly magnetic markers [12]. Figure 5 presents a block diagram of the actual control hardware architecture. The current implementation of the framework is based on Linux as its underlying real-time operating system. The component-based software selected for the proposed software framework is GenoM (Generator of Modules) [15], which is an environment for description and implementation of software components.

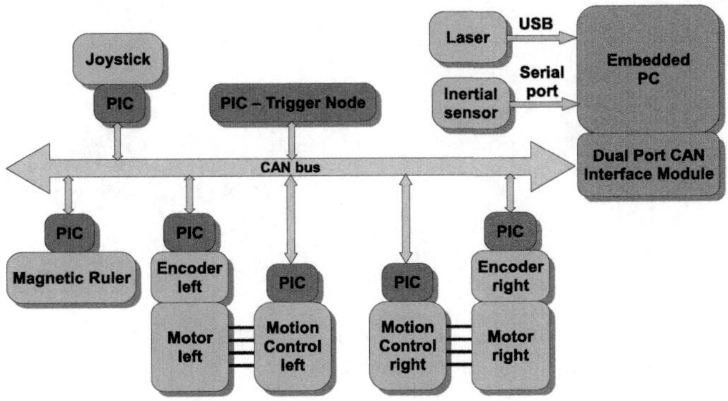

Fig. 5. RobChair hardware architecture

An embedded PC is responsible for giving some degree of intelligence to the robot. This computer is connected to distributed devices through fieldbuses. The platform is connected to external devices through a wireless link. This connection allows the implementation of a distributed architecture, which exhibits the possibility and capability to extend our single robot to other perspectives, like multi-robot cooperation, its integration in intelligent environments, etc.

All the distributed devices, connected through CAN, use a base printed circuit board, containing a microchip micro-controller (μC), as described in [14]. A custom communication protocol, based on the time-triggered protocol paradigm, was designed and implemented. All events are synchronized by a message, sent from a Synchronization Micro-Controller Unit (Trigger Node, in Fig. 5), that synchronizes the other Micro-Controller Units, and defines the control loop time reference.

The odometric data provided by the wheel encoders is fused with the data from magnetic markers. The extended Kalman filter (EKF) was chosen for the fusion process [17]. This navigation technology, based on sensing magnetic markers, is well suited

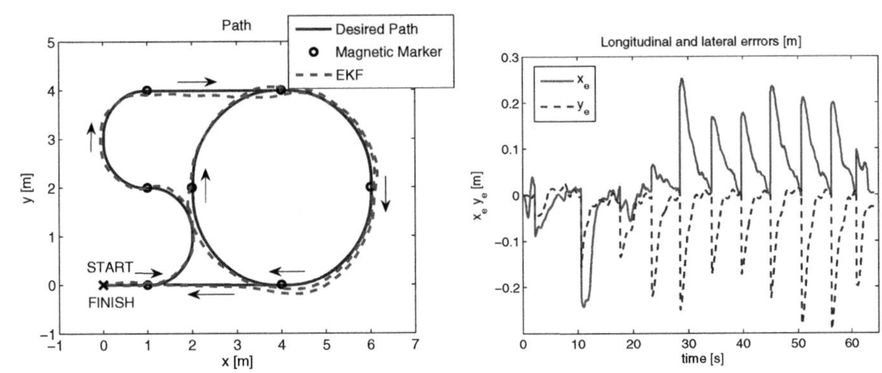

Fig. 6. A) Experimental sliding-mode trajectory-tracking control using an EKF-based fusion in the on-line pose estimation. B) Longitudinal and lateral errors for SM-TT control under magnetic-markers navigation (case B).

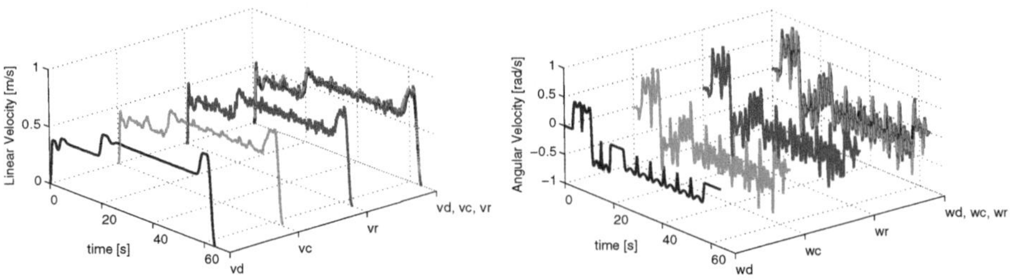

Fig. 7. Desired (v_d, ω_d), command (v_c, ω_c) and real (v_r, ω_r) linear/angular velocities for SM-TT control under magnetic-markers navigation (case B)

when high precision navigation and robustness is required, and it can be used to complement other navigation systems, such as GPS.

The inertial sensor RGA300CA-100 (Crossbow) was used for measuring the wheelchair accelerations in three orthogonal directions.

Several experiments has been performed in outdoor environment (see Fig. 9). Experimental results of the SM-TT controller using the planned path presented in Fig 1, are shown in Figs. 6- 7. Figure 7 shows desired, command and real linear and angular velocities for SM-TT control under magnetic-markers navigation. Corrections in the pose after each magnetic marker detection provokes an error signal that is efficiently dealt by the SM-TT controller, and rapidly the tracking errors converge to zero (see Fig. 6.B).

The analysis of user comfort is made in three different situations:

- case A: SM-TT control under odometry navigation;
- case B: SM-TT control under magnetic-markers navigation (odomeric data is fused, using an EKF-based fusion, with absolute position data from magnetic markers detection);
- case C: SM-TT control with purposely-incorrectly-tuned parameters.

The experimental data of all three cases are summarized in Table 3. Three experimental trials were executed in cases A and B. The table shows the vibration dose value (VDV),

Table 3. Experimental results

Case	No	VDV_x RobChair	VDV_x Head	T_r	RMS accel. RobChair	Max. accel RobChair	RMS accel. Head	Max. accel Head
A	I	2.9219	1.3826	0.4732	0.3287	1.5686	0.1656	0.7286
	II	2.7634	1.3429	0.4859	0.3191	1.4988	0.1589	0.7307
	III	2.7177	1.2802	0.4710	0.3187	1.5590	0.1510	0.8210
average		2.8010	1.3352	0.4767	0.3222	1.5421	0.1585	0.7601
B	I	2.9481	1.4408	0.4887	0.3538	1.2853	0.1797	0.7100
	II	3.2946	1.5707	0.4767	0.3663	1.7635	0.1856	0.7988
	III	3.0318	1.4860	0.4901	0.3629	1.4136	0.1828	0.7755
average		3.0915	1.4992	0.4852	0.3610	1.4875	0.1827	0.7614
C	-	11.4623	5.4607	0.4764	1.5185	4.1708	0.7039	2.2210

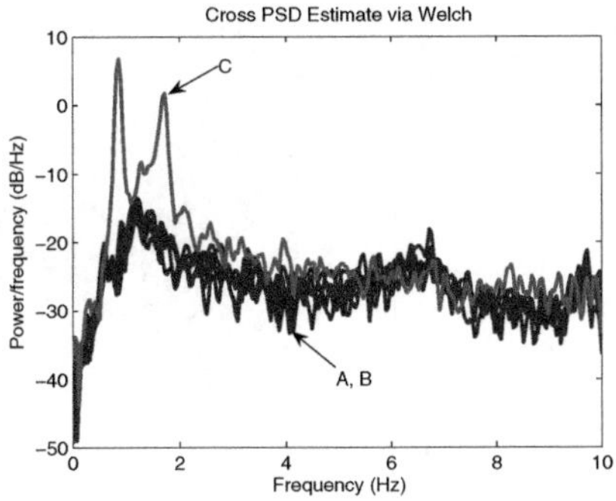

Fig. 8. Cross-spectal density functions for all experiments

transmissibility (T_r), root mean square accelerations (RMS) and maximum values (Max). The results of columns "RMS accel. Robchair" and "Max. accel. RobChair" concern the acceleration results obtained by the inertial sensor; and the "RMS accel. Head" and "Max. accel. Head" were obtained from the model of head-neck complex (10). The overall rms acceleration of head (along the x axes) in cases A and B are in range of "not uncomfortable", but in case C is in range of "uncomfortable" (see Table 1 and equation (4)).

Each experiment was made for the same trajectory (see Fig. 1). The time domain VDV values obtained in cases A and B are below the limit value of $9.1ms^{-1.75}$, only in case C, VDV values are above that limit. As can be observed from Table 3, the transmissibility tends to be under unity, suggesting that the vibrations are attenuated.

Figure 8 shows cross-spectral density values for all experiments. The maximum of H_{CSD} magnitude occurs in case C, and the corresponding frequencies are between $0.8 - 1.8Hz$. When the magnitude of H_{CSD} increases, the user comfort decreases.

To evaluate the quality of the estimated transfer function from the measured signals, the following coherence function is determined

Fig. 9. Snapshots concerning real experiments performed in an outdoor environment

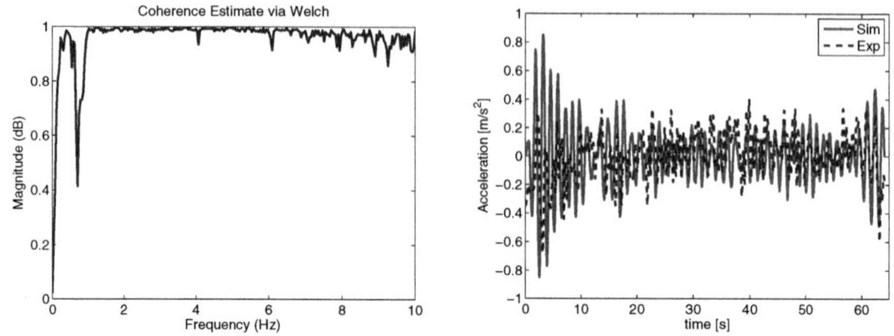

Fig. 10. A) The coherency functions for one experiment. B) Comparison of the simulated (Sim.) and experimental (Exp.) outputs for one experiment in the time domain.

$$h(f) = \frac{|CSD_{in-out}(f)|^2}{CSD_{in-in}(f) \cdot CSD_{out-out}(f)} \tag{14}$$

The coherence functions revealed that the input and output of the HNC were well correlated, such that it was acceptable to use a linear model for the HNC (see Fig. 10.A). For this reason a second inertial sensor was applied to measure the HNC accelerations. Coherence is a frequency dependent function with values in the interval $[0, 1]$. A value of coherence close to unity usually gives promise of a successful identification. If the system is linear and the input and output signals are not corrupted by noise, then the value of coherence will be unity.

An accelerometer was used for measuring the user's head movement accelerations. This sensor can detect three-dimensional accelerations on X, Y and Z-axis. To validate the results, we compared simulation (using eq. 10) and experimental results (from the accelerometer on the head) for one experiment in the time domain (Fig. 10.B).

As future work we intend to investigate on a better HNC model from experimental results using two accelerometers.

5 Conclusions

Wheelchair is exposed to vibration coming not only from a variety of different road surface but also from the command of the wheelchair (manually - using Joystick or automated - using different type of controllers). The paper analyzes the comfort of wheelchair users when a SM-TT controller is used. The user comfort is examined not only in time domain (using the fourth power VDV), but also in frequency domain (using the cross-spectral density method). Outdoor experiments, using RobChair with SM-TT controller were performed (see Fig. 9). The experimental tests presented in this paper are representative of the average performance of the controllers. We had summarized our acquired experience in general observations that can be useful guidelines for implementation of the same control strategies in other type of mobile robots.

Acknowledgements. This work was supported in part by ISR - Coimbra and Portuguese Foundation for Technology and Science (FCT), under contract NCT04: POSC/EEA/SRI/ /58016/2004. R. Solea acknowledges a PhD research fellowship from FCT.

References

1. ISO-2631: Mechanical vibration and shock - Evaluation of human exposure to whole body vibration - Part 1: General requirements. International Organization for Standardization (1997)
2. Atapourfard, M., Ishihara, T., Inooka, H.: The influences of trunk horizontal vibration to the head-neck complex. In: SICE 2002, Osaka, pp. 1053–1058 (2002)
3. Atapourfard, M., Ishihara, T., Inooka, H.: Identification of the head-neck complex in response to trunk horizontal vibration. Biological Cybernetics 90(6), 418–426 (2004)
4. Gurses, S., Dhaher, Y., Hain, T.C., Keshner, E.A.: Perturbation parameters associated with nonlinear responses of the head at small amplitudes. CHAOS 15(2), 23905 (2005)
5. Kitazaki, S., Griffin, M.J.: A modal analysis of whole-body vertical vibration using a finite element model of the human body. Journal of Sound and Vibration 200(1), 83–103 (1997)
6. Yang, J.M., Kim, J.H.: Sliding mode control for trajectory tracking of nonholonomic wheeled mobile bobots. IEEE Transactions on Robotics and Automation 15(3), 578–587 (1999)
7. Chwa, D.: Sliding-mode tracking control of nonholonomic wheeled mobile robots in polar coordinates. IEEE Transactions on Control Systems Technology 12(4), 637–644 (2004)
8. Chwa, D., Hong, S.K., Song, B.: Robust posture stabilization of wheeled mobile robots in polar coordinates. In: 17th International Symposium on Mathematical Theory of Networks and Systems, Kyoto, pp. 343–348 (2006)
9. Solea, R., Nunes, U., Filipescu, A.: Trajectory planning and sliding-mode control for WMR trajectory-tracking and path-following respecting human comfort travel. In: 7th Portuguese Conference on Automatic Control - CONTROLO 2006, Lisbon (2006)
10. Solea, R., Nunes, U.: Trajectory planning and sliding-mode control based tra-jectory-tracking for cybercars. Integrated Computer-Aided Engineering 14(1), 33–47 (2007)
11. Gao, W., Hung, J.C.: Variable structure control of nonlinear systems: A new approach. IEEE Transactions on Industrial Electronics 40(1), 45–55 (1993)
12. Lopes, A.C., Moita, F., Nunes, U., Solea, R.: An outdoor guidepath navigation system for AMRs based on robust detection of magnetic markers. In: 12th IEEE International Conference on Emerging Technologies and Factory Automation, Patras, pp. 989–996 (2007)
13. European-Parliament and the Council of the European Union: On the minimum health and safety requirements regarding the exposure of workers to the risks arising from physical agents (vibration). Official Journal of the European Communities, Directive 2002/44/EC (2002)
14. Maia, R.: Movimento de robots moveis com rodas de tracao diferencial: modelacao e controlo do sistema motriz. Universidade de Coimbra (in portuguese), Coimbra (2004)
15. Fleury, S., Herrb, M., Chatila, R.: GenoM: A tool for the specification and the implementation of operating modules in a distributed robot architectures. In: IEEE of the International Conference on Intelligent Robots and Systems, Grenoble, pp. 842–848 (1997)
16. Pires, G., Nunes, U.: A wheelchair steered through voice commands and assisted by a reactive fuzzy-logic controller. Journal of Intelligent and Robotic Systems 34, 301–314 (2002)
17. Bento, L.C., Nunes, U., Moita, F., Surrecio, A.: Sensor fusion for precise autonomous vehicle navigation in outdoor semi-structured environments. In: 8th IEEE International Conference on Intelligent Transportation Systems (ITSC 2005), Vienna, pp. 245–250 (2005)

Calibration and Comparison of Visual Tracking on the Ground for Multi Camera Tracking

Jorge Raul Gomez, Jose J. Guerrero, and Elias Herrero-Jaraba

Aragon Institute for Engineering Research, University of Zaragoza
Maria de Luna 1, Zaragoza, Spain
jrg@unizar.es, jguerrer@unizar.es, jelias@unizar.es

Abstract. Tracking is an important field in visual surveillance systems. Trackers have been applied traditionally in the image, but a new concept of tracking has been used gradually, applying the tracking on the ground map of the surrounding area. The purpose of this article is to compare both alternatives and prove that this new usage makes possible to obtain a higher performance and a minimization of the projective effects. Moreover, it provides the concept of *multi-camera* as a new tool for mobile object tracking in surveillance scenes, because a common reference system can be defined without increasing complexity. An automatic camera re-calibration procedure is also proposed, which avoids some practical limitations of the approach.

1 Introduction

Real-time object tracking is recently becoming more and more important in the field of video analysis and processing. Applications like traffic control, user computer interaction, on-line video processing and production and video surveillance need reliable and economically affordable video tracking tools. In the last years this topic has received an increasing attention by researchers. However, many of the key problems are still unsolved. The surveillance tracking community in particular has studied target tracking techniques for a number of years, mainly in the context of finding efficient methods to track missiles, aircrafts etc. and tracking targets of unknown motion. Their work has been used for a variety of applications.

There have been previous contributions in order to improve such systems. Tissainayagam and Suter [1] proposed a tracking method using a model switching. Other authors, as Isler [2], use the technique of multiple or distributed sensors, assigning sensors to track targets so as to minimize the expected error in the resulting estimation for target locations. In most of the cases, the sensors used for these tasks are inherently limited, and individually incapable of estimating the target state, and they only can be used for a unique task. This limitation disappears with the use of cameras. In this way, Lee et al. [3] suggested to establish a common coordinate frame and to capture image signals from several cameras arranged in a particular environment. This last idea is used in this paper in order to demonstrate that this solution contributes a better solution to the tracking problem.

We propose a simple tracker based on the Kalman Filter. This tracker is used in two different ways (on the image plane and on the ground plane), making a comparative

J.A. Cetto et al. (eds.): Informatics in Control, Automation and Robotics, LNEE 37, pp. 127–138.
springerlink.com

between both. Theoretically, the perspective effects must disappear in the second one, and therefore, the tracking must involve better. This paper shows this event such in laboratory conditions as in real environments.

This paper is organized as follows: Section 2 briefly details the transformation between the image and the ground. After that, an automatic re-calibration procedure that avoids some of the practical limitations of the approach is proposed. Section 3 provides details of the tracking on the floor, showing the different stages of the proposed tracking system: detection, tracking, uncertainty transformation, and tuning. In Section 4 we provide our results in a particular environment (our laboratory) under stable conditions. Finally, some results for a more complex environment (a football match) are shown also in section 4.

2 From Image to the Ground

In order to transform the coordinates from the image to the planar ground, a plane projection transformation is used. At the moment no distortion of the camera lens is assumed. A point in the projective plane is represented by three coordinates, $\mathbf{p} = (x_1, x_2, x_3)^T$, which represents a ray through the origin in the 3D space [4]. Only the direction of the ray is relevant, so all points written as $\lambda\mathbf{p} = (\lambda x_1, \lambda x_2, \lambda x_3)^T$ are equivalent. The classical Cartesian coordinates of the point (x, y) can be obtained intersecting the ray with a special plane perpendicular to x_3 axis and located at unit distance along x_3. This is equivalent to scale $\mathbf{p}$ as, $\mathbf{p} = (x, y, 1)^T$. Projected points in an image and real points in a planar ground are both represented in this way.

A projective transformation between two projective planes (1 and 2) can be represented by a linear transformation $\mathbf{p}_2 = \mathbf{T}_{21}\mathbf{p}_1$. If the transformation is represented in Cartesian coordinates it results non-linear. Since points and lines are dual in the projective plane, the transformation for the line coordinates is also linear, being $\left(\mathbf{T}_{21}^{-1}\right)^T$ the corresponding transformation matrix for lines.

2.1 Computing the Transformation to Calibrate the Camera

Let it be $\mathbf{p}_c = (x_i, y_i, 1)^T$ the coordinates of a point i in the camera reference system. Let it be (x_i^g, y_i^g) the coordinates of the corresponding point in a reference system of the planar ground obtained from the plane of the building, and therefore let it be $\mathbf{p}_g = (x_i^g, y_i^g, 1)$ its homogeneous coordinates.

We obtain the projective transformation $\mathbf{T}_{gc}$ up to a non-zero scale factor, for points, $\mathbf{p}_g = \mathbf{T}_{gc}\mathbf{p}_c$. For each couple i of corresponding points, two homogeneous equations to compute the projective transformation are considered. They can be written as, $(\lambda_i x_i^g, \lambda_i y_i^g, \lambda_i)^T = \mathbf{T}_{gc}(x_i, y_i, 1)^T$. Developing them in function of the elements of the homography matrix, we have

$$\begin{pmatrix} x_i & y_i & 1 & 0 & 0 & 0 & -x_i^g x_i & -x_i^g y_i & -x_i^g \\ 0 & 0 & 0 & x_i & y_i & 1 & -y_i^g x_i & -y_i^g y_i & -y_i^g \end{pmatrix} \mathbf{t} = \begin{pmatrix} 0 \\ 0 \end{pmatrix}$$

where $\mathbf{t} = (t_{11}\, t_{12}\, t_{13}\, t_{21}\, t_{22}\, t_{23}\, t_{31}\, t_{32}\, t_{33})^T$ is a vector with the elements of the homography matrix $\mathbf{T}_{gc}$.

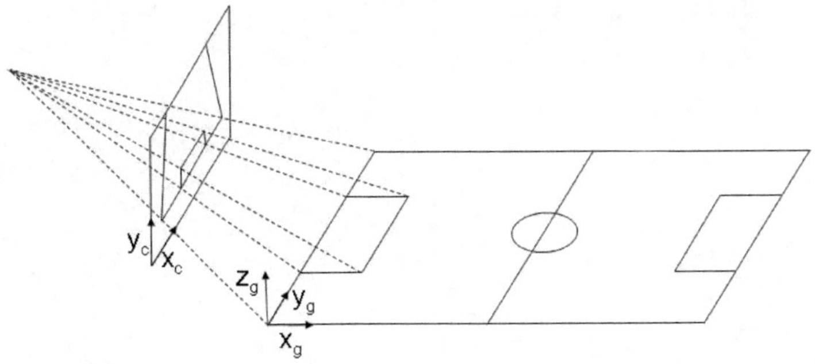

Fig. 1. Diagram depicting the transformation of coordinates from image to the ground

Using four pairs of corresponding points (no three of them being collinear), we can construct a 8x9 matrix $\mathbf{M}$, where $\mathbf{Mt} = \mathbf{0}$. Then, the solution $\mathbf{t}$ corresponds with the eigenvector associated to the least eigenvalue (in this case the null eigenvalue) of the matrix $\mathbf{M}^T$ $\mathbf{M}$, which can be easily solved by singular value decomposition (svd) of matrix $\mathbf{M}$. In order to have a reliable transformation, more than the minimum number of point correspondences must be considered, solving in a similar way [5].

It is known that a previous normalization of data is suitable to avoid numerical computation problems [6]. We have transformed the coordinates of the points (in the image and in the ground) before the computation of the homography to reference systems located in the centroid of the points and scaled in such that the maximum distance of the points to its centroid is 1. After computation of the homography, it is inversely transformed by simple matrix computation to express the homography in the desired reference systems.

2.2 Automatic Camera Re-calibration

Once we have calibrated the camera using at least 4 pairs of corresponding points in the image and in the ground, it cannot be moved, which is the main limitation of this proposal. In practice, due for example to the flexibility of the camera support, the orientation of the camera changes. A little change of orientation has a great influence in the image coordinates of a point, and therefore invalidates previous calibration. However if the camera is not changed in position, or position change is small with respect to the depth of the observed scene, the homography can be re-calibrated automatically with high robustness and without 3D computations. As camera position changes suppose main reconfiguration of the surveillance system, but orientation changes are usual, the automatic re-calibration procedure presented below eliminates the limitation in practice. Besides that, this re-calibration procedure can also be used for changes in zoom lens or motions in pan-tilt cameras demanded by the user.

The re-calibration can be made using features extracted in the image like points and/or lines. We propose to do it using lines because they are plentiful in man made environments and have other advantages. The straight lines have a simple mathematical representation, they can be extracted more accurately than points being also easier to match them and they can be used in cases where there are partial occlusions.

Fig. 2. Two images of a football match before and after an unexpected camera motion in a real application. The automatic robust line correspondences has been superimposed to the images.

After extracting the lines, automatic computation of correspondences and homographies is carried out, as previously presented in [7], which uses robust estimation techniques. Thus, initially the extracted lines are matched to the weighted nearest neighbor using brightness-based and geometric-based image parameters.

With the coordinates of at least four pairs of corresponding lines we can obtain an homography that transforms both images. As usually we have many more than four line correspondences, an estimation method can be used to process all of them, getting better results. The least squares method assumes that all the measures can be interpreted with the same model, which makes it to be very sensitive to wrong correspondences. The solution is to use robust estimation techniques which detect the outliers in the computation. From the existing robust estimation methods, we have chosen the least median of squares method [8].

In figure 2 we can see an example of two images before and after an unexpected camera motion. The automatic robust matching of lines that allows to compute the camera re-calibration has been superimposed to the images. The initial matching has about 20% of wrong correspondences, but the robust computation of the homography allows to reject wrong matches and also to search more matches according to it in a subsequent step. From the line correspondences the homography to recalibrate the camera is accurately obtained.

3 Tracking on the Ground

3.1 Detection and Tracking

A widely used technique for separating moving objects from their backgrounds is based on background subtraction [9]. In this approach, an image $I_B(x, y)$ of the background is stored before the introduction of a foreground object. Then, given an image $I(x, y)$ from a sequence, feature detection of moving objects are restricted to areas of $I(x, y)$ where:

$$|I(x, y) - I_B(x, y)| > \sigma \tag{1}$$

where σ is a suitable chosen noise threshold.

But this approach exhibits poor results in most real image sequences due to four main problems:

- Noise in the image.
- Gray level similarity between background and moving objects, even if the color is different.
- Continuous or quick illumination changes in the scene.
- Variation of the static objects in the background.

These problems can be partially solved by an appropriate selection of the threshold value. Some authors [10] [11] have proposed a region-based motion segmentation using adaptive thresholding, according to illumination changes. In addition to this, morphological filters have to be used to eliminate noisy pixels and to fill the moving regions poorly segmented. However, in spite of these improvements, the results that can be found in real situations are far away from a satisfactory solution.

To detect the moving objects, we present an approach in motion detection, based on difference, introducing two procedures, *Neighborhood-Based Detection* and *Overlapping-Based Labelling*, in order to obtain a more robust segmentation in real scenes. The first one uses a local convolution mask, instead of a punctual one, to compute difference and obtain a more reliable difference image. The second one uses an overlapping criterion between two difference image to classify blobs in two different types: static or dynamic. This last characteristic makes a distinction between moving objects and shadows or illumination changes.

After the motion detection, a Kalman tracker [12] with a constant velocity model [13] is used for tracking, using the center of each detected object as measure data. Internally, the tracker has a state with 4 elements: 2 for the position and 2 for the velocity.

The Kalman filter is divided in two main parts: prediction and estimation. Between them, a matching procedure associates the measures obtained in the motion detection with the prediction of the tracking. It select the nearest-neighbor if it is close enough in function of the covariance of the innovation.

3.2 Uncertainty Transformation

The measure in the Kalman tracker is the position (x, y) of the mobile object. The measure noise has pixel units, but in order to do the tracking in the ground, it must be transformed according to the homography to metric units. To transform the covariance matrix from image to ground, as proposed in [14], a transformation in three steps is required: change to homogeneous coordinates, transformation of coordinates from image to ground and transformation to inhomogeneous coordinates again.

Given a covariance matrix, which express the uncertainty location in image coordinates:

$$\Lambda_{\mathbf{x_c}}^{2\times 2} = \begin{pmatrix} \sigma_x^2 & \sigma_{xy} \\ \sigma_{xy} & \sigma_y^2 \end{pmatrix} \tag{2}$$

the correspondent homogeneous one is obtained:

$$\Lambda_{\mathbf{x_c}} = \begin{pmatrix} \Lambda_{\mathbf{x}}^{2\times 2} & \mathbf{0} \\ \mathbf{0}^{\top} & 0 \end{pmatrix} \tag{3}$$

The change of homogeneous coordinates from image to the ground is made with the homography matrix $\mathbf{T}_{gc}$. Therefore, the covariance matrix is transformed as:

$$\Lambda_{\mathbf{x_g}} = \mathbf{T}_{gc}\Lambda_{\mathbf{x_c}}\mathbf{T}_{gc}^{\top} \tag{4}$$

Once we have the uncertainty in the ground in homogeneous coordinates, we need to transform to non-homogeneous coordinates in order to have the measurement. We will use ∇f as a first-order approximation of the relationship between homogeneous and inhomogeneous coordinates. If $\mathbf{X}_g = (X, Y, W)^{\top}$

$$\nabla f = 1/W^2 \begin{pmatrix} W & 0 & -X \\ 0 & W & -Y \end{pmatrix} \tag{5}$$

Therefore, the covariance matrix of the measurements noise in ground coordinates is

$$\Lambda_{\mathbf{x_g}}^{2\times 2} = \nabla f \Lambda_{\mathbf{x_g}} \nabla f^{\top} \tag{6}$$

This transformation allows to have a noise model which considers the influence of the perspective effect when we made the tracking in the ground.

3.3 Tuning in Practice

The constant velocity model only can be considered locally valid. In practice, there are velocity changes that we model in the process noise. If we consider that the goal have an acceleration which is modelled as a white noise with zero mean and covariance q_i, the state noise matrix $\mathbf{Q}_i$ for each coordinate $i = x_g, y_g$ is [13]:

$$\mathbf{Q}_i = q_i \cdot \begin{bmatrix} \frac{dt^4}{4} & \frac{dt^3}{2} \\ \frac{dt^3}{2} & dt^2 \end{bmatrix} \tag{7}$$

where dt is the time interval.

If we consider the tracker in the ground, the tuning has a well known meaning, because $\sqrt{q_i}$ represents directly the acceleration of the mobile. On the other hand, the classical tracker in the image needs an empirical tuning in pixel units, that depends of the perspective effect.

The *measure noise matrix* $\mathbf{R}$ is defined in the image for both trackers and transformed to the ground using the homography matrix for the tracker on the ground, as seen in section 3.2.

4 Experiments

4.1 Description

The objective of these experiments is to compare the performance of a tracker on the ground versus a tracker on the image. Two Kalman trackers will be compared, using the same constant velocity model, although each one may have a different, but equivalent, tuning, since coordinates in the image and coordinates in the ground represent different magnitudes.

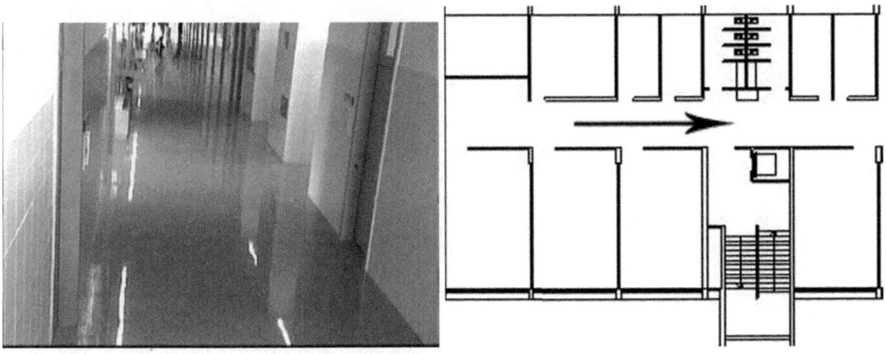

Fig. 3. First image of the sequence (left) and building plane (right) where the tests have been performed. The remote-controlled car moves along the corridor in the tests, as the arrow shows.

The three first tests performed compare the precision of the predictions of both trackers in a sequence of 550 frames, recorded with a still camera located in a corridor at 2.5 meters high. The target is a remote-controlled car moving at nearly constant velocity though a corridor, as it can be seen in figure 3.

After obtaining the position of the car in each frame, a tracker on the image has been applied. At the same time, the homography previously calculated, allows us to obtain the corresponding points in the ground for each point in the image, making possible to track the same object with another independent similar tracker on the ground.

To compare the performance of these two trackers, the mean difference between the prediction points and their corresponding measures has been computed. This difference is only taken into account if the measure is considered to belong to the object being tracked. Predictions from these two trackers are going to be compared at three levels: short, medium and long-term.

4.2 Comparative Analysis

Test 1: The first test compares the precision of short-term predictions. In the original sequence, the object is moving away the camera, with occasional lateral movements. The sequence will be tested also in reverse mode, starting from the last frame to the first, making the target go towards the camera. Mean distances obtained between prediction and measures, denoted as d_{pm}, are shown in table 1. All distances are measured in the ground, using Euclidean distance.

Table 1. Mean distances in mm. between predictions and measures for the image tracker and the ground tracker with the sequence processed forwards and backwards.

d_{pm}	Image	Ground
Forwards	54,90	18,89
Backwards	44,02	20,01

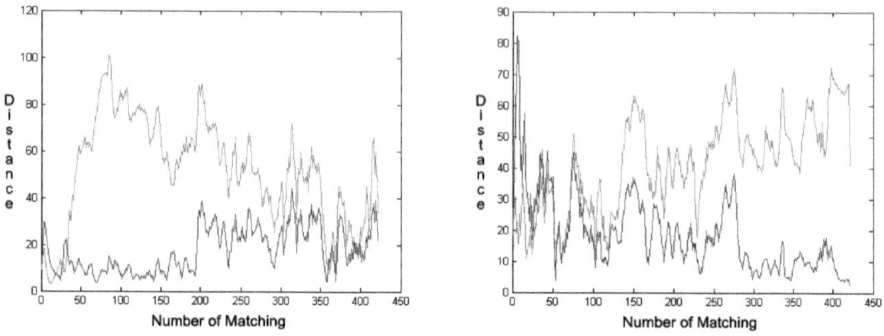

Fig. 4. Distance between prediction and measure with tracking on the ground (dark line) and tracking on the image (light line). (left) Target moving away the camera (right) Target moving to the camera.

In this first test, a tracking on the ground has a great advantage over a tracking on the image, since the effect of the perspective deformation is avoided. Distances between predictions and measures are represented in figure 4.

As it can be seen in figure 4, a tracker on the ground obtains better results if the moving object is near the camera, because the velocity of the object in the image is more changeable. However, a tracker on the image obtains opposite results, since only measure noise causes this error. In any case, even when the moving object is far from the camera, the tracker on the ground obtains better results.

It must be noticed that two kinds of source of noise could be considered in the measure: one generated in the detection and other caused by the errors in the homography. At the moment, this second noise source has not been modelled, considering that its effect is small, because all the trajectories are inside the points used to compute the homography matrix in the calibration phase.

Test 2: To compare the accuracy of medium-term predictions, another test will be carried out, in the same sequence, consisting in decimating the number of measures. After the decimation, we will have $550/f$ measures, where f is the decimation factor. The same values for $\mathbf{Q}$ and $\mathbf{R}$ are used. Mean distance d_{pm} will be used again to measure the precision in each case. Obtained results can be seen in table 2, also in mm.

Table 2. Mean distances in mm. between prediction and measure applying different grades of decimation.

f	Image	Ground
1	54,9	18,89
2	78,49	22,89
3	97,75	26,58
5	132,01	31,06
8	180,02	40,33
12	248,08	55,93

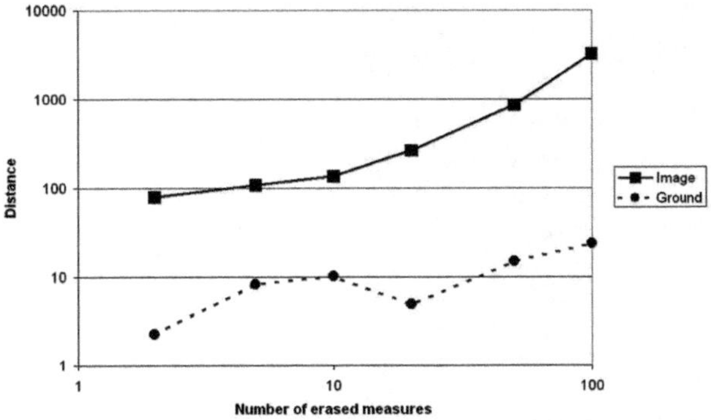

Fig. 5. Distances obtained after erased blocks of measures of different lengths (in logarithmic scale)

Mean distances grow as decimation factor increases, since the movement of the object is not totally predictable. The tracker on the image has an extra error originated by the deviation in velocity produced by the perspective effect.

Test 3: To test the precision of long-time predictions, a determined number of consecutive measures will be erased for both trackers. Also the same values of $\mathbf{Q}$ and $\mathbf{R}$ are used. This test evaluates the possibility of recovering the object after an occlusion.

Different numbers of measures have been erased in each test, from 2 to 100. To measure the accuracy of each tracker, the distance in the ground between the measure and the prediction after the erased block of measures has been used. The results can be seen in figure 5.

Here the differences between both trackers are higher. The ground tracker does not lost the measure even after an occlusion of about 100 frames, maintaining short distances between prediction and measure. However, the image tracker easily lost the measure after an occlusion of about 10 frames, giving very bad predictions.

4.3 Using in Practice

Performance in Real Scenarios. Once tested the superiority of the ground tracker in the laboratory, it has been confirmed in real-life videos of a football match. The test consists of a video sequence, as the frame in figure 6.a, where two football players are going to be tracked. Both players are running in parallel trajectories, but at different distances from the camera. In this test, the necessity of re-tuning of each trackers when trying to track different objects will be evaluated.

Both trackers will be configured with different equivalent tunings for the distant player. Using the homography, we can determine the relationship between a pixel in the image and the scale of the field to tune both trackers in a equivalent way. The measure noise $\mathbf{R}$ is fixed, defined for the image tracker, and translated using the homography matrix for the ground tracker. This tuning will be used on the nearby player to check its validity.

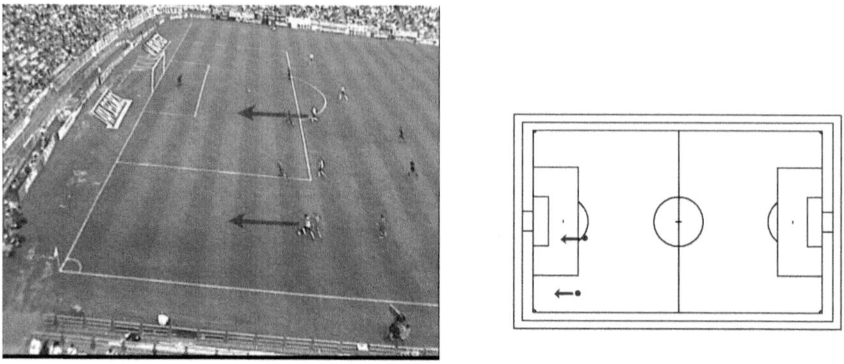

Fig. 6. Frame of the video (left) and plane of the football field (right) of the video used. The arrows represent the direction of the players that have been used for the test.

The results can be seen on figure 7. As the two trajectories are not equivalent, and may have different accelerations and noises, the results cannot be directly compared. In any case, the figure shows that the number of matchings obtained for the distant player is similar for the two trackers, while the ground tracker obtains more matchings than the image tracker for the nearby player using the same tuning.

Although the number of matchings always can be increased with higher values of the the matrices **R** and **Q**, the prediction error and the possibility of crossing with other measures would increase as well. Hence, the application for a ground tracker is more important if multiple objects are attempted to be tracked.

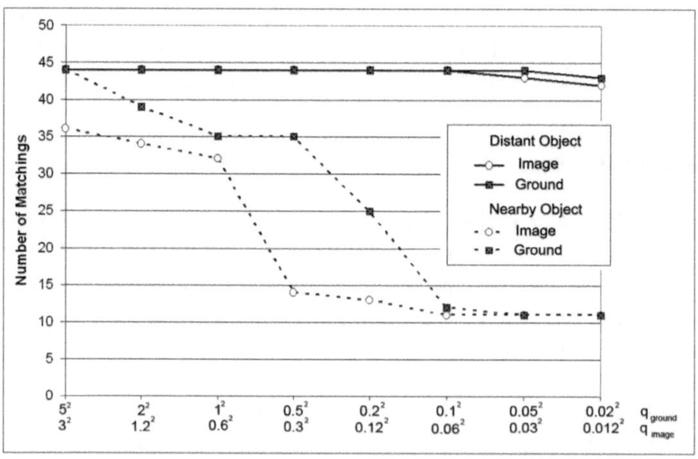

Fig. 7. Number of matchings for different state noises. The two scales are equivalent for the position of the distant player.

Using a Multi Camera Tracker. One of the main applications of a tracking on the ground is to be able to use a multi camera tracking algorithm using a common reference. The measures from different cameras are transformed to the ground and then a matching algorithm is used to decide the correspondences among measures and trackers.

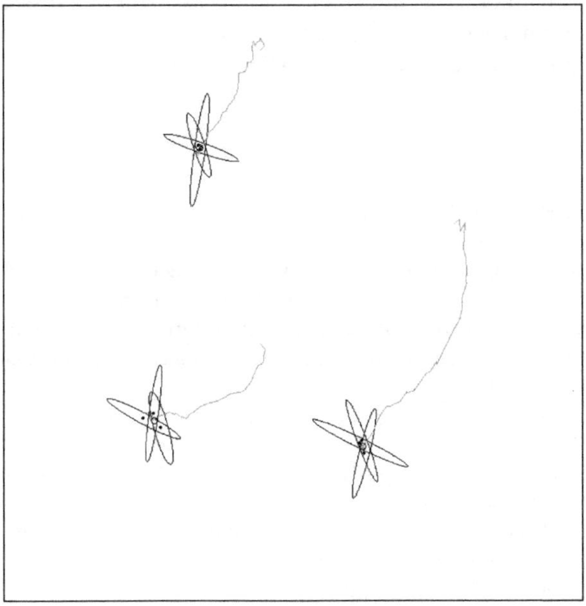

Fig. 8. Tracking of 3 players using 3 cameras. All the measures are transformed to the ground and represented as a point and a ellipsoidal error around the central point. The trajectory of the trackers is also shown in gray.

Using the method described in this paper, the measures of 3 different cameras are drawn in the ground, as seen in figure 8. Each one of the cameras has a circular measure error around the position of the measure in the camera, that is also transformed to the ground as an ellipse. In the figure we can see how the different cameras produce slanted ellipsoidal errors in different angles, depending on the relative position between the camera and the player.

5 Conclusions

In this paper we compared a tracker on the image versus a tracker on the ground. A plane projective transformation allows to make the tracking in real coordinates which facilitates the tuning of the tracker, gives measures in real coordinates and allows to relate different cameras in a common reference system. Experimental results from laboratory test and from real environments proved empirically that the tracker on the ground achieves better results. We have also shown some preliminary results for the automatic re-calibration of the camera which avoids some of the practical limitations of the approach. The continuation of this work will be focused to the usage of multiple cameras, having the plane of the surroundings as a common reference for the tracking.

Acknowledgements. This work was supported by project OTRI 2005/0388 "UZ - Real Zaragoza - DGA Collaboration Agreement for development of a research project in the sport performance improvement based on the image analysis", and it establishes the grounding for taking real measures and statistics over the ground.

References

1. Tissainayagam, P., Suter, D.: Visual Tracking with Automatic Motion Model Switching. Arch. Rat. Mech. Anal. (2001)
2. Isler, V., Khanna, S., Spletzer, J., Taylor, C.J.: Target tracking with distributed sensors: The focus of attention problem. Computer Vision and Image Understanding 100, 225–247 (2005)
3. Lee, L., Romano, R., Stein, G.: Monitoring Activities from Multiple Video Streams: Establishing a Common Coordinate Frame. IEEE Transactions on Pattern Analysis and Machine Intelligence 22(8) (2000)
4. Mundy, J.L.S., Zisserman, A.: Geometric Invariance in Computer Vision. MIT Press, Boston (1992)
5. Hartley, R., Zisserman, A.: Multiple View Geometry in Computer Vision. Cambridge University Press, Cambridge (2000)
6. Hartley, R.I.: In defense of the eight-point algorithm. IEEE Trans. on Pattern Analysis and Machine Intelligence 19(6), 580–593 (1997)
7. Guerrero, J.J., Sagüés, C.: Robust line matching and estimate of homographies simultaneously. IbPRIA. In: Perales, F.J., Campilho, A.C., Pérez, N., Sanfeliu, A. (eds.) IbPRIA 2003. LNCS, vol. 2652, pp. 297–307. Springer, Heidelberg (2003)
8. Rousseeuw, P.J., Leroy, A.M.: Robust Regression and Outlier Detection. John Wiley, Chichester (1987)
9. Herrero, E., Orrite, C., Senar, J.: Detected motion classification with a double-background and a neighborhood-based difference. Pattern Recognition Letters 24, 2079–2092 (2003)
10. Durucan, E., Ebrahimi, T.: Change detection and background extraction by linear algebra. Proceedings of the IEEE 89(10), 1368–1381 (2001)
11. Moscheni, F., Bhattacharjee, S., Kunt, M.: Spatiotemporal segmentation based on region merging. IEEE Transactions on Pattern Analysis and Machine Intelligence 20(9), 897–915 (1998)
12. Kalman, R.E.: New Approach to Linear Filtering and Prediction Problems. Transactions of the ASME–Journal of Basic Engineering 82, Series D, 35–45 (1960)
13. Bar-Shalom, T., Fortmann, T.E.: Tracking and Data Association. Academic Press Inc., London (1988)
14. Criminisi, A., Reid, I., Zisserman, A.: A plane measuring device. Image and Vision Computing (IVC) 8(17), 625–634 (1999)

Mobile Robot Teleoperation via Wireless Multihop Networks - Parameter Tuning of Protocols and Real World Application Scenarios

Florian Zeiger, Nikolaus Krämer, Markus Sauer, and Klaus Schilling

University of Wuerzburg, Department of Robotics and Telematics
Am Hubland, Wuerzburg, Germany
zeiger@informatik.uni-wuerzburg.de, kraemer@informatik.uni-wuerzburg.de,
sauer@informatik.uni-wuerzburg.de, schi@informatik.uni-wuerzburg.de

Abstract. Currently, the use of wireless networks is very common in the field of networked robotics and can be considered as a key issue for capable multi robot systems with a high grade of mobility. Nevertheless, this mobility requests for special features of the communication infrastructure, which leads to the integration of mobile robots into wireless ad-hoc networks. Since the late nineties, more than 80 ad-hoc routing protocols were developed and nowadays some of them are implemented and ready to use in real world applications. A comparison of four ad-hoc routing protocols (AODV, DSR, OLSR, and BATMAN) showed shortfalls of the default parameter settings not allowing a reliable teleoperation of mobile robots while using AODV, OLSR, or BATMAN. This work is focused on the parameter tuning of the routing protocols to use them in wireless ad-hoc networks of mobile robots. The time required for route reestablishing, as well as the packet loss during rerouting is investigated in hardware tests of a network with dynamic network topology consisting of mobile robots. It could be demonstrated, that an appropriate parameter setting of OLSR and BATMAN allow the teleoperation of mobile robots in outdoor environments via a wireless ad-hoc network. Finally a small application is presented to show the interaction of the different protocol layers in order to provide a communication network for services and applications to increase the capabilities of networked mobile robots.

1 Introduction

Currently, more and more research is done in the field of teleoperation of mobile robot teams via wireless networks. As now a larger number of mobile robots are developed which are capable to operate in impassable or hazardous environments with little or no communication infrastructure, the communication infrastructure is set up by the robots itself on demand. Within these wireless ad-hoc networks, different types of nodes might be present: human workers or rescue personnel equipped with modern communication devices, mobile robots, or even some stationary nodes. All of them are able to act as data source, data sink, and communication relay and must support a dynamic network topology. In the field of networked robotics, several approaches are using wireless ad-hoc networks in many different areas of robot teleoperation. Multi-robot exploration with robots using wireless networks [1] or a mobile robot team connected via wireless network which performed localization and control tasks [2]. Also in the field of rescue robotics [3], or for integrating UAVs into IP based ground networks [4], the use of wireless networks is quiet common nowadays.

J.A. Cetto et al. (eds.): Informatics in Control, Automation and Robotics, LNEE 37, pp. 139–152.
springerlink.com

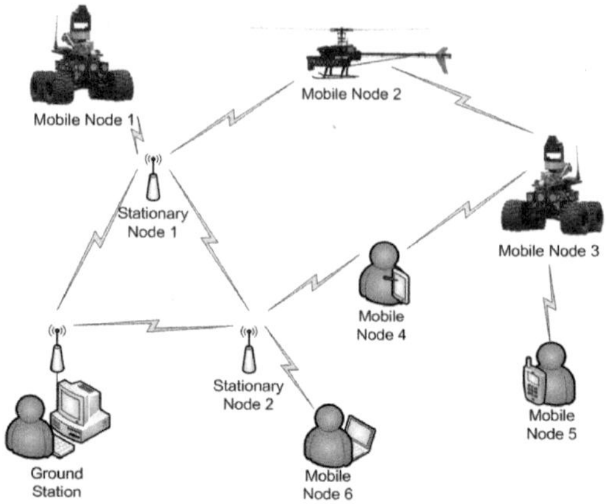

Fig. 1. Future scenario of a heterogeneous network of mobile robots and human

An example for the network topology of these future scenarios is given in Figure 1. The network consists of several stationary nodes or ground stations and several mobile nodes which can be ground vehicles, aerial vehicles, or humans equipped with communication devices. All these nodes are connected by an ad-hoc wireless network which should guarantee a transparent any-to-any communication. Nevertheless, wireless communication always implies unpredictable communication delays, packet loss, or in worst case the loss of the link which makes the provision of the required quality a challenging task [5]. To avoid the loss of communication, research focused on a dynamic setup of the required telecommunication infrastructure by placing relay nodes on demand [6][7] or using mobile robots as relay nodes [8][9]. These approaches are using communication relays in wireless ad-hoc networks to setup communication networks with dynamic topologies. In these wireless networks no fixed infrastructure exists, and each mobile node not only works as host but also as router for data packets of other nodes. These dynamic topologies of wireless communication networks have advantages like providing direct and indirect any-to-any communication of each network node, redundant communication links in larger networks, no central administration, and a distribution of the traffic load in large networks. Of course, these advantages can only be used with rather complex and special routing protocols providing each node the necessary information about the network topology. The nodes itself are working as routers and must store the routing information of the complete network locally. In the field of wireless telecommunication, more than 80 ad-hoc routing protocols for wireless networks were developed [10][11][12][13]. Also some simulations for performance evaluations for larger scale telecommunication networks were done in the past [14][15][16]. [17] compared several ad-hoc routing protocols in a simulation study and [18] gives a survey of currently existing real-world implementations of ad-hoc routing protocols, including some real-world scenario tests.

This work is based on the results of a former publication [19], which compares several ad-hoc routing protocols with respect to mobile robot teleoperation. The standard parameter settings of the routing protocols AODV, OLSR, DSR, and BATMAN were investigated. Unfortunately, only DSR showed to be an appropriate solution for mobile robot teleopreation and the performance other three routing protocols had to be improved by parameter tuning. This work is focused on the parameter tuning of OLSR, AODV, and BATMAN. The objective is an acceptable packet loss and time for rerouting in a highly dynamic network topology. Therefore, existing protocol implementations of AODV and DSR (http://core.it.uu.se/core/index.php) from Uppsala University and the University of Basel, OLSR (http://www.olsr.org) and BATMAN (https://www.open-mesh.net/batman) are used in real-world test scenarios where mobile robots are teleoperated in an outdoor environment.

The presented work is structured as follows. in Section 2, the used hardware and the real-world test scenarios are described. Section 3 gives a brief summary of a comparison of AODV, DRS, OLSR, and BATMAN with respect to mobile robot teleoperation with standard parameter settings. The results of this work – the parameter tuning of ad-hoc routing protocols to enable mobile robot teleoperation via wireless ad-hoc networks – are presented in Section 4. By presenting an adaptive traffic shaping mechanism which is used for video transmission of a mobile robot in an outdoor environment Section 5 shows how the application layer can be used to provide useful services for mobile robot teleoperation via ad-hoc networks. A conclusion is given in Section 6.

2 Hardware and Test Setup

2.1 Hardware

For the performed tests, several mobile nodes were used. One node is a PC for the operator. Up to 4 MERLIN robots (standard version) were used as stationary communication relay nodes, and one Outdoor MERLIN was used (cf. Figure 2) [20].

All MERLIN robots have a C167 micro controller for low-level operations and sensor data processing, as well as a PC-104 for more complex and computationally more intensive tasks. The PC-104 uses a Linux operating system and all nodes are equipped with 802.11b standard WLAN equipment (Atheros chip). More information on the MERLIN robots is given in [20]. For steering the mobile robot, the operator's PC is running an application which generates command packets of a size between 6 and 22

Fig. 2. The Teleoperated OutdoorMERLIN Robot

bytes of payload. These packets are sent via UDP over the wireless network to the mobile robot. The onboard software of the mobile robot generates a UDP packet stream of packets with variable size containing the sensor data.

2.2 Test Setup

The scenario is set up in a way that the rerouting procedure will start with the mobile robot being at a certain location. Therefore, a large building is used as obstacle (cf. Figure 3). Relay nodes are placed at the corners of the building, such that they have always the neighbor nodes at the next and previous corner of the building within their communication range. As soon as the mobile robot is moved out of the line-of-sight of one node, the rerouting procedure is initialized. This scenario represents a worst case in terms of link redundancy, as only one route between operator PC and mobile robot is available. Relevant measurement categories are the packet loss and the duration of a communication drop-out during rerouting.

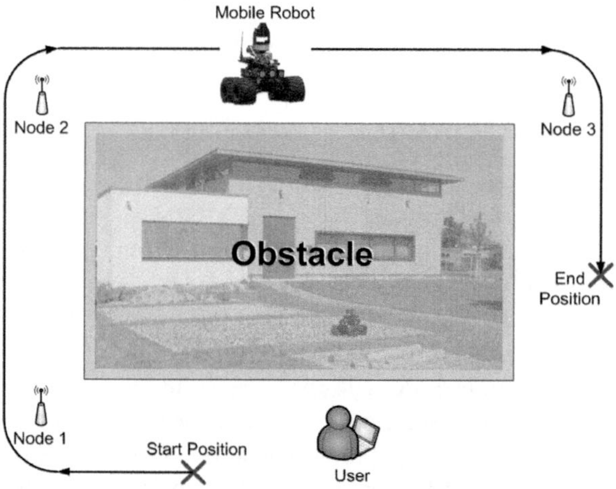

Fig. 3. Test Setup

3 Ad-Hoc Routing and Teleoperation

3.1 Investigated Protocols

This work investigates the parameter settings for different ad-hoc routing protocols: Ad-hoc On-demand Distance Vector (AODV), Dynamic Source Routing (DSR), and Optimized Link State routing (OLSR).

AODV [12] [13] is a reactive routing protocol and determines required routes on-demand. To discover a route to an unknown destination, a Route Request (RREQ) message is broadcasted. Each intermediate node which is not the destination and without a route to the destination receiving a RREQ broadcasts it further. In case the RREQ is received more than once, only the first reception will result in a broadcast. To avoid uncontrolled dissemination of RREQs, each has a certain time to live (TTL) after which it is discarded. When the destination receives a RREQ message a Route Reply (RREP) message is generated and sent back to the source in unicast hop by hop fashion along the route which was determined by the RREQ message. After generating a RREP message, the RREQ message is discarded at this node. As the RREP propagates, each intermediate node creates a route to the destination. After the source receives the RREP, it records the route to the destination and begins sending data. In case the source receives multiple RREPs, the route with the shortest hop count is chosen. The status of each route is maintained in the local routing table and timers are used to determine link failures. In case a certain node is part of an active route, Hello messages are used to obtain the route status. These Hello messages are broadcasted periodically to all neighbors. If a neighbor does not send a Hello message within a specified time a link loss is detected and the node is deleted from the routing table. In addition, a Route Error message (RRER) is generated. More detailed information on AODV is given in [12]. In the test scenarios of this work, AODV-UU version 0.9.5 from Uppsala University (Sweden) is used (http://core.it.uu.se/core/index.php/AODV-UU).

DSR is also a reactive ad-hoc routing protocol which works similar to AODV but without using Hello messages for route maintenance. However, it uses source routing [21]. DSR does not use any periodic routing advertisement, link status sensing, or neighbor detection packets, and does not rely on these functions from any underlying protocols in the network. DSR is composed of two main mechanisms that work together to allow the discovery and maintenance of source routes in the ad-hoc network. In case source node (S) wants to send data to an unknown destination host (D), S initiates the route discovery mechanism. S broadcasts a route request message which identifies the source and destination of the route discovery to all neighbors. A route request also contains a record listing the address of each intermediate node which forwarded this particular copy of the route request. A node which receives this route request without being the destination looks up for a source route to the requested destination in its own route cache. Without any source route present in its route cache, the node appends its own address to the route record and broadcasts the route request message. In case this request message was received more than once, it is simply discarded. As soon as the route request message arrives at the desired destination D, a route reply message to S is created which contains an accumulated route record of the route request. After S receives this route reply, it caches the corresponding route in its route cache and S is ready to transmit data. Of course, there exist mechanisms to omit flooding of the network with route requests. A hop limit was introduced and every time a route request is forwarded, the hop limit is decremented by one. As soon as it reaches zero, the request is discarded. Also mechanisms for avoiding infinite recursion of route discoveries are implemented. A more detailed description of this protocol is given in [10] [22]. The presented work uses DSR-UU version 0.2 from Uppsala University (Sweden) (http://core.it.uu.se/core/index.php/DSR-UU).

OLSR is a table-driven pro-active routing protocol for mobile ad-hoc networks. It uses hop-by-hop routing (each node uses its local information to route packets). OLSR minimizes the overhead from flooding of control traffic by using only selected nodes called Multipoint Relays (MPR) to retransmit control messages. Each node in the network selects a set of nodes in its neighborhood, which may retransmit its messages. This set of selected neighbor nodes is called the MPR set of that node. The neighbors of node N which are not in its MPR set, receive and process broadcast messages but will never retransmit broadcast messages received from node N. The MPR set is selected such, that every node in the 2-hop neighborhood of N has a link to the MPRs of N. OLSR continuously maintains routes to all destinations in the network by distributing link and neighborhood information (periodically exchange Hello messages). These messages are also used for link sensing and for checking the connectivity. More details on OLSR are given in [23]. The scenario tests in the present work are performed with OLSR version 0.5.3 (http://www.olsr.org/index.cgi?action=download).

BATMAN (Better approach to mobile ad-hoc networking) is a new approach to ad-hoc routing. Unlike other algorithms that exist right now, BATMAN does not calculate routes. It continuously detects and maintains the routes by receiving and broadcasting packets from other nodes. Instead of discovering the complete route to a destination node, BATMAN only identifies the best single-hop neighbor and sends a message to this neighbor. These messages contain the source address, a sequence number, and a time-to-live (TTL) value that is decremented by 1 every time before the packet is broadcasted. A message with a TTL value of zero is dropped. The sequence number of these messages is of particular importance for the BATMAN algorithm. As a source numbers its messages, each node knows whether a message is received the first time or repeatedly. More details on BATMAN are given in [24]. In the test scenarios of the presented work, BATMAN version 0.2 is used (https://www.open-mesh.net/batman).

3.2 Rerouting Time and Packet Loss with Standard Parameter Settings

In [19], four different ad-hoc routing protocols were investigated with respect to mobile robot tele-operation. A mobile robot was commanded in a test scenario which forced the routing protocols to increase the number of participating nodes in the communication link while the robot was moved around an obstacle. The four compared ad-hoc routing protocols were all used with the standard parameter settings and behaved quite different. While it was not possible to accomplish the scenario with BATMAN, the other protocols at least allowed a teleoperation of the mobile robot – often with only very limited performance. AODV was originally designed for highly dynamic networks. Routes are established on demand. In some cases this rerouting took only a very short time (cf. Table 1), but sometimes, the communication drop-out duration was longer than 30 seconds. This is by far too long for the telecommand of a mobile robot. Compared to

Table 1. Packet Loss & Times for Route Reestablishing from [19]

Protocol	Packet loss during test run	Time for re-routing	
		min.	max.
AODV	29.2%	$2.1s$	$> 30s$
OLSR	14.2%	$10.1s$	$> 30s$
DSR	11.2%	$2.4s$	$2.7s$
BATMAN	conn. lost	–	–

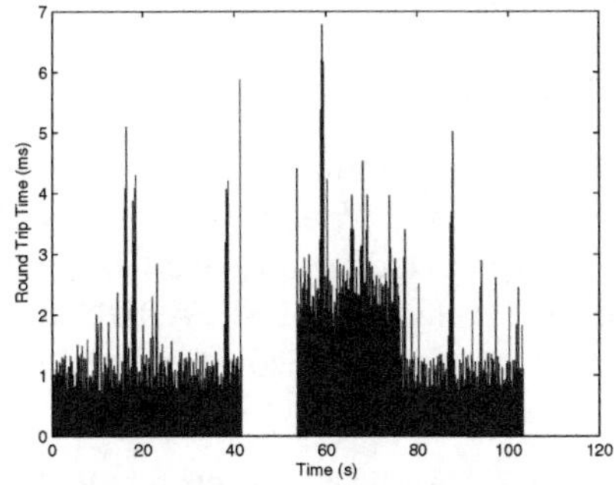

Fig. 4. Example for OLSR round trip times for scenario 1 with default parameter setting

AODV, the minimum of the required re-routing time of OLSR is less. Rarely, also communication drop outs were observed. With only half of the packet loss, OLSR showed a slightly better performance than AODV. Although OLSR worked more reliable than BATMAN or AODV, the observed minimum time for re-routing of 10.1 seconds is quite high with respect to teleoperation and will not be appropriate for any kind of control via this network. DSR showed to be the most reliable and the fastest protocol which was tested. A packet loss of about 11% and a re-routing time between 2.4 and 2.7 seconds make this protocol suitable for reliable telecommand of a mobile robot. With respect to the test scenario, it was expected that DSR performs best, as only one node (the robot) is mobile and all other nodes are stationary. Here, DSR discovers the topology quiet fast and only the changes due to the robots movement result in routing messages. The used test scenario also represents a worst case in the means of route redundancy due to the availability of only one possible route between controller and mobile robot. This could be the reason for the relatively poor performance of AODV and OLSR. Originally, these protocols were developed to handle much larger networks with higher node mobility and a higher grade of meshing as in the current test scenario. Nevertheless, the presented test scenario is quiet typical with respect to teleoperation. In Table 1, also the average packet loss during route reestablishing is given. Again DSR showed the best performance (11.2% packet loss) compared to OLSR (14.2% packet loss) and AODV (29.2% packet loss). BATMAN was not able to establish a new route via additional relay nodes.

Figure 4 displays an example of the round trip times for test scenario 1 and shows a clear communication drop out between 40 and 50 seconds test time. In Figure 5 the behavior of DSR in the same scenario is showed. Here, the communication drop-outs were significantly shorter. The compared ad-hoc routing protocols followed different principles (e.g. proactive vs. reactive) and were designed by different inspirations and for different application areas. Thus, they have also different parameters with different default settings, too. Nevertheless, this investigation showed that a better performance is required for mobile robot teleoperation. Based on these experiments, protocol parameters are identified and tuned in order to increase the quality of the communication. The results are presented in the following sections.

4 Parameter Tuning

4.1 Variable Protocol Parameters

This section summarizes the default parameter settings of the investigated ad-hoc routing protocols OLSR (cf. Table 2), AODV, and BATMAN. In [19], these settings were chosen as a trade-off between generated routing overhead and fast topology discovery. Unfortunately, these parameter settings for OLSR, AODV, and BATMAN showed to be not suitable to use these ad-hoc routing protocols for mobile robot teleoperation.

Table 2. Variable parameters for OLSR

Name:	Default Value
Willingness	dyn. calc.
LinkQualityLevel	2
LinkQualityWinSize	10
Pollrate	0.05 sec
TcRedundancy	0
HelloInterval	2 sec
HelloValidity	6.0 sec
TCInterval	5 sec
TCValidity	15.0 sec

For AODV, relevant parameters are "force gratuitous", "local repair", and "no wait on reboot" which are disabled by default. For BATMAN the "originator interval" will be used for optimization. By default, "originator interval" is set to 1000 milliseconds. OLSR uses "HelloInterval", "HelloValidity", "TcInterval", and "TcValidityTime" (cf. 3) as parameters.

4.2 OLSR

To increase the performance of OLRS in order to use it for mobile robot teleoperation, the four parameters were changed as shown in Table 3. As the parameter settings are

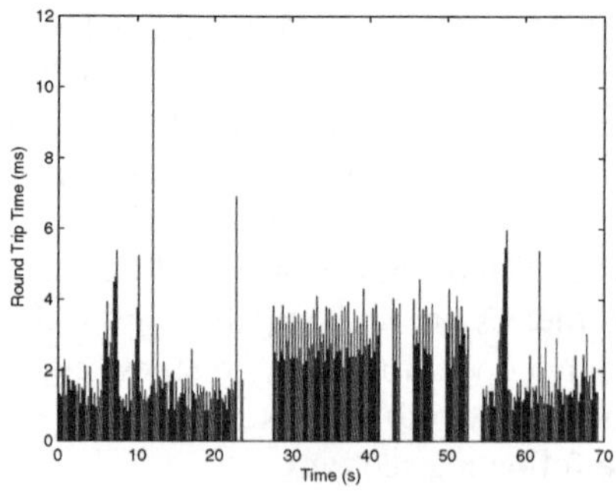

Fig. 5. Example for DSR round trip times for scenario 1 with default parameter setting

Table 3. Tuned parameters for OLSR

Name:	Default Value:	New Value:
HelloInterval	2 sec	0.5 sec
HelloValidity	6 sec	1.5 sec
TcInterval	5 sec	2.5 sec
TcValidityTime	15 sec.	5.0 sec

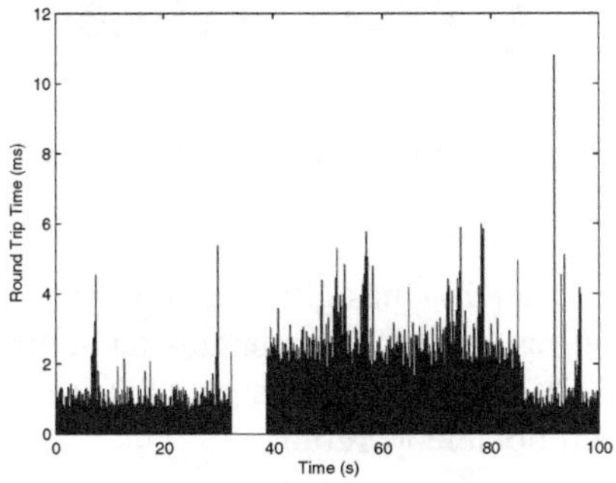

Fig. 6. Example for OLSR round trip times for scenario 1 with HelloInterval=0.5, HelloValidity=1.5, TCInterval=2.5 and TCValidity=5.0

interdependent, some simple rules must be followed as not all combinations of values are useful. The hello- and tc-intevals have to be smaller than half the corresponding validity times. Of course, the traffic for routing will be increased due to more hello and tc-messages but will not cause decrease the throughput significantly.

The rerouting time was reduced from more than 10.1 to an average value of 5.96 seconds (with a minimum of 5.2 and a maximum of 7.4) for including the first relay node into the communication link. The packet loss during rerouting was reduced from 14.2% to 4%. Figure 6 shows an example of the behavoir of the round trip time during a rerouting process with OLSR. As the rerouting is initiated at 32.4 seconds, a short communication drop out occurred until the link is reestablished at 37.2 seconds simulation time. Thus, the OLSR performance could be increased but still, the observed performance of DSR during the comparison in [19] could not be reached.

4.3 BATMAN

As already mentioned, the number of variable parameters for BATMAN is very small. A suitable parameter to be tuned is the "originator interval", which is the time to wait sending one message and before the batman daemon sends the next message (default value is 1000 milliseconds). In [19], all BATMAN test runs ended up in a communication loss. Now, the value for "originator interval" is decreased which should result in a faster response on topology changes by the routing protocol. Figure 7 shows an Example of a test run with "originator interval" set to 125 ms. Of course, this setting increases the routing overhead but as a result, BATMAN now can handle the rerouting

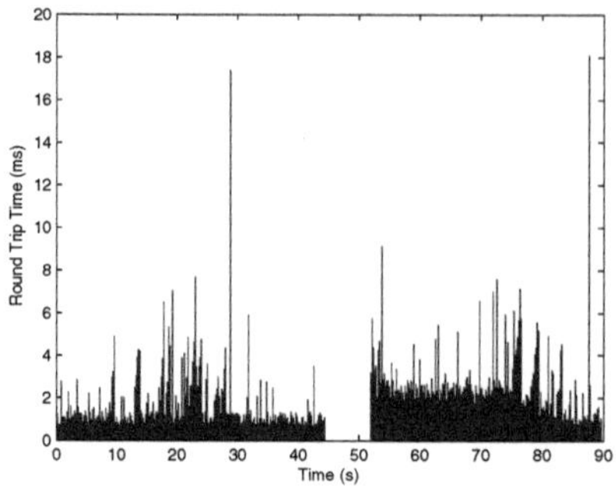

Fig. 7. Example for BATMAN with "originator interval" set to 125 ms

without losing the communication. During the tests, the average rerouting duration is 7.78 seconds with a packet loss of 8%. This parameter setting significantly improved the performance of BATMAN but still, the duration of the rerouting procedure takes too long to be used for mobile robot teleoperation.

4.4 AODV

For AODV, it is possible to tune several parameters. The performed tests showed, that only an combined change of several parameters might have the opportunity to show a positive effect with respect to teleoperation. Unfortunately, it was not possible to find a suitable parameter setting – all combinations made the protocol behaving very unstable. For example, enabling "force gratuitous", "local repair", "no wait on reboot", and setting "treat node as neighbor" to 2 (cf. Figure 8) lead to many communication losses.

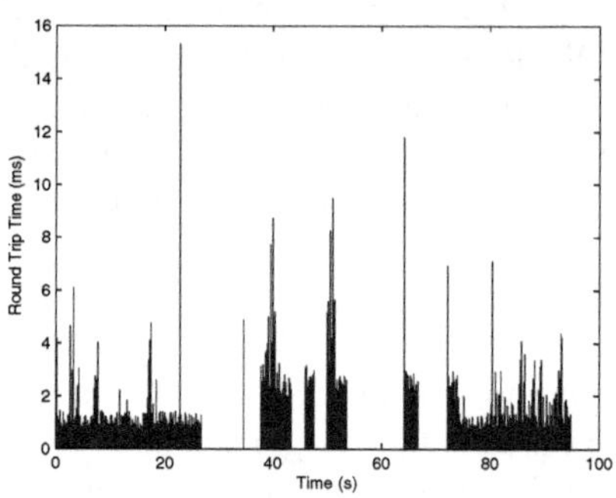

Fig. 8. Example for AODV with "force gratuitous", "local repair", and "no wait on reboot" enabled and "treat node as neighbor" = 2

5 Application Layer

The above presented results allow for the application of wireless ad-hoc networks for mobile robot teleoperation. So far, this work presents how parameter tuning of lower protocol layers can improve the teleoperation capabilities of mobile robots by faster route reestablishing. Nevertheless, as soon as additional services are provided or applications use the network for bandwidth consuming data exchange, communication failures and instable communication links may occur due to link overload. This section is focused on the application layer and gives an example of a traffic shaping mechanism to show how applications can use the above presented ad-hoc networks efficiently. [25] presents a traffic shaping mechanism which is used in an ad-hoc network to provide a video stream for mobile robot teleoperation. The mechanism uses a feedback which is generated at the different nodes and gives information about the link status inside the network. Based on this feedback, the traffic source can now shape its data stream in order to avoid overload situations of the network. The video traffic shaping is basically done by adapting the amount of transmitted data by changing the quality of the video images according to the link status. In case the link quality decreases, the shaping mechanism starts to transmit video images with a reduced quality. Thus, the operator of a mobile robot receives a video stream with the quality degrading stepwise. This turned out to be more convenient to the operator than a highly variable frame rate (cf. [25]). By using a real-world scenario where a mobile robot is teleoperated via a wireless ad-hoc network the efficiency of this mechanism is presented. The control commands as well as the video stream are transmitted and several relay nodes are used for the communication. In case the mobile robot is teleoperated without traffic shaping, some links are overloaded and the frame rate varies significantly. As a result, a very bad video in terms of constant frame rates is received at the operator which makes teleoperation based on the video data impossible. This result is also shown in Figure 9. As soon as traffic shaping is active, the amount of transmitted data is adaptively reduced and the frame rate stays at a constant level (cf. Figure 10). This small example shows how the provided service benefits from a seamless interaction of higher and lower protocol layers and

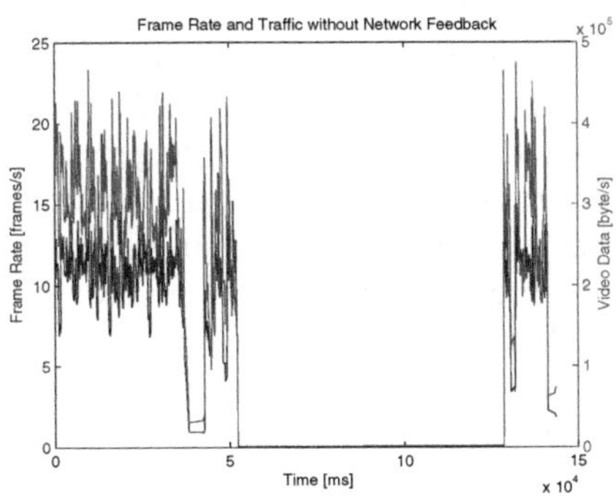

Fig. 9. Frame rate and received amount of video data at the receiver without shaping the video stream (from [25])

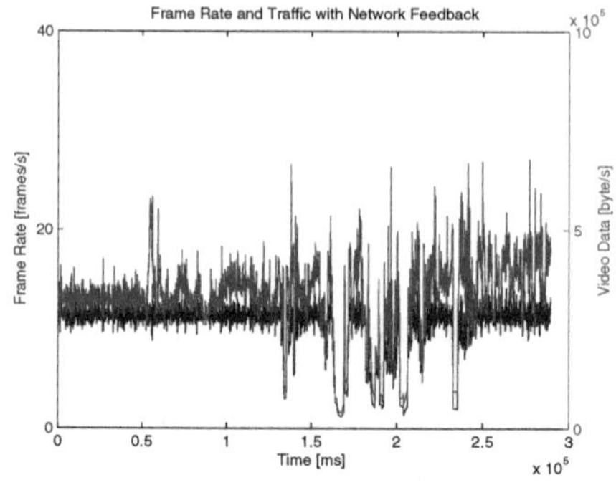

Fig. 10. Frame rate and received amount of video data at the receiver (from [25])

how complex ad-hoc networks can easily be used to increase the capabilities of mobile robots. For more information on the described traffic shaping mechanism please refer to [25].

6 Conclusions

This work uses results of a comparison of four ad-hoc routing protocols as a basis where the default parameter setting showed to be not appropriate for mobile robot teleoperation. Here, results of a study for parameter tuning of real implementations of the ad-hoc routing protocols OLSR, AODV, and BATMAN are presented. Real hardware tests of a mobile robots teleoperation scenarios were performed and the behavior of the communication link was analyzed with respect to mobile robot teleoperation. It is shown, that a tuning of a combination of relevant parameters for OLSR can improve its performance. Also the reduction of "originator interval" of BATMAN results in a more reliable network than experienced with the default parameter settings. For AODV, sometimes a better reaction in terms of required time for rerouting could be observed. The evaluated protocol implementations are suitable for some teleoperation approaches – e.g. systems with local autonomy functions to ensure a defined behavior of the mobile robot during the rerouting times. Nevertheless, the behavior of all tested ad-hoc routing implementations was less than expected. The previously mentioned simulation results showed much shorter rerouting times (in a magnitude of some milliseconds). These short rerouting times were never observed in the presented real hardware tests. Future work will now be focused on a detailed analysis of the differences between the simulation studies and the effects described in this work to further improvement the usability of AODV, OLSR, DSR, and BATMAN for mobile robot teleopertation or remote control approaches via wireless multi-hop networks. In addition, the required interaction between applications (e.g. teleoperation interfaces or control algorithms) and the network status via the lower protocol layers must be analyzed.

References

1. Rooker, M.N., Birk, A.: Multi-robot exploration under the constraints of wireless networking. Control Engineering Practice 15(4), 435–445 (2007)
2. Das, A., Spletzer, J., Kumar, V., Taylor, C.: Ad Hoc Networks for Localization and Control. In: Proceedings of the 41st IEEE Conference on Decision and Control (CDC 2002), vol. 3, pp. 2978–2983 (2002)
3. Rooker, M.N., Birk, A.: Combining Exploration and Ad-Hoc Networking in RoboCup Rescue. In: Nardi, D., Riedmiller, M., Sammut, C., Santos-Victor, J. (eds.) RoboCup 2004. LNCS (LNAI), vol. 3276, pp. 236–246. Springer, Heidelberg (2005)
4. Zeiger, F., Selbach, C., Ruderisch, B., Schilling, K.: An Application Protocol to Integrate a Small Size Helicopter into an IP based Ad-Hoc Network. In: Proceedings of ROBOCOMM 2007, Athens, Greece (2007)
5. Hsieh, M.A., Cowley, A., Kumar, V., Taylor, C.: Towards the deployment of a mobile robot network with end-to-end performance guarantees. In: Proceedings of the IEEE International Conference on Robotics and Automation (ICRA 2006), Orlando, FL (USA), pp. 2085–2090 (2006)
6. Nguyen, H.G., Pezeshkian, N., Gupta, A., Farrington, N.: Maintaining Communication Link for a Robot Operating in a Hazardous Environment. In: Proceedings of ANS 10th International Conference on Robotics and Remote Systems for Hazardous Environments, Gainesville, Florida, March 28-31 (2004)
7. Pezeshkian, N., Nguyen, H.G., Burmeister, A.: Unmanned Ground Vehicle Radio Relay Deployment System for Non-Line-Of-Sight Operations. In: Proceedings of the 13th IASTED International Conference on Robotics and Applications, Wuerzburg, Germany (2007)
8. Nguyen, H.G., Pezeshkian, N., Raymond, M., Gupta, A., Spector, J.M.: Autonomous Communication Relays for Tactical Robots. In: The 11th International Conference on Advanced Robotics, Proceedings of ICAR 2003, Coimbra, Portugal, June 30-July 3 (2003)
9. Pezeshkian, N., Nguyen, H.G., Burmeister, A.: Unamnned Ground Vehicle Non-Line-of-Sight Operations Using Relaying Radios. In: Proceedings of the 12th IASTED International Conference on Robotics and Applications, Honolulu, Hawaii, USA (2006)
10. Johnson, D.B., Maltz, D.A.: Dynamic Source Routing in Ad-Hoc Wireless Networks. Mobile Computing 353 (1996)
11. Redi, J., Welsh, R.: Energy-Conservation for Tactical Mobile Robots. In: Proceedings of Military Communications Conference, MILCOM (1999)
12. Das, S., Perkins, C.E., Belding-Royer, E.M.: Ad hoc On-Demand Distance Vector (AODV) Routing. IETF RFC 3561 (2003)
13. Chakeres, I.D., Belding-Royer, E.M.: AODV Routing Protocol Implementation Design. In: Proceedings of the International Workshop on Wireless Ad hoc Networking (WWAN), Tokyo, Japan, pp. 698–703 (2004)
14. Broch, J., Maltz, D.A., Johnson, D.B., Hu, Y.C., Jetcheva, J.: A Performance Comparison of Multi-Hop Wireless Ad Hoc Network Routing Protocols. In: Mobile Computing and Networking, pp. 85–97 (1998)
15. Das, S.R., Perkins, C.E., Royer, E.M., Marina, M.K.: Performance Comparison of Two On-demand Routing Protocols for Ad hoc Networks. IEEE Personal Communications Magazine special issue on Ad hoc Networking, 16–28 (February 2001)
16. Dyer, T.D., Boppana, R.V.: A comparison of TCP performance over three routing protocols for mobile ad hoc networks. In: Proceedings of the 2nd ACM International Symposium on Mobile Ad Hoc Networking & Computing, Long Beach, CA, USA, pp. 56–66 (2001)

17. Johansson, P., Larsson, T., Hedman, N., Mielczarek, B., Degermark, M.: Scenario-based Performance Analysis of Routing Protocols for Mobile Ad-hoc Networks. In: Proceedings of Mobicom 1999, Seattle, Washington (1999)
18. Kiess, W., Mauve, M.: A Survey on Real-World Implementations of Mobile Ad-Hoc Networks. Elsevier's Ad Hoc Networks 5(3), 324–339 (2007)
19. Zeiger, F., Kraemer, N., Schilling, K.: Commanding Mobile Robots via Wireless Ad-Hoc Networks - A Comparison of Four Ad-Hoc Routing Protocol Implementations. In: Proceedings of the IEEE International Conference on Robotics and Automation (ICRA 2008) (2008)
20. Eck, D., Stahl, M., Schilling, K.: The Small Outdoor Rover MERLIN and its Assistance System for Tele-Operations. In: Proceedings of International Conference on Field and Service Robotics (FSR), Chamonix, France (2007)
21. DARPA Internet Program: RFC 791, Internet Protocol Specification (1981)
22. Hu, Y.C., Johnson, D.B., Maltz, D.A.: The Dynamic Source Routing Protocol (DSR) for Mobile Ad Hoc Networks for IPv4. IETF MANETWorking Group, IETF RFC 4728 (2004)
23. Clausen, P.J.T.: RFC 3626, Optimized link state routing protocol (OLSR). IETF, Network Working Group (2003)
24. B.A.T.M.A.N. (better approach to mobile ad-hoc networking): 2007/09/07 (2007), https://www.open-mesh.net/batman
25. Zeiger, F., Sauer, M., Schilling, K.: Intelligent shaping of a video stream for mobile robot teleoperation via multihop networks in real-world scenarios. In: Proceedings of IEEE/RSJ International Conference on Intelligent Robots and Systems (IROS 2008) - Workshop on Network Robot Systems: human concepts of space and activity, integration and applications (2008)

Safe Test Flights for Small Rotorcrafts

Nikos Vitzilaios and Nikos Tsourveloudis

Intelligent Systems & Robotics Laboratory
Department of Production Engineering & Management
Technical University of Crete
73300 Chania, Greece
{vitzilaios,nikost}@dpem.tuc.gr

Abstract. In this paper we present an experimental test bed for the development and evaluation of unmanned helicopters control. Main component of the suggested test bed is a flying stand which permits all possible movements but, also, prevents helicopters' damage because of crashing. A variety of sensors mounted on the stand, monitor and feedback rotorcraft's attitude and operational states. Based on the sensory information, a fuzzy logic controller has been developed and tested on a real rotorcraft, using the developed test bed. This controller proved able to perform hovering and altitude control. Experimental results are presented for various test cases.

Keywords: Unmanned helicopters, Experimental test bed, Altitude control, Fuzzy control, Aerial robotics.

1 Introduction

Unmanned helicopters are the most flexible flying machines among the variety of UAVs (Unmanned Aerial Vehicles), since they have the ability to take off and land vertically as well as to perform aggressive maneuvers and hovering, which gives them the advantage of effective observation from various positions. These advantages along with the continuous development of robotic vehicles' technology have led to the use of unmanned helicopters and rotorcrafts in many applications [1]. During the last years, small scale (about 1500 mm in length) helicopters are preferred for development and experimentation due to their low cost and expendability.

Although small scale unmanned helicopters offer as experimentation platforms the advantages of low cost and easy operation, the development of autonomous navigation systems for such vehicles is a difficult and dangerous procedure that may increase this overall cost, since except from the equipment needed (helicopter, sensors, telemetry systems etc) one should add the cost of crashes and damages that may occur during experimentation. For this reason, the development of an autonomous navigation controller begins with numerous tests in a software-based simulation environment. In this environment, controllers are evaluated for their ability to control efficiently the helicopter. If the simulation results are encouraging, the controller may be tested on the real vehicle.

The simulation procedure has drawbacks as well. At first, the simulation environment cannot imitate helicopter's navigation in detail with all possible

J.A. Cetto et al. (eds.): Informatics in Control, Automation and Robotics, LNEE 37, pp. 153–166.
springerlink.com

environmental disturbances. Therefore, a controller that seems to work satisfactorily in the simulation may be insufficient for the navigation of the real vehicle in a real environment. Moreover, independently of any simulation evaluation, first/initial tests with a real vehicle generally are the most dangerous, since a lot of unexpected problems may arise at this time. As a result, it would be desirable to test the controller on a real vehicle but in a safe environment, without having the danger of crashing and destroying the equipment or harm people that monitor the flight.

In the past years, there have been proposed ways of testing controllers on a real vehicle safely. Normally there is a mechanical construction where a real helicopter (or a simplified model of it) can fly indoors without crashing or harming the humans involved in the experimentation.

At first, in the literature we meet constructions that simulate a real helicopter [2-4]. These setups involve custom helicopter-like constructions whose degrees of freedom are reduced, and are used for the design and evaluation of flight stabilization controllers. Further in the literature, we meet systems that use real helicopters for the experiments. In [5] and [6], a mechanical construction holds the helicopter in a stable position allowing only small and safe movements. Using mechanical limitations, the helicopter is able to move in only one or two axes and within limits. As a result the helicopter cannot take any dangerous orientation or collide to the ground.

The drawback of the work presented in the above references, is that either a helicopter emulation construction is used, or a real helicopter with reduced degrees of freedom. In both cases, the developed controller partially covers the control of the vehicle in one or two axes and it is not sufficient to fully control a helicopter in real conditions. The motivation of this paper is the construction of a laboratory test bed where small helicopters can be safely (for both humans and the equipment involved) used indoors for experimental validation without limitations in helicopter's movement. Indoor flying gives the ability for continuous tests regardless of weather conditions. Moreover, the suggested setup minimizes the need for experienced helicopter pilots within the research group. Flying small helicopters requires pilot training which stems research efforts towards autonomous helicopter flights.

In this paper we also propose a fuzzy controller for the altitude and hovering control of an unmanned helicopter. The controller is developed using the proposed test bed and is able to stabilize the helicopter in desired positions (each position is defined by horizontal and vertical coordinates). Except from hovering at a desired altitude, the tasks of autonomous take-off and landing are also considered here.

In the literature there is previous work on the autonomous altitude control of unmanned helicopters. Usually altitude control is a part of an autonomous navigation controller [7, 8], where a subsystem dedicated to altitude control cooperates with other subsystems in order to navigate the helicopter. In [9] an adaptive approach is proposed for altitude control for an unmanned helicopter which utilizes rotor RPM to track altitude commands. Significant work has been done also in the field of autonomous landing problem for unmanned helicopters [10, 11].

2 Experimental Test Bed

The laboratory test bed consists of three basic elements; a customized flying stand, a customized helicopter and a ground control station (Fig. 1).

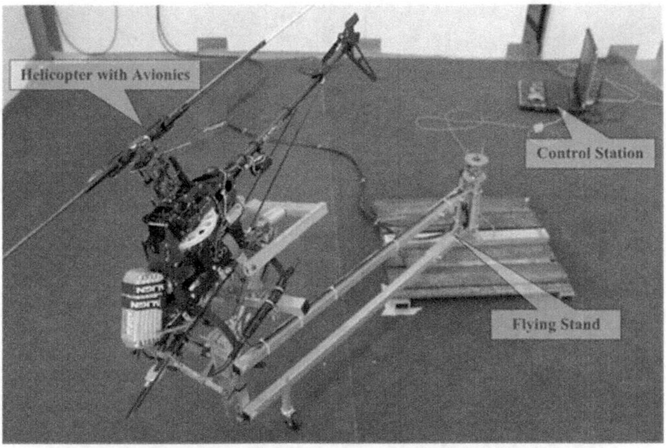

Fig. 1. View of the experimental test bed

2.1 Helicopter Flying Stand

Helicopter flying stand is a mechanical construction able to hold the helicopter, allowing full movements (6 degrees of freedom) while protecting it from damaging and crashing. It is a customized construction based on the commercially available Whiteman flying stand [12] that it is used by inexperienced pilots for flight training. The stand allows the helicopter move naturally without any constraint around a 2.1m diameter circle (Fig. 7), flying forwards, backwards or sideways. A gas strut is used to counterbalance the weight of the stand. As a result the helicopter does not lift any extra weight. In Fig. 2, rotations as well as the Euler angles of the helicopter are presented.

Since the test bed will be used for indoor experiments, a positioning system must be developed in order to know helicopter's position during testing. To avoid high cost indoor positioning and localization vision systems [13], we utilize the rotary movement of the central shaft of the stand. The stand and consequently the helicopter move around a circle (planar rotation at Fig. 2, Fig. 7) with a rotation angle which may easily be monitored. For this reason, we put a rotation encoder on the central shaft of the stand (Fig. 3). The encoder initializes its position to zero and then gives signed numbers that denote the current position relative to the initial position. Positive numbers denote rotation to the left while negative numbers denote rotation to the right side. The rotation encoder gives the planar position of the helicopter at each time instant.

Moreover, we need to know the altitude in which the helicopter flies. The flying stand gives the ability to the helicopter to fly at a maximum height of 60cm. An infrared sensor is used to monitor the actual value of altitude. The sensor is mounted at the lower part of the bracket that holds the helicopter, as it is shown in Fig. 3. The accuracy of the altitude readings is less than 1cm, which is far better than the accuracy of outdoor altimeters or GPS.

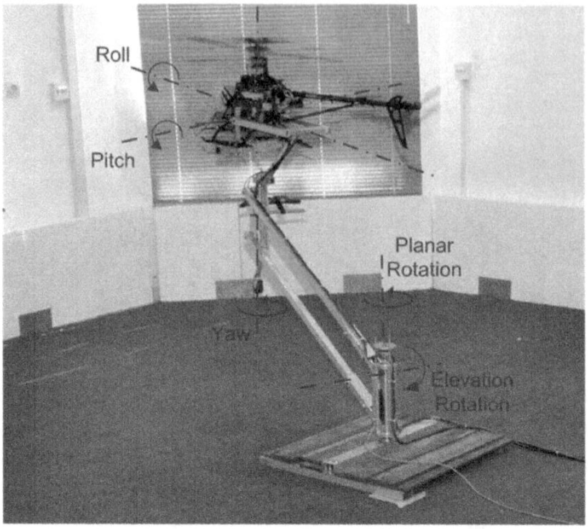

Fig. 2. Euler angles and rotation axes

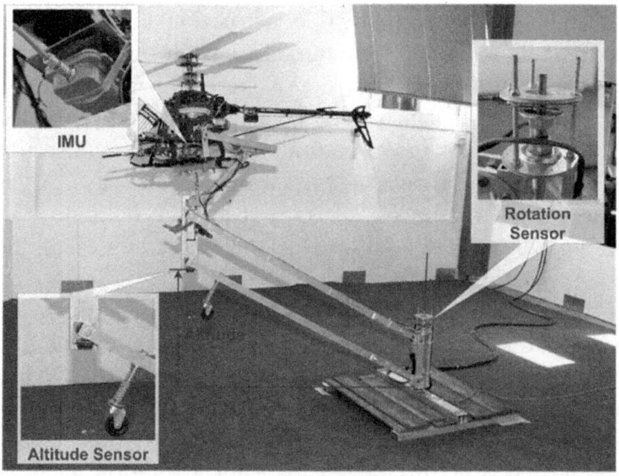

Fig. 3. Positioning sensors

The final position of the helicopter can be monitored during the experiments using the altitude and the rotation sensors. Orientation (Euler angles) of the helicopter is what needed in order to have a complete reference system and make calculations during the experiments. For this reason, an Inertial Measurement Unit (IMU, Fig. 3) is used. This sensor is mounted on the helicopter and provides its orientation at each time instant. Further information about this unit is provided in Section 2.2.1.

2.2 Helicopter and Avionics

The VTOL (=Vertical Take-Off and Landing UAV), that we use in our test bed is a customization of a 50-size (1200 mm length, 405 mm height, 1350 mm main rotor diameter) commercially available electric powered RC helicopter. An important characteristic of this helicopter is that it has electric motor so there is no need for fuel gas, and therefore it does not produce any exhaust gasses during its operation, which is important for indoor testing. This helicopter has been heavily customized in order to be ready for experimental use. In what follows we describe the additional equipment and avionics we have put on board.

2.2.1 Inertial Measurement Unit (IMU)

This unit gives the orientation of the helicopter. The unit consists of 3D gyroscopes, accelerometers and magnetometers and outputs the 3 Euler angles (roll, pitch and yaw). The IMU used is the commercial MTi model of Xsens Motion Technologies. For the communication between IMU and control station a USB-serial data and power cable is used.

2.2.2 Digital Switch

This is the interface that manages the switching from manual to autonomous flight. Manual flight is controlled remotely by a human operator, while autonomous flight is supervised by a Central Processing Unit (CPU). Switching between manual and autonomous flight is an important operation because it allows the human tester to regain manual control at any time instant during experimentation, which is very useful in case of failure or insufficient controller behaviour.

2.2.3 Servo Driver/Controller

RC servos are the actuators used to control the motion of the helicopter. In manual operation, the onboard receiver forwards the transmitter commands to servos by sending appropriate PWM signals. In order to send such signals from the control station to the servos, a servo driver is needed. For that reason a PIC microcontroller is used, which translates control signals from the ground station to RC PWM servo signals and drives the servos. Further, the PIC reads the input from the localization system (x-y position, altitude) and transmits it to the control station.

2.2.4 Communication System

A wireless communication system has been established between the control station and the PIC microcontroller. Having 2 receiver/transmitter units (one on the helicopter and one on the ground station) and by using the Bluetooth protocol, we obtain two-way communication between the serial port of the PIC and the serial port of the control station.

2.2.5 Power System

The electric helicopter has high power consumption. During hovering, the electric motor needs about 50A current of 25V. Normally in these helicopters, LiPo batteries are used that have high capacity and the ability to sustain big currents. With this consumption and with a high capacity LiPo battery, the helicopter can perform hovering for about 15 minutes. To overcome this limitation in the duration of experiments, the test bed is provided with constant power supply of 24V that gives continuous current to the helicopter.

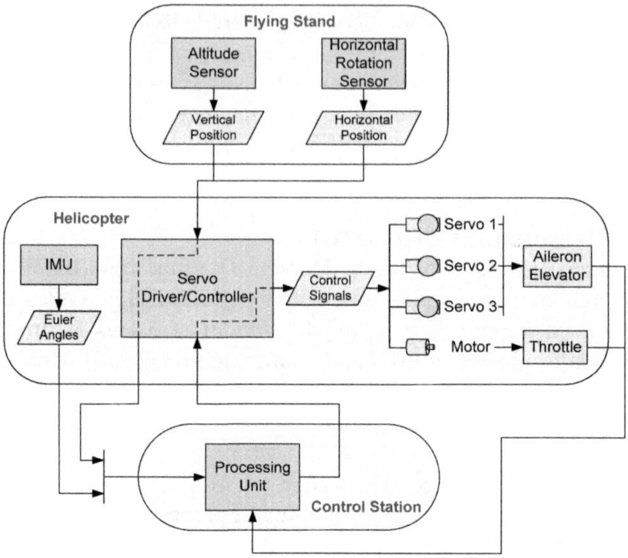

Fig. 4. System Architecture

2.3 Ground Control Station

Since our test bed works indoor and we can have all the signals through wireless communication (expect from the IMU), there is no need to put any processor unit onboard. For this reason we use portable CPU which serves as the "control station". Because of this solution, the helicopter has fewer payloads to lift, while the control station has increased processing power able to run control algorithms at high speeds.

In Fig. 4, a block diagram presents the connections of the equipment and the data transmission through these connections, for each subsystem (flying stand, helicopter and control station).

2.4 Safety

Safety is a very important issue that must be carefully considered when handling model helicopters. A model helicopter is potentially a very dangerous piece of equipment since it's rotor blades may spin at over 1.700 rpm. Even the most experienced pilots might make a mistake or experience various failures that may cause severe property or physical damage and in the worst case serious injuries. When a helicopter crashes, several parts break and spread all over the crash area. The potential hazard is even bigger when using unmanned helicopters for autonomous navigation. The use of a control system in the phase of development involves risks of insufficient control that may result in unexpected behavior, and even if a safety pilot is standby to take control of the vehicle, accidents may be difficult to avoid. For this reason, safety was carefully taken into account in the development of the test bed.

Fig. 5. View of the safety cage that contains the test bed

In our case, the experimental test bed is designed to work indoors. Even though the helicopter is attached on the flying stand, which holds it in safe positions, we want to increase test bed's safety by securing appropriately the test bed area. For this reason a safety cage was built around the test bed area, so as to protect the human operator and people monitoring the experiments (Fig. 5). The cage is made of unbreakable glass that permits clear view of the test bed area. In the unlikely event of an accident or a malfunction where the stand fails to hold securely the helicopter, the safety cage will prevent any helicopter part to get outside the cage area.

3 Altitude and Hovering Control

The controller developed and tested in the test bed is a fuzzy controller for altitude and hovering control. The objective of the controller is to hold stable the helicopter at a predefined horizontal position and altitude.

3.1 Fuzzy Controller

A fuzzy controller of the Mamdani type has been designed and implemented (Fig. 6) in the MATLAB environment. The objective of this controller is to keep the helicopter "hovering" at predefined positions subject to wind and other disturbances. As shown in Fig. 6, the inputs of the fuzzy controller are the *roll* and *pitch* angles of the helicopter at every time instant, as well as the *position error*, the *change of position error* and the *altitude error*. In Fig. 7 we show the representation of the *position error* input, which is defined as the difference between the current horizontal position and the target horizontal position.

As *position error* represents how far the helicopter is from the target point, the *change of position error* represents the way that position error changes and if the helicopter reaches the target point or moves away from it. The *altitude* error is also calculated as the difference between the current and the target altitude. The outputs of the controller are the change of the roll and pitch angles (*aileron* and *elevator* variables respectively), as well as the change in the *throttle* of the helicopter.

In what follows, we present in detail the above input and output variables.

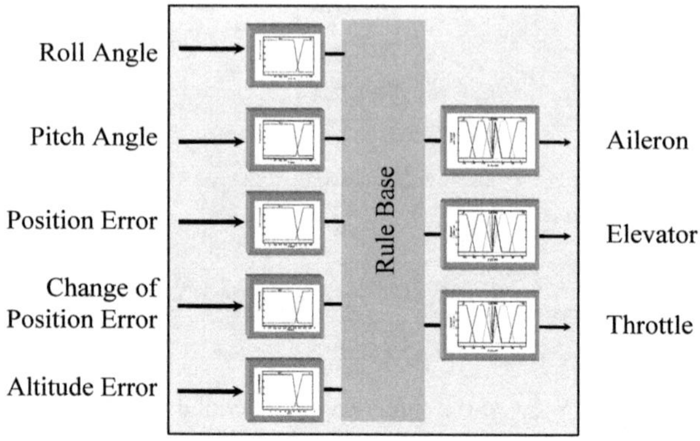

Fig. 6. Hovering and Altitude Fuzzy controller

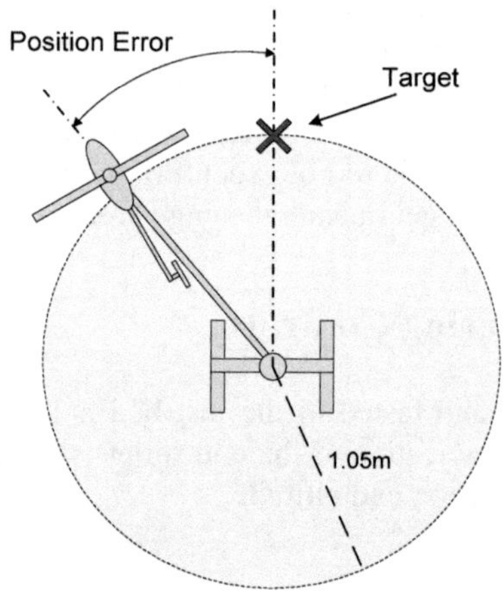

Fig. 7. Position error representation

Roll angle is given by the IMU in real time. Although the flying stand permits roll angles from -30° to 30°, the flight control system takes as input degrees from -90° to 90°. The linguistic variables that represent the *roll* angle are: *left big* (*LB*), *left* (*L*), *zero* (*ZERO*), *right* (*R*), *right big* (*RB*), and their membership functions are shown in Fig. 8.

The second input variable is the *pitch* angle of the helicopter. The linguistic variables for this input are: *back big* (*BB*), *back* (*B*), *zero* (*ZERO*), *front* (*F*), *front big* (*FB*), with membership functions also presented in Fig. 8.

The third input variable is the *position error*, which is defined as the difference between the current and the desirable position. Since for safety reasons we do not want the stand to rotate out of its limits (-180° to 180° which corresponds to -30 to 30 in odometer units) we set the range of the *position error* variable to be between -30 to 30 (in odometer units). The linguistic variables for these inputs are: *negative big* (*NB*), *negative* (*N*), *zero* (*ZERO*), *positive* (*P*), *positive big* (*PB*) (Fig. 8).

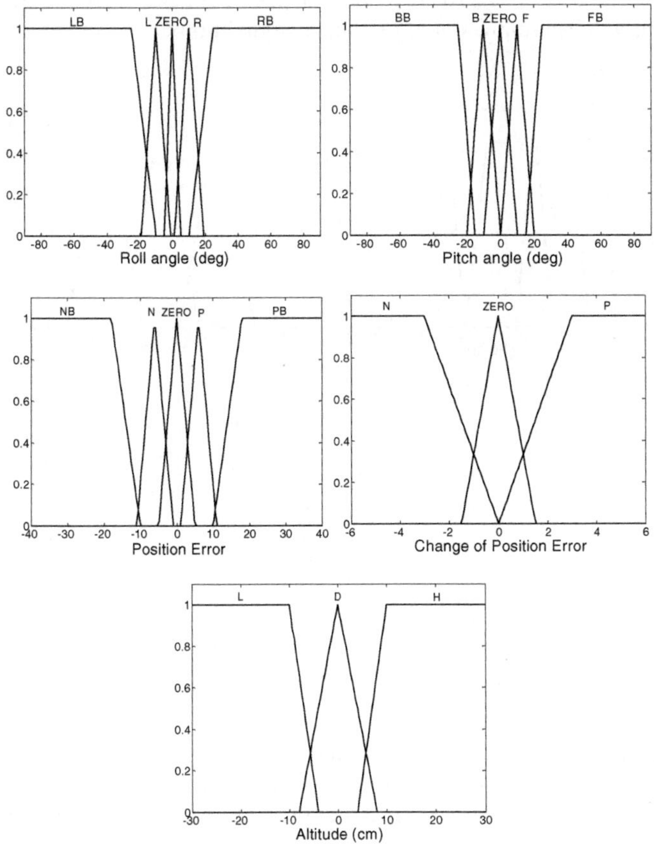

Fig. 8. Membership functions for input variables

The next input in the fuzzy controller is the *change of position error*. While position error shows how far the helicopter is from the desire position, *change of position error* shows how fast the vehicle is moving towards or away from the desired point. This input is defined as the difference (in odometer units) between the previous position error and the current position error, and it is represented by the linguistic variables: *negative* (*N*), *zero* (*ZERO*), *positive* (*P*) (Fig. 8).

The last input is the *altitude* error input. This input represents the difference in cm between actual and desired altitude by counting if the helicopter is placed lower or higher than the desired position. The linguistic variables for this input are: *lower* (*L*), *desired* (*D*), *higher* (*H*) (Fig. 8).

The outputs of the fuzzy controller are the changes of roll and pitch angles (*Aileron* and *Elevator* movements respectively) and *Throttle* change. The membership functions of *aileron, elevator* and *throttle*, are presented in Fig. 8. The linguistic variables for *aileron* are *left big* (*LB*), *left* (*L*), *left small* (*LS*), *zero* (*ZERO*), *right small* (*RS*), *right* (*R*) and *right big* (*RB*). The linguistic variables for *elevator* are *back big* (*BB*), *back* (*B*), *zero* (*ZERO*), *front* (*F*) and *front big* (*FB*). Both *aileron* and *elevator* output values are presented in control signal units.

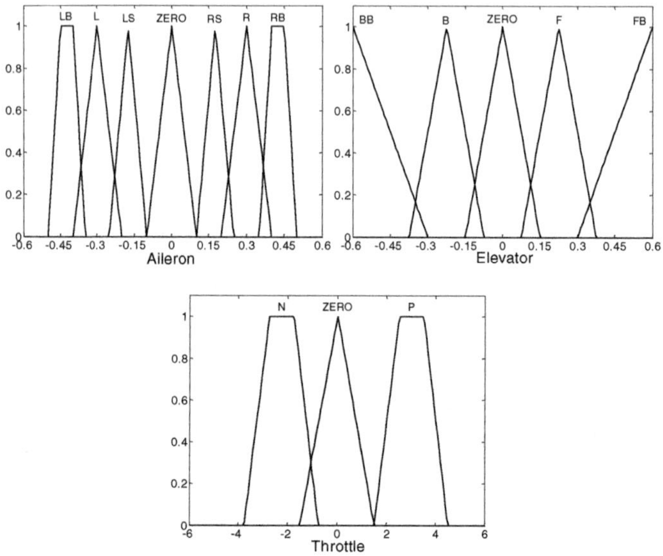

Fig. 9. Membership functions for output variables

The linguistic variables for *throttle* change are *negative* (*N*), *zero* (*ZERO*) and *positive* (*P*). The values of throttle output are also presented in control signal units. Negative output reduces throttle of the helicopter while positive output increases it.

3.2 Control Rules

The control objective in the experiments performed was the stabilization of helicopter at a certain point (defined by horizontal and vertical target coordinates). The transition between the states of the controller is presented in Fig. 10. After take-off, the controller has as a target to hover the helicopter. Then checks actual horizontal position and drives the helicopter to the desired one. The next step is checking of actual altitude in order to drive the helicopter to the desired one. After some iterations where the helicopter hovers in the target point, the controller lands it.

For the implementation of this scheme, three sets (rule bases) of fuzzy IF-THEN rules were used. The one was responsible for the control of the pitch angle. The target was to keep the pitch angle always close to zero as this is what needs to be done when the helicopter hovers. The second rule base contains rules that lead the helicopter towards the desired point as they tend to minimize the distance between the helicopter's horizontal position at each moment and the desired one. The third rule base is responsible for handling the throttle of the helicopter. The policy we follow here is that the changes in the throttle of the helicopter occur only when the helicopter is in stable hovering attitude on the desired horizontal position (roll and pitch angles are close to zero, change of position error is close to zero) or when the altitude becomes higher than a top safety limit.

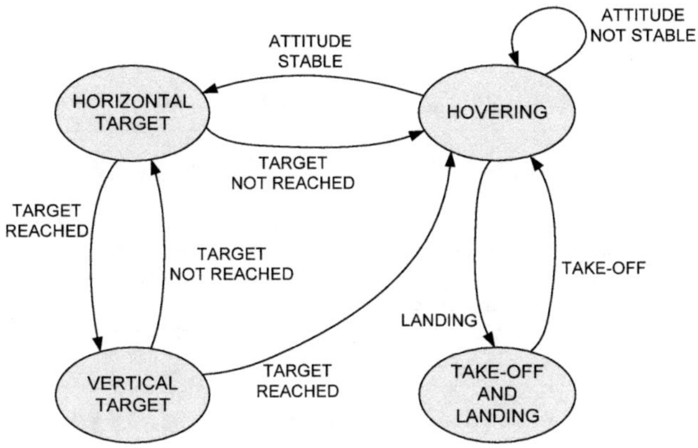

Fig. 10. Controller state transition

4 Experimental Results

Experimental results for two test cases may be seen in Fig. 11 and Fig. 12. In these figures *Roll* and *Pitch* values are measured in degrees, while *Position Error* and *Change of Position Error* are measured in odometer units (here, 1 odometer unit corresponds to 6 degrees) and *Altitude* is measured in centimetres. *Elevator, Aileron* and *Throttle* values are measured in control signals. The initial altitude of the helicopter (when the flying stand is on the ground) is 10cm.

In **test case 1** (Fig. 11) the ability of the controller to perform autonomous take-off and keep the helicopter in a hovering state, is evaluated. The helicopter is placed on the desired horizontal position and the autopilot takes over with a target altitude of about 22 cm. The controller keeps roll and pitch angles close to zero and gradually increases throttle in order to increase the altitude and reach the target one (Fig. 11). When the target altitude is reached few oscillations around the horizontal position occur but the controller manages to hold the helicopter in hovering. In the beginning, it is clear that position error tends to be a small positive number, which means that the helicopter always drifts to the left. This is explained by the position of the test bed area which is close to the walls of the building. Air flow from the main rotor of the helicopter circles through the walls and return to the helicopter. This air flow gives a small drift to helicopter to the left. The developed controller seems to recognize the disturbance and successfully stabilize the helicopter in the desired position.

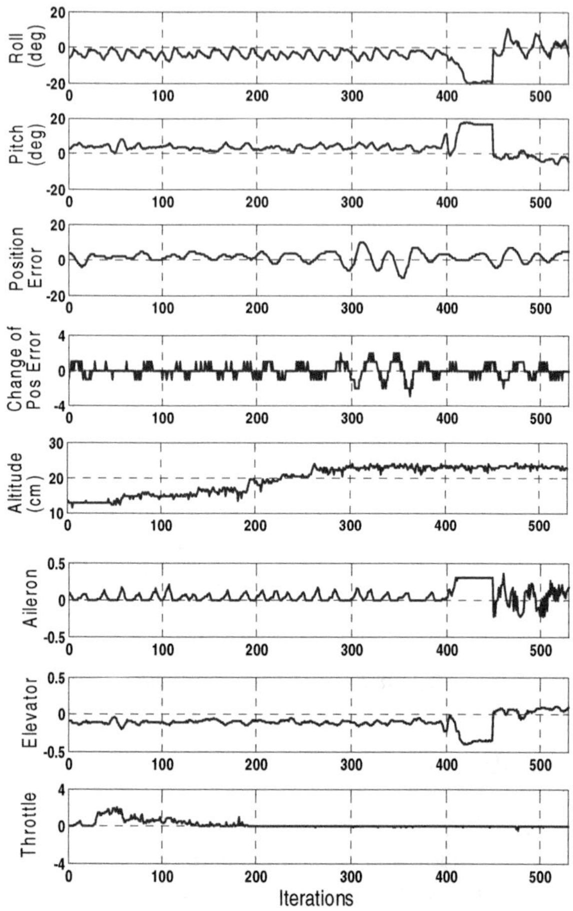

Fig. 11. Experimental results for **test case 1**

In Fig. 12 we present the results of **test case 2**. In this test, the initial position of the helicopter is different from the desired one and the controller objective is to drive the helicopter to the desired position and then land it autonomously. The helicopter is placed manually to a random position and then the fuzzy autopilot gains control of the helicopter. As one may see in the *Position Error* plot of Fig. 12, the helicopter moves manually from its initial position to a random position. At time instant 50, the autopilot gains control of the vehicle. The target of the autopilot is to move the helicopter to the initial position and in 20 cm altitude. It is clear that the autopilot drives the helicopter to the target point by moving it to the desired horizontal position at first and then by raising the altitude until the targeted one has been reached. After a few iterations that the target position has been reached, the controller reduces the throttle and lands the helicopter. Small oscillations occur while the autopilot tries to keep the helicopter in stable position. It is also clear, as in test case 1, that we face the air disturbance that causes small drift in the helicopter in this test case too.

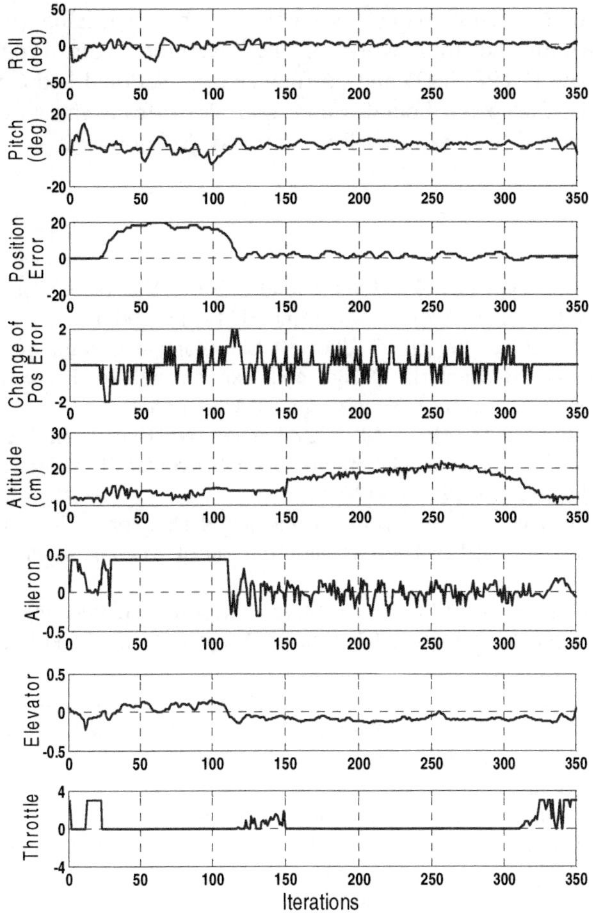

Fig. 12. Experimental results for **test case 2**

5 Conclusions

In this paper we presented a fuzzy controller for hovering and altitude control of a small-scale helicopter. The controller was developed and tested on a custom made laboratory experimental test bed, where tests on unmanned helicopters can be performed with safety. The test bed works indoors, is independent of power supply and can be used for continuous tests. The development of the controller is done on a real helicopter and not in simulation, so we can have direct and reliable results. Special attention has been paid to the safety of the test bed. The experimental results show that this setup works well. Experimental results from the evaluation of the altitude fuzzy controller were presented.

Future work, involves development of other kinds of controllers which will be tested and evaluated on the test bed. This work will lead to a comparison of controllers based on their efficiency and ability to control successfully an unmanned helicopter.

Acknowledgements. This paper is part of the 03ED365 research project, implemented within the framework of the "Reinforcement Program of Human Research Manpower" (PENED) and co-financed by National and Community Funds (75% from E.U.-European Social Fund and 25% from the Greek Ministry of Development-General Secretariat of Research and Technology).

References

1. Spanoudakis, P., Doitsidis, L., Tsourveloudis, N.C., Valavanis, K.P.: The Market for VTOL UAVs. Unmanned Systems Magazine 21, 14–18 (2003)
2. Andrievsky, B., Peaucelle, D., Fradkov, A.L.: Adaptive Control of 3DOF Motion for LAAS Helicopter Benchmark: Design and Experiments. In: Proceedings of the 2007 American Control Conference, New York City, USA (2007)
3. Kutay, A.T., Calise, A.J., Idan, M., Hovakimyan, N.: Experimental results on adaptive output feedback control using a laboratory model helicopter. IEEE Transactions on Control Systems Technology 13, 196–202 (2005)
4. Tanaka, K., Ohtake, H., Wang, H.O.: A practical design approach to stabilization of a 3-DOF RC helicopter. IEEE Transactions on Control Systems Technology 12, 315–325 (2004)
5. Dzul, A., Lozano, R., Castillo, P.: Adaptive control for a radio-controlled helicopter in a vertical flying stand. International Journal of Adaptive Control and Signal Processing 18, 473–485 (2004)
6. Mancini, A., Caponetti, F., Monteriu, A., Frontoni, E., Zingaretti, P.: Longhi, S.: Safe flying for an UAV Helicopter. In: Proceedings of the15th Mediterranean Conference on Control & Automation, Athens-Greece (2007)
7. Kim, H.J., Shim, D.H.: A flight control system for aerial robots: algorithms and experiments. Control Engineering Practice 11, 1389–1400 (2003)
8. Shin, J., Nonami, K., Fujiwara, D., Hazawa, K.: Model-based optimal attitude and positioning control of small-scale unmanned helicopter. Robotica 23, 51–63 (2005)
9. Kim, N., Calise, A., Corban, J., Prasad, J.: Adaptive output feedback for altitude control of an unmanned helicopter using rotor RPM. In: Proceedings of the AIAA Guidance Navigation and Control Conference and Exhibit, Rhode Island, USA, pp. 2004–5323 (2004)
10. Merz, T., Duranti, S., Conte, G.: Autonomous Landing of an Unmanned Helicopter based on Vision and Inertial Sensing. In: Experimental Robotics IX: The 9th International Symposium on Experimental Robotics, pp. 343–352 (2006)
11. Saripalli, S., Sukhatme, G.S., Montgomery, J.F.: An Experimental Study of the Autonomous Helicopter Landing Problem. Experimental Robotics VIII (2003)
12. Whiteman, R.: Training Apparatus for Remote Control Model Helicopters. United States Patent No. 4917610 (1990)
13. Valenti, M., Bethke, B., Fiore, G., How, J.P.: Indoor Multi-Vehicle Flight Testbed for Fault Detection, Isolation, and Recovery. In: AIAA Guidance, Navigation, and Control Conference and Exhibit, Keystone, Colorado (2006)

Particle Filter with Temporal Smoothing for Mobile Robot Vision-Based Localization

Walter Nisticò and Matthias Hebbel

Institute for Robot Research
Technische Universität Dortmund
Otto-Hahn-Str.8, 44227 Dortmund, Germany
forename.surname@uni-dortmund.de

Abstract. Particle filters based on the Sampling Importance Resampling (SIR) algorithm have been extensively and successfully used in the field of mobile robot localization, especially in the recent extensions (Mixture Monte Carlo) which sample a percentage of particles directly from the sensor model. However, the Markov assumption on which these methods rely is frequently violated, due to "ghost percepts" and undetected collisions, and this can be troublesome especially when working with small particle sets, due to limited computational resources and real-time constraints. In this paper we present an extension of Monte Carlo localization which relaxes the Markov assumption by tracking and smoothing the changes of the particles' importance weights over time, and limits the speed at which the samples are redistributed over the state space after a single resampling step. We present the results of experiments conducted on vision based localization in an indoor environment for a legged-robot, in comparison with state of the art approaches.

1 Introduction

Vision-based localization is becoming very popular for autonomous mobile robots, and particle filters are a successful technique to integrate visual data for localization. Due to size, cost and power consumption limits, many robots use a camera as their only exteroceptive sensor; although range finders such as laser scanners would provide more accurate measurements, vision can often provide unique features such as *landmarks* to speed up global localization.

1.1 The Platform

This work has been developed on the Sony Aibo ERS-7 robot [1], which has been one of the most popular complete standard platforms adopted for robotic applications. The robot is equipped with a 576MHz 64bit RISC CPU, 64MB of main memory, and a low-power CMOS camera sensor with a maximum resolution of 416 × 320 pixel. The camera is mounted on the robot head, with 3 degrees of freedom in the neck; it is severely limited in terms of resolution, field of view (56.9° horizontally), and is affected by a significant amount of noise. The Aibo production has been recently discontinued by Sony, but several new commercially available robotic kits are being introduced, such as the humanoid robot "Nao" from Aldebaran Robotics[1]. These platforms have similar

[1] *http://www.aldebaran-robotics.com/*

J.A. Cetto et al. (eds.): Informatics in Control, Automation and Robotics, LNEE 37, pp. 167–180.
springerlink.com

characteristics in terms of size and power, and are often equipped with embedded RISC CPUs or PDAs and inexpensive compact flash cameras.

1.2 Related Work

Practical applications of particle filters for mobile robot localization began with the introduction of the Sampling Importance Resampling (SIR) filter proposed in [2, 3] as an important extension of the Sequential Importance Sampling (SIS) filter [4]. The SIS filter requires a huge number of particles to work, since it does not make an efficient use of them, as the particle location is not influenced by sensor data in any way; the resampling step introduced in the SIR filter acts as a kind of "survival of the fittest" strategy that increases the particle density in the areas of highest probability of the posterior distribution. However, in the context of mobile robot localization the SIR filter suffers of two major problems: difficulty to recover in the "kidnapped robot" case (i.e. the robot is moved by an external agent to a new location); and *filter degeneracy*, where the filter performance starts to drop instead of improving as the sensors become "too" accurate, so that it becomes increasingly difficult to find particles close enough to the peaks of the likelihood function. Both problems can be (partially) eased by increasing the number of particles used, but this increases the computational cost; a much better solution has been proposed in [5] which samples a certain percentage of particles directly from the sensor model ("Sensor Resetting Localization" (SRL)), ensuring that no measurement gets lost due to a lack of particles. A more formal representation of the sensor resetting idea has been described in [6] with the "Mixture Monte Carlo" (MMCL) algorithm, where a small percentage of particles are sampled from the sensor model and receive importance weights proportional to the process model, thus (unlike SRL localization) the belief converges to the true posterior for an infinite number of samples. In the context of vision based localization on autonomous mobile robots, in [7] the authors presented an approach to deal with high levels of noise/uncertainty while using a small particle set (100 samples). Their approach keeps track of the individual importance weights of different landmark classes, and uses this information to constrain the maximum change of likelihood of the particle set in a single iteration of the algorithm, thus easily rejecting outliers in the measured data. The proposed approach is not formally correct, as we will show, however it provides surprisingly good results in practice and it is very popular on the Aibo platform, thanks also to the availability of its source code[2]. Here we show how to solve the problems of this approach and introduce a new resampling strategy; this paper is an extended version of the work we presented in [8].

2 Particle Filters

The *Particle Filter* [9] is a non-parametric implementation of the general *Bayes Filter* [10]. Given a time series of *measurements* $z_{1:t}$, *control actions* $u_{1:t}$ and an *initial belief* $p(x_0)$, the Bayes Filter is a recursive algorithm that calculates the belief or *posterior*

[2] *http://www.germanteam.org/GT2005.zip*

$bel(x_t)$ at time t of the state x_t of a certain process, by integrating sensor observations z_t and control actions u_t over the belief of the state at time $t-1$. The Bayes Filter is based on the Markov assumption or *complete state* assumption which postulates the conditional independence of past and future data given the current state x_t. This can be expressed in terms of conditional independence as follows:

$$\begin{aligned} p(z_t|x_t, z_{1:t}, u_{1:t}) &= p(z_t|x_t) \\ p(x_t|x_{t-1}, z_{1:t-1}, u_{1:t}) &= p(x_t|u_t, x_{t-1}) \end{aligned} \tag{1}$$

A Particle Filter represents an approximation of the posterior $bel(x_t)$ in the form of a set of samples randomly drawn from the posterior itself; such a representation has the advantage, compared to closed form solutions of the Bayes Filter such as the Kalman Filter [11], of being able to represent a broad range of distributions and model non-linear processes, whereas parametric representations are usually constrained to simple functions such as Gaussians. In the context of robot localization and object tracking, particle filters are often referred to as *Monte Carlo Localization*. Given a set of N samples or *particles* $\chi_t := x_t^1, x_t^2, \ldots, x_t^N$, at time t each particle represents an hypothesis of the state of the observed system; in case of robot localization, the state space is usually represented by the (x, y) Cartesian coordinates in the plane, and the heading θ. Obviously, the higher the number of samples N, the better the approximation, however [12] has shown how to dynamically adjust N. In this work we do not consider dynamic adjustments in the number of samples used, since we are interested in keeping the run-time of the algorithm approximately constant, due to the real-time constraints. An estimate of $p(x_t|u_t, x_{t-1})$ is called *Process Model*, while $p(z_t|x_t)$ is known as

---

Algorithm 1. SIR Particle Filter.

Require: particle distribution χ_{t-1}, control action u_t, measurement observation z_t
for $i = 1$ to N **do**
1. **Process update**: update the particles' state as the result of the control action u_t: $\overline{x}_t^i \sim p(x_t|u_t, x_{t-1}^i)$
2. **Measurement update**: calculate the particle importance factors $w_t^i = p(z_t|\overline{x}_t^i)$ from the latest observation
Add $\langle \overline{x}_t^i, w_t^i \rangle$ to the temporary set $\overline{\chi}_t$
end for
3. **Resampling**: create χ_t from $\overline{\chi}_t$ by drawing the particles x_t^i in number proportional to their importance w_t^i. All the importance factors w_t in χ_t are reset to 1.

Sensor Model. $\overline{\chi}_t$ before the measurement update step is a sampled representation of the *prior distribution* $\overline{bel}(x_t)$. If we omit the Resampling step in Algorithm 1we obtain the SIS filter. Under the Markov assumption, current measurements are statistically independent from past measurements, as such the importance factors during the Measurement update step can be calculated as:

$$w_t^i = p(z_t|x_t^i) \cdot w_{t-1}^i \tag{2}$$

The SIS filter for localization generally performs poorly, in fact since the importance weights are updated multiplicatively, a single percept is sufficient to set the importance of a given particle close to zero. It has been formally demonstrated that the variance of the importance weights in the SIS filter can only increase (stochastically) over time [13]. The SIR filter performs much better, because after the resampling step, the particles of low importance are discarded while the particles of high importance are replicated and all the importance weights are reset to 1, i. e. all particles carry the same importance. Consequently the distribution approximates the true posterior $bel(x_t) = \eta p(z_t|x_t)\overline{bel}(x_t)$.

2.1 Sampling from Observations: The Kidnapped Robot Problem

The situation when a mobile robot is physically moved ("teleported") by an external agent in a new location is known as the "Kidnapped Robot" problem. This situation is challenging because the robot has no information about such external action and its sensor and control data is not in accordance with the new state. The SIR filter as it has been presented has serious problems dealing with this situation, because it is unlikely to find a particle in the proximity of the location where the robot has been teleported, due to the effect of resampling which tends to concentrate the particle distribution in the area of high likelihood of the posterior at time $t - 1$, which in this case corresponds to the robot's location prior to the teleporting action. An efficient solution to such problem is to reverse the Monte Carlo process (*"Dual Monte Carlo"*) by drawing a certain percentage of the samples directly from the measurement model instead of the prior belief distribution:

$$\overline{x}_t^i \sim p(z_t|x_t) \tag{3}$$

and calculate the importance weights from the prior distribution (integrating the process model):

$$w_t^i = \int p(\overline{x}_t^i|u_t, x_{t-1})bel(x_{t-1})dx_{t-1} \tag{4}$$

Dual Monte Carlo performs better than normal MCL with very precise sensors, and can quickly recover in case the robot is teleported, since the samples are drawn directly from the last observations. Conversely, it is extremely sensitive in case of high sensor noise and performs poorly especially with ghost percepts; for these reasons, Mixture Monte Carlo uses a small percentage of samples which are updated using the Dual MCL approach, and the rest are following the normal MCL algorithm, thus combining the benefits of both approaches. The main implementation problems of the Dual MCL approach is that it is often difficult to sample from the observation model (Equation 3), when such model does not have a closed form, and it is both difficult and computationally expensive to calculate the importance weights from the prior distribution (Equation 4).

3 Temporal Smoothing Particle Filter

Particle filters can generally deal efficiently with sensor noise, provided that there is a sufficient amount of particles (most authors suggest to use at least 1000 particles) and that the measurement model is accurate enough. However a limit of this approach is that in practice the Markov assumption can often be violated: unmodeled dynamics in the environment, such as the presence of other mobile agents, "ghost" percepts (for example, echoes in sonar data, or incorrectly classified objects in a vision system), inaccuracies in the probabilistic sensor and process model (slippage, collisions), approximation errors due to the sampled representation, etc.

3.1 Sensor Smoothing

In [7] the authors proposed an approach which can deal with highly noisy visual measurements while using a sample set of just 100 particles, which can then be executed in real-time on severely constrained (in terms of processing power) mobile robots. The idea is to limit the effect of a single measurement (which might be an outlier) on the particle population; this is achieved by a filtering method which constrains the importance weight w_t^i for a given particle at time t to differ at most by a fixed $\Delta \in (0..1]$ from the value that the particle had at time $t-1$.

Let $\bar{w}_t^i$ be the importance weight of particle i at time t **calculated from the measurement update step** of the particle filter, w_t^i can be calculated from $\bar{w}_t^i$, w_{t-1}^i, and a pair of values $\Delta_{up}, \Delta_{dn} \in (0..1]$ as:

$$w_t^i = \begin{cases} (w_{t-1}^i + \Delta_{up}) \; if(\bar{w}_t^i > w_{t-1}^i + \Delta_{up}) \\ (w_{t-1}^i - \Delta_{dn}) \; if(\bar{w}_t^i < w_{t-1}^i - \Delta_{dn}) \\ \qquad \bar{w}_t^i \qquad\qquad\quad else \end{cases} \tag{5}$$
$$(\Delta_{up} = 0.01, \Delta_{dn} = 0.005)$$

Equation 5 describes this approach as presented in [7]; the authors chose for their particular application values of $(\Delta_{up} = 0.01, \Delta_{dn} = 0.005)$, which means that it takes a series of 100 consistent percepts to raise the importance weight of a given particle from 0 to 1, and twice as much to lower it from 1 to 0; as such it is evident that the influence of a single percept is very limited. The visual features used for localization are separated in a set of different percept classes (field lines, borders and goals are the recognized objects in [7]), the perceived measurements of different classes are assumed to be statistically independent, and each particle carries a different importance weight per percept class. Consequently, the importance weight of a particle is calculated as the product of the weights associated to each percept class.

So, if we define a set of *percept classes* $\Gamma = \{fieldlines, fieldborders, goals\}$, the importance weight w_t^i at time t of a particle i, by virtue of the assumed conditional independence of the percept classes, can be expressed as a function of the individual importance weights $w_t^{i,j}$ associated with the percept classes $j \in \Gamma$ as:

$$w_t^i = \prod_{j \in \Gamma} w_t^{i,j} \tag{6}$$

From a theoretical point of view, the smoothing performed in Equation 5 is a violation of the Markov assumption, in fact past and current measurements for objects in the same class are correlated by this update method. On the other hand, this process can be seen as an incremental estimate of the importance weight of a given particle with respect to a series of measurements of a certain object, rather than considering each measurement independently. The result of this is a smoothing of the effects of the measurement update, which can easily filter noise as well as limit the effects of outliers, where Δ represents an upper bound to the negative effect of a wrong measurement to the particle population. Another useful effect of this process is that the particle filter behavior becomes less dependent on the accuracy of the measurement model, since a too peaked likelihood function gets smoothed as well. Δ can be set according to the frequency of detection of a given percept class and the magnitude of the noise and occurrence of outliers.

3.2 Reducing Incorrect Correlations

Despite its effectiveness in practice, this approach as presented in [7] has some problems:

1. having $\Delta_{up} \neq \Delta_{dn}$ introduces an unnecessary bias in the importance weights (in [7] such bias was exploited to control a component of the process model which was importance-weight dependent);
2. measurements relative to different objects are put in the same class, and the particles keep their "memory" indefinitely, so this mechanism incorrectly correlates measurements which should be independent.

While the first problem is trivial, the same is not true about the second; the correct solution would be to assign each perceived object to a different percept class, so that they can be treated as statistically independent. This however is not always possible, because in some cases the robot is observing different objects which are indistinguishable among themselves. Practical examples of this are walls or doors in an office environment (for vision-based localization), or range measurements obtained with active sensors such as laser scanners or sonars. This problem can be eased under an assumption of *temporal coherence* of a series of measurements, i. e. we expect "bursts" of successive measurements relative to a certain object to appear within a short time span. Under this assumption, we can here ([8]) introduce an "*aging law*" to progressively fade out the memory of past measurements in the importance weights, thus reducing the incorrect correlations among percepts of the same class.

So, given a set Γ of percept classes, let $w_t^{i,j}$ be the importance weight for the percept class $j \in \Gamma$ relative to the i-th particle at time t. If $w_t^{i,j} = 1$ as can be seen in Equation 6 the percept class j does not contribute to the global weight, which is the desired behavior in case of a percept which was never seen in a long time or at all, so we want to use an aging law which fades the weights toward 1. So, let $\alpha^j \in [0..1]$ be an aging factor for class j, $\hat{w}_t^{i,j}$ the result of the aging process, we can here introduce the following law:

$$\hat{w}_t^{i,j} = w_t^{i,j} + (1 - w_t^{i,j}) \cdot \alpha^j \tag{7}$$

This function asymptotically fades an importance factor toward 1 with a speed controlled by the parameter α^j, which can be set according to the frequency of independent observations of a given percept class. α^j represents a percentage of the distance

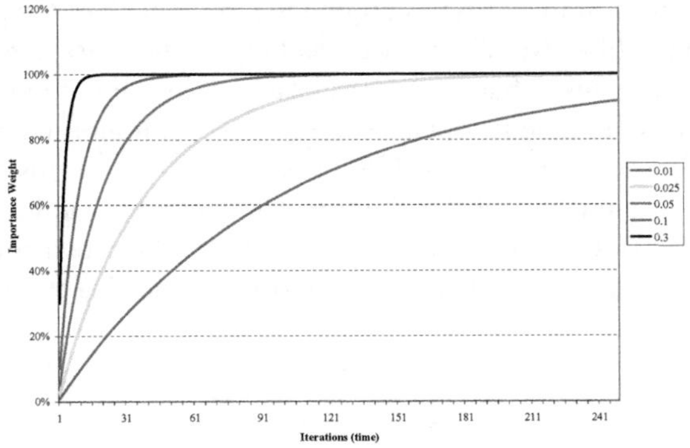

Fig. 1. Aging function, for different values of α

between $w_t^{i,j}$ and 1 and we chose the law presented in Equation 7 because it fades a certain percentage of any percept in constant time. So for example, for $\alpha = 0.05$ (Figure 1) if a certain percept set the importance weight to 0, it takes 30 iterations to fade out 80% of its effect (final weight 0.8), and it takes the same 30 iterations to fade out the 80% of the effect of another percept from 0.8 to 0.96.

This aging process is particularly beneficial in case a series of past measurements for a given class resulted in a very low weight, which without aging would affect negatively the global weight of the particle, regardless of the values of the other percept classes. Finally, after the *aging step* (Equation 7), given a $\bar{w}_t^{i,j}$ calculated from the measurement model, we can apply the *temporal smoothing* step:

$$\begin{aligned} \hat{w}_{t-1}^{i,j} &= w_{t-1}^{i,j} + (1 - w_{t-1}^{i,j}) \cdot \alpha^j \\ w_t^{i,j} &= \begin{cases} (\hat{w}_{t-1}^{i,j} + \Delta^j) \; if(\bar{w}_t^{i,j} > \hat{w}_{t-1}^{i,j} + \Delta^j) \\ (\hat{w}_{t-1}^{i,j} - \Delta^j) \; if(\bar{w}_t^{i,j} < \hat{w}_{t-1}^{i,j} - \Delta^j) \\ \qquad \bar{w}_t^{i,j} \qquad\qquad\qquad else \end{cases} \\ w_t^i &= \prod_{j \in \Gamma} w_t^{i,j} \end{aligned} \tag{8}$$

3.3 "Lazy" Resampling

The main problem of the approach described in [7] is the resampling step: particles are copied in the new distribution χ with a frequency proportional to their importance weight w_t^i, as in the Step 3 of Algorithm 1, but the weights $w_t^{i,j}$ are not reset to 1 afterward, this because otherwise it would not be possible to use them to filter the measurements at the following iterations of the algorithm. So, supposing that a particle with importance w_t^i is copied $n^i > 1$ times in the target distribution χ, the probability density in that location will increase by a factor of n^i. Consequently, after the resampling step the particle distribution does not approximate the true posterior anymore: the probability density will artificially increase in areas of high likelihood, while it will further decrease in areas of low likelihood. The result is that this filter converges very quickly and tends to focus most of the particles around the main peak of the probability distribution; in practice the filter works very well as long as the noise can be filtered by its sensor smoothing mechanism, but when there is a high level of ambiguity in the sensor data, for example if no unique landmark is observed in a long time, then the particles will start to quickly oscillate among different peaks of the true posterior distribution, rather than assume a multi-modal distribution. This problem can be fixed however, if we know the number of times n^i that a certain particle i has been copied. This can be used to normalize the weights in the target distribution, so that it again approximates the posterior:

$$\tilde{w}_t^i = \frac{w_t^i}{n_t^i} \Rightarrow \frac{\prod\limits_{j \in \Gamma} w_t^{i,j}}{\prod\limits_{j \in \Gamma} \nu_t^i} \Rightarrow \prod_{j \in \Gamma} \frac{w_t^{i,j}}{\nu_t^i} \tag{9}$$

$$\nu_t^i = \sqrt[c]{n_t^i}; c = \|\Gamma\|$$

where, given n^i and c (the number of percept classes) we can normalize the global weight of the particle $\tilde{w}_t^i$ dividing its importance weights by the coefficient ν_t^i. As the basis for our new resampling algorithm, we use the *low variance sampler* [10]. The low variance

Algorithm 2. Low Variance Sampling.

Require: proposal particle distribution $\bar{\chi}_t$, importance weights W_t, number of particles N
$\chi_t = \emptyset$
$r = rand(0; N^{-1})$
$c = w_t^1$
$i = 1$
for $m = 1$ to N **do**
  $U = r + (m-1) \cdot N^{-1}$
  while $U > c$ **do**
    $i = i + 1$
    $c = c + w_t^i$
  end while
  add x_t^i to χ_t
end for
return target distribution χ_t

sampler has several advantages over the independent random sampler described in the basic SIR filter algorithm: it has $O(N)$ complexity (instead of $O(Nlog(N))$), it guarantees that if all particles have the same importance, χ_t is the same as $\bar{\chi}_t$, and most importantly for us, it covers the set of samples systematically, making it much easier to track the values n^i to use for normalization. At this point, we can introduce a new technique, that here we will call *"lazy resampling"*, to enforce temporal coherence in the particle distribution: we can clip the number of new samples in the target distribution χ which can be generated from a single particle in the proposal distribution $\bar{\chi}$ to be at most a chosen $\bar{n}_{max} \in [1..N]$. The idea is to reduce the effect of a single resampling step on the particle distribution: if a robot is well localized and it is not teleported, we would expect that the peaks of the posterior distribution at time t are close to the position they had at time $t-1$, factoring in the odometry data in the process update step. Especially when drawing some samples from the measurement model, as in the Sensor Resetting / Mixture Monte Carlo approaches described, a few bad measurements would move many particles away from the area of high likelihood, degrading the performance of the filter; by setting a limit to the number of times that a single particle can be copied as a consequence of resampling, we are reducing the effects of outliers on the particle distribution. The downside is that, in case the robot is really teleported, our filter would be slower in relocating most of the particles in the new area of high likelihood: the robot would still relocate its estimate quickly, because of the update in the particle importance weights, but there would be initially less resolution in the area where the robot has been teleported. Thus, it is necessary to set a value of $\bar{n}_{max}$ as a compromise dependent on the frequency of robot teleports and the amount of sensor noise and ambiguity. For $\bar{n}_{max}=1$, a particle in $\bar{\chi}$ is copied unaltered in χ, as such the SIS filter can be seen as a special case of our lazy resampling; at the other end of the spectrum, setting $\bar{n}_{max}=N$ results in no limitations in the effects of resampling, so our filter behaves exactly like a SIR filter. Algorithm 3 describes a possible implementation, r is used to calculate how many times a certain particle should be copied in the target distribution based on its importance weight; such number is clipped to a maximum resample limit of $\bar{n}_{max}$. Particles of low weight ($n<1$) are copied in the target distribution starting from the end of the list (indexed by j), so that they can be overwritten by the more important particles, which are copied starting from the beginning of the list (indexed by i).

4 Experiments and Results

In the following experiments, we will evaluate the performance of our new approach (from now on referred to as Temporal Smoothing Monte Carlo “TSMCL”) which uses our *sensor smoothing with aging* (Equations 7,8) and the *lazy resampling* strategy described in Algorithm 3. We compare it with a SIR filter with Sensor Resetting, and the approach described in [7] (here for convenience referred to as Sensor Smoothing Monte Carlo “SSMCL”).

4.1 Test Environment

The robot is placed in a soccer field whose size is $6m \times 4m$, where it can recognize 4 unique landmarks: 2 goals (1 yellow, 1 blue) and 2 cylindrical beacons (yellow on blue, and blue on yellow). The position of the landmarks on the map is known to the robot. Additionally, the robot can recognize non unique features such as the white lines on the field, and the intersections where the lines meet (see Figure 2).

Since the environment is color coded, the vision system, similar to the approach described in [14], uses color based image segmentation. The vision and localization systems run at 30 frames per second.

As can be seen in Figure 3, the measurement error is high (especially the measured distance error grows almost exponentially with the distance to the observed object), due to the low resolution of the camera and the fact that the camera pose in space can

Algorithm 3. "Lazy" Resampling.

Require: proposal particle distribution $\bar{\chi}_t$, importance weights W_t, number of particles N, resample limit $\bar{n}_{max}$

$\chi_t = \emptyset$

$i = 1$

$j = N$

$r = \frac{N}{\sum_i w_t^i}$

for $m = 1$ to N **do**

  $n = min(\lfloor r \cdot w_t^m \rfloor; \bar{n}_{max})$

  if $n > 0$ **then**

    normalize $\tilde{w}_t^m = \frac{w_t^m}{n}$

    for $z = 1$ to n **do**

      add $x_t^m, \tilde{w}_t^m$ to $\left\langle \chi_t, \tilde{W}_t \right\rangle$ in position i

      $i = i + 1$

    end for

  else

    if $j > i$ **then**

      add x_t^m, w_t^m to $\left\langle \chi_t, \tilde{W}_t \right\rangle$ in position j

      $j = j - 1$

    end if

  end if

end for

return target distribution $\left\langle \chi_t, \tilde{W}_t \right\rangle$

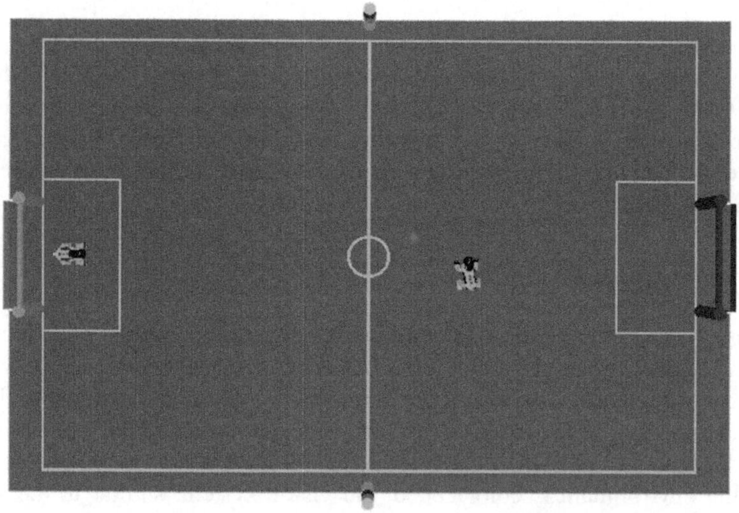

Fig. 2. Test environment: a yellow and a blue goal, two colored beacons (on the middle line), two robots and an orange ball

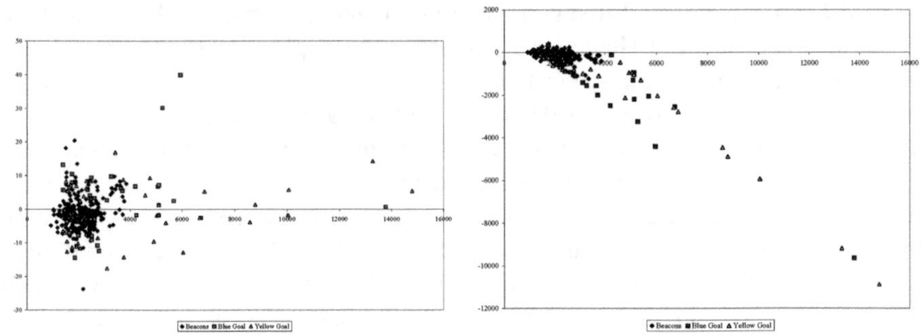

(a) Bearing Error (b) Distance Error

Fig. 3. Landmark Measurement error. The bearing (degrees) and distance (mm) errors are plotted as a function of the perceived distance.

only be roughly estimated. In fact the camera is mounted on the head of the robot, it has 3 degrees of freedom and due to the narrow field of view, it has to constantly scan around at high angular velocity. Moreover, the height of the neck on which the camera is mounted is not constant since the robot uses legged locomotion, and it can only be approximately estimated through kinematic calculations from the servos' encoders in the robot legs. For these reasons our sensor model calculates the measurement error in terms of horizontal angle (*bearing*) and vertical angle (*depression*) to the observed object. This has the advantage that, while the distance error grows more than linearly with the real distance to the object, the vertical angle error variance is approximately constant. We assume the horizontal and vertical angle error components to be independent, and use Gaussian likelihood functions to model each component, with different constants $\sigma^j_{horizontal}, \sigma^j_{vertical}, j \in \Gamma$ for each percept class. Field lines and their intersections (crossings) due to their small size can only be recognized reliably up to a maximum distance of approximately $800mm$; for these non-unique features we adopt a closest point matching model (with pre-computed lookup tables) as described in [7]. Due to the different measurement errors (see Table 1) and frequency of detection, we distinguish our percepts in 3 different classes: $\Gamma = \{Beacons, Goals, Lines\}$. All three localization algorithms use overall 100 particles, draw a small percentage of them directly from observations, and are executed in parallel in real-time on our test platform. The locations for such samples are calculated from landmarks via triangulation or using two bearings and two distances; for this purpose, the location of seen landmarks

Table 1. Measurement error. ρ represents the distance error, in expressed in *mm*; α the bearing error, in *deg*.

Percept Class	μ_ρ	σ_ρ	μ_α	σ_α
Beacons	162.7	178.5	3.2	3.1
Goals	708.5	1294.6	4.6	4.8
Lines	122.6	215.8	3.1	3.5

is retained for 5 seconds and updated using odometry information. Additionally, line crossings are classified as T-shaped or L-shaped, and samples are drawn from all possible locations which match the seen shape (8 at most, for L-shaped crossings, hence we use a mixture with 8% of particles drawn from observations), and two further possible poses are found if the center circle is seen. To provide ground truth to evaluate the localization performance, we use an external vision system based on a ceiling-mounted camera which is able to track the robot position at 25 frames per second with a maximum error of 40mm (position) and 2° degrees (heading). The *process model* updates the particle positions based on odometry data which is provided by the robot's motion module; the odometry error is modeled as a bi-dimensional Gaussian with the major axis parallel to the direction of movement, and an independent Gaussian represents the heading error.

4.2 Experiments

In the static localization case, with the robot having unlimited time to reach a specified position on the field, all three approaches are able to localize the robot with a position error below 7 cm and heading error below 3 degrees, and in this situation it is not possible to make an analysis of the relative performance, due to measurement errors in the ceiling camera system providing ground truth and the fluctuations derived from the randomized nature of the localization algorithms employed. So in the following experiments, we will analyze the accuracy of localization while the robot runs around the field: in the first half of each experiment the robot will be allowed to look around to see unique landmarks, in the second half it will be chasing a ball looking only at it, hence seeing landmarks only occasionally, having to rely on non-unique features such as lines and crossings for localization. In the first experiment, we want to illustrate the effects of our *lazy resampling*, by comparing the performance of our algorithm for different values of the *resample limit* $\bar{n}_{max}$: 8 (the experimentally derived optimum for our application), 15 and 30 (for all practical purposes, equivalent to "unlimited", given a set of 100 particles). As can be seen in Figure 4, all three versions localize about at the same time ($\approx$ 180 frames or 6 seconds), then they perform equally well, as

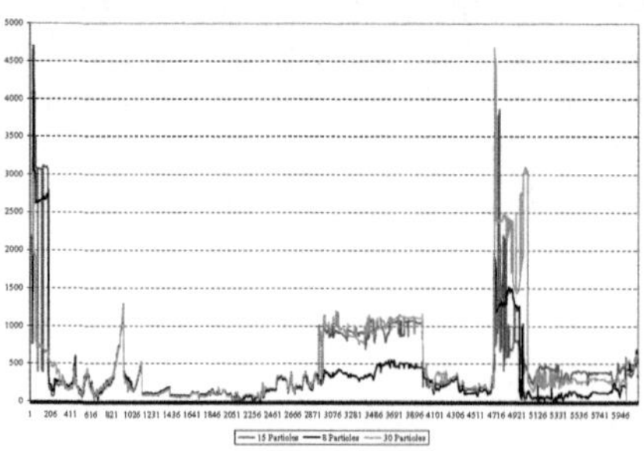

(a)

	First Half			Overall		
	μ	σ	max	μ	σ	max
$\bar{n}_{max} = 8$	177.5	157.7	1298	261.0	281.4	1940
$\bar{n}_{max} = 15$	183.5	154.6	1263	396.9	553.1	3875
$\bar{n}_{max} = 30$	190.7	164.8	1281	482.3	588.2	4689

(b)

Fig. 4. Resample limit comparison (Test 1). x-axis: time (frames); y-axis: position error. At frames 2900-3950 and 4700-5100 the scarcity of unique features and the odometry errors make the localization jump to a wrong location.

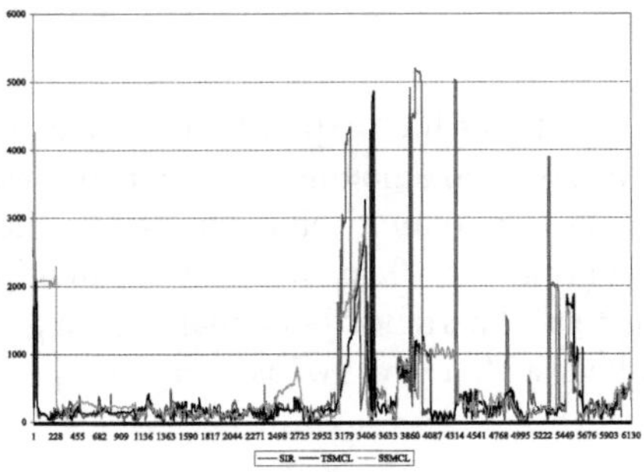

(a)

	First Half			Overall		
	μ	σ	max	μ	σ	max
TSMCL	144.8	62.8	430	323.6	503.0	4864
SSMCL	201.9	130.3	813	444.1	553.1	2875
SIR	146.2	76.1	548	436.7	897.1	5195

(b)

Fig. 5. Localization algorithm comparison (Test 2). x-axis: time (frames); y-axis: position error.

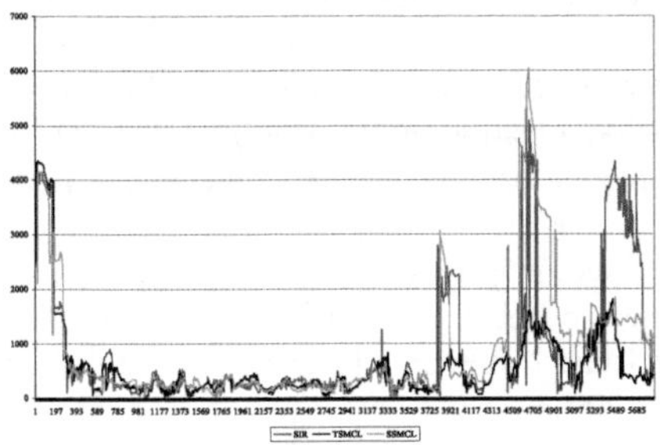

(a)

	First Half			Overall		
	μ	σ	max	μ	σ	max
TSMCL	277.5	160.1	847	432.7	355.0	1811
SSMCL	276.3	128.6	867	745.8	998.1	6051
SIR	294.2	169.0	1261	706.9	993.7	5567

(b)

Fig. 6. Localization algorithm comparison (Test 3). x-axis: time (frames); y-axis: position error.

long as the robot receives reasonably good percepts. However, in the situations where the robot looks only at the ball for a long time, the version with $\bar{n}_{max} = 8$ performs much better, being more effective in smoothing the noise and resolving ambiguities by keeping more particles in the areas where the robot was previously localized. In the second experiment, we compare our TSMCL algorithm with $\bar{n}_{max} = 8$ to the state of the art approaches. In the first half of the experiment, all systems perform equally well, with TSMCL and SIR nearly identical and SSMCL slightly worse but within the limits due to random factors. When the robot starts to chase the ball however (after frame 3000), SIR and SSMCL start to oscillate much more (Figure 5). Our third test is similar to the second, but it represents a worse scenario since this time the beacons (which are the best landmark, in terms of measurement error and ease of detection) are removed. The results presented in Figure 6 show our approach clearly outperforming the others. The run-time of all 3 algorithms is $3 - 5ms$ per frame, depending on the number and type of percepts seen. We could not measure significant differences in the time taken to globally localize and recover from kidnapping ($2 - 6s$ in all cases).

5 Future Work

We presented an extension of Monte Carlo Localization which exploits temporal coherence to improve localization accuracy in presence of high amounts of noise and sensor ambiguity. In the future we plan to apply this approach to multi-object tracking, where the use of temporal coherence could allow our algorithm to "keep memory" of multiple modes in the distribution more effectively than normal particle filters, making it a good candidate to compete with banks of multiple filters with degrees of association.

References

1. Sony Corporation: OPEN-R SDK Model Information for ERS-7. Sony Corporation (2004)
2. Gordon, N.J., Salmond, D.J., Smith, A.F.M.: Novel approach to nonlinear/non-gaussian bayesian state estimation. Radar and Signal Processing, IEE Proceedings F 140(2), 107–113 (1993)
3. Dellaert, F., Fox, D., Burgard, W., Thrun, S.: Monte carlo localization for mobile robots. In: IEEE International Conference on Robotics and Automation (ICRA 1999) (May 1999)
4. Geweke, J.: Bayesian inference in econometric models using monte carlo integration. Econometrica 57(6), 1317–1339 (1989)
5. Lenser, S., Veloso, M.: Sensor resetting localization for poorly modelled mobile robots. In: Proceedings of the IEEE International Conference on Robotics and Automation, 2000 (2000)
6. Thrun, S., Fox, D., Burgard, W., Dellaert, F.: Robust monte carlo localization for mobile robots. Artificial Intelligence 128(1-2), 99–141 (2001)
7. Röfer, T., Jüngel, M.: Vision-Based Fast and Reactive Monte-Carlo Localization. In: Proceedings of the 2003 IEEE International Conference on Robotics and Automation, ICRA 2003, Taipei, Taiwan, September 14-19, 2003, pp. 856–861. IEEE, Los Alamitos (2003)
8. Nisticò, W., Hebbel, M.: Temporal smoothing particle filter for vision based autonomous mobile robot localization. In: Proceedings of the 5th International Conference on Informatics in Control, Automation and Robotics (ICINCO), vol. RA-1, pp. 93–100. INSTICC Press (2008)
9. Fox, D., Hightower, J., Liao, L., Schulz, D., Borriello, G.: Bayesian Filtering for Location Estimation. PERVASIVE computing, 10–19 (July-September 2003)
10. Thrun, S., Burgard, W., Fox, D.: Probabilistic Robotics. MIT Press, Cambridge (2005)
11. Kalman, R.E.: A new approach to linear filtering and prediction problems. Transactions of the ASME - Journal of Basic Engineering 82, 35–45 (1960)
12. Fox, D.: Adapting the sample size in particle filters through kld-sampling. I. J. Robotic Res. 22(12), 985–1004 (2003)
13. Doucet, A., Godsill, S., Andrieu, C.: On sequential monte carlo sampling methods for bayesian filtering. Statistics and Computing 10, 197–208 (2000)
14. Nisticò, W., Röfer, T.: Improving percept reliability in the sony four-legged league. In: Bredenfeld, A., Jacoff, A., Noda, I., Takahashi, Y. (eds.) RoboCup 2005. LNCS, vol. 4020, pp. 545–552. Springer, Heidelberg (2006)

Oscillatory Controllers with Recurrent Neural Networks

Guillaume Jouffroy

Laboratoire d'Informatique Avance de Saint-Denis (LIASD)
University Paris 8, France
gj@ai.univ-paris8.fr

Abstract. In the robotics field, a lot of attention is given to the complexity of the mechanics and particularly to the number of degrees of freedom. Also, the oscillatory recurrent neural network architecture is only considered as a black box, which prevents from carefully studying the interesting features of the network's dynamics. In this paper we describe a generalized teacher forcing algorithm, and we build a default oscillatory recurrent neural network controller for a vehicle of one degree of freedom. We discuss a weak teacher forcing heuristic method to improve the stability of the oscillator. We then build a feedback system as a constraint method for the joint. We show that with the default oscillatory controller the vehicle can however behave correctly, even in its transient time from standing to moving, and is robust to the oscillatory controller's own transient period and its initial conditions. We finally discuss how the default oscillator can be modified, thus reducing the local feedback adaptation amplitude.

Keywords: Joint constraint method, oscillatory recurrent neural network, generalized teacher forcing, feedback, adaptive systems.

1 Introduction

Central Pattern Generators (CPG) are biological periodic oscillatory neural networks responsible for a wide range of rhythmic functions. They can be made of endogenous oscillatory neurons connected to non oscillatory ones or from the sole interaction between non oscillatory neurons.

Particularly, they are a great source of inspiration in the robotics field, for the control of joints in locomotion. In general, an oscillatory network controls a joint angle in both directions, and the phase relationships needed between all joints arise from the coupling between the different networks.

The needed parameters for an artificial Recurrent Neural Network (RNN) to have a periodic oscillatory behavior cannot be measured experimentally. This network is most of the time a relatively simplified model of its biological counterpart when available. Only clinical temporal data of joints kinetics and kinematics can be of use, where however it is difficult to isolate the real control of a particular joint from the influence of the others.

In the case of non endogenous oscillatory neurons, in the literature, parameters are thus mainly determined empirically or with genetic algorithms [1, 2, 3, 4, 5, 6], comparatively to relatively few learning methods [7, 8, 9]. Though this is useful with

J.A. Cetto et al. (eds.): Informatics in Control, Automation and Robotics, LNEE 37, pp. 181–190.
springerlink.com

large networks in complex mechanical models, there are two drawbacks. It is very difficult in general to isolate the resulting dynamics of the different networks, and to understand their interaction to each other and with the mechanical system dynamics. Also, it is not clear how to modify such networks in an adaptive context, e.g. in the case of a permanent constraint change on a joint due to injury.

Based on this considerations, we apply a *generalized* formulation of the so called *teacher forcing* gradient descent-based learning algorithm, to create an oscillatory RNN as a *torque controller* for an interesting vehicle with one single degree of freedom, the Roller Racer. The RNN is put in a closed loop with the Roller Racer, such that the vehicle can be freely controlled, where the RNN can be modified permanently.

The paper is structured as follows. In section 2, we briefly present the Roller Racer model and we show how it can be controlled with a torque input. In section 3 we describe the control system. The subsection 3.1 presents the generalized formulation of the teacher forcing learning algorithm with which we build the oscillatory RNN as a basic torque controller for the Roller Racer vehicle. In subsection 3.2 we describe the local feedback control that can be built so that the vehicle direction can be controlled, limiting the effect of its transient state. We discuss how the basic oscillatory system can be In section 4, we give concluding remarks and discuss how the basic oscillatory system can be modified to better fit the needs of the Roller Racer, thus reducing the local feedback adaptation amplitude.

2 Vehicle Model

The Roller Racer is a toy-vehicle with one single degree of freedom which is the handlebar. The direction wheels are shifted back from the the axis. Thus, oscillating the handlebar from side to side, a component of the reaction force on the ground which points backward is created, moving forward the vehicle.

In [10], we revisited and synthesized a mathematical model of the Roller Racer from the original work of Krishnaprasad and Tsakriris [11]. The input control was the angle of the axis. Here, we will describe the torque input control formalization.

Recall the state of the Roller Racer vehicle is $\mathbf{x} \triangleq (\theta_r, x_r, y_r, p, \theta_c, \dot{\theta}_c)^T \in \mathbb{R}^6$, and its dynamics is $\dot{\mathbf{x}} = \mathbf{f}(\mathbf{x}, u)$, where

$$\mathbf{f}(\mathbf{x}, u) \triangleq \begin{pmatrix} \frac{1}{\Delta(\theta_c)}\Big(\sin\theta_c p - \delta(\theta_c)\dot{\theta}_c\Big) \\ \frac{\cos\theta_r}{\Delta(\theta_c)}\Big(\chi(\theta_c)p - \gamma(\theta_c)\sin\theta_c\dot{\theta}_c\Big) \\ \frac{\sin\theta_r}{\Delta(\theta_c)}\Big(\chi(\theta_c)p - \gamma(\theta_c)\sin\theta_c\dot{\theta}_c\Big) \\ \Big[A_1(\theta_c)\dot{\theta}_c - C_1(\theta_c)\Big]\,p + \\ \qquad \Big[A_2(\theta_c)\dot{\theta}_c + C_2(\theta_c)\Big]\,\dot{\theta}_c \\ \dot{\theta}_c \\ u \end{pmatrix}, \tag{1}$$

θ_c is the angle of the handlebar, p is the momentum of the vehicle and (x_r, y_r) are the rear coordinates respectively to the global reference space. Friction constraint is built in the model through the functions $C_1(\theta_c)$ and $C_2(\theta_c)$. Thus one does not need to deal with mechanical aspects, leaving focus on the control strategy, and on the learning aspects of the RNN.

Here we control the Roller Racer using the torque input control T_c with the following equation

$$\ddot{\theta}_c = u = B_1(\theta_c)\dot{\theta}_c p + B_2(\theta_c)\dot{\theta}_c^2 + B_3(\theta_c)T_c \tag{2}$$

The right hand side of the equation replaces u in (1). The parameters are defined as

$$B_1 \triangleq -\frac{A_2(\theta_c)}{\Delta_1(\theta_c)}$$
$$B_2 \triangleq \frac{m_1\gamma(\theta_c)\sin\theta_c}{\Delta(\theta_c)\Delta_1(\theta_c)}[\gamma(\theta_c)\cos\theta_c + d_1\delta(\theta_c)]$$
$$B_3 \triangleq \frac{\Delta(\theta_c)}{\Delta_1(\theta_c)},$$

where

$$\Delta_1 \triangleq I_1 I_2 \sin^2\theta_c + m_1(I_1 d_2^2 + I_2 d_1^2\cos^2\theta_c),$$

and the other parameters are as defined in [10].

3 Design of the Oscillatory Controller

3.1 The Oscillatory Recurrent Neural Network Torque Controller

Let us consider the RNN system

$$\dot{\mathbf{x}} = \mathbf{f}(\mathbf{x}, \mathbf{W}), \tag{3}$$

with $\mathbf{x} \in \mathbb{R}^n$ is the state vector of the network, $\mathbf{W} \in \mathbb{R}^{n\times n}$ is the matrix of the weight connections, w_{ij} to be considered as the weight from the neuron i to the neuron j. We consider a *fully* connected RNN, which means all neurons are interconnected and self-connected ($w_{ii} \neq 0$). For the neuron model we use the rate based neuron model of the simplest form

$$\mathbf{f}(\mathbf{x}, \mathbf{W}) = (\mathbf{I}\tau^{-1})(-\mathbf{x} + \mathbf{W}^T s(\mathbf{x})), \tag{4}$$

with $s(x)$ a squashing function such as $\tanh(x)$. $\mathbf{I}$ is the identity matrix and $\tau \in \mathbb{R}^n$ the time constant vector of the system.

Each component x_i^* of the teacher vector $\mathbf{x}^*$ is of the form

$$x_i^* = \sin(t + \phi_i) \tag{5}$$

The learning is achieved when a global error criterion E, $E \in \mathbb{R}^n$ is less or equal than a minimum $\epsilon \in \mathbb{R}$, $\epsilon \simeq 0$

$$E = \frac{1}{2}(\mathbf{x} - \mathbf{x}^*)^{\mathbf{T}}(\mathbf{x} - \mathbf{x}^*) < \epsilon, \tag{6}$$

The weight matrix $\mathbf{W}$ is adjusted according to the following gradient rule

$$\dot{w_{pq}} = -\eta \sum_{i=1}^{n} \frac{\partial E}{\partial x_i} z_{pq}^i, \tag{7}$$

with $\eta \in \mathbb{R}$ is the learning rate. $\mathbf{z} \in \mathbb{R}^{n \times n^2}$ is the sensitivity of the state of the system with respect to a weight w_{pq}. For convenience $\mathbf{z}$ is of the form

$$\begin{bmatrix} z_{11}^1 & \cdots & z_{nn}^1 \\ \vdots & \ddots & \vdots \\ z_{11}^n & \cdots & z_{nn}^n \end{bmatrix} \tag{8}$$

which can be written in the matrix form

$$\frac{d\mathbf{z}}{dt} = (\mathbf{I}\tau^{-1})(\mathbf{J}_{\mathbf{f_x}}(M)\mathbf{z} + \mathbf{J}_{\mathbf{f_w}}(M)), \tag{9}$$

where $\mathbf{J}_{\mathbf{f_x}}(M)$ and $\mathbf{J}_{\mathbf{f_w}}(M)$ are the jacobian matrices of the function $\mathbf{f}$ respectively to $\mathbf{x}$ and $\mathbf{w}$ at the point M. $\mathbf{J_{fx}} \in \mathbb{R}^{n \times n}$, and $\mathbf{J_{fw}}$ is of the same form as $\mathbf{z}$, i.e. $\mathbf{J_{fw}} \in \mathbb{R}^{n \times n^2}$. In the teacher forcing special case (9) reduces to

$$\frac{d\mathbf{z}}{dt} = (\mathbf{I}\tau^{-1})(-\mathbf{I}\mathbf{z} + \mathbf{J}_{\mathbf{f_w}}(M)), \tag{10}$$

Providing a target signal(s) x_i^* only for some neuron(s) i, letting $\mathbf{J}_{\mathbf{f_x}} = \mathbf{A}$, the sensitivity equation (9) can be written as

$$A_{ij} = -a_{ij} + b_{ij} w_{ji} \frac{\partial s(x_j)}{\partial x_j}, \tag{11}$$

with $a_{ij} = 1$ when $i = j$, 0 otherwise, $b_{ij} = 1$ when i is not a forced neuron, 0 otherwise. $\mathbf{J}_{\mathbf{f_w}} = \mathbf{B}$ is of the same form as $\mathbf{z}$ and its elements are such that $B_{pq}^i = 0$ when $i \neq q$, $B_{ij} = s(x_p^*)$ if p is a forced neuron, $B_{ij} = s(x_p)$ otherwise.

With this algorithm we build a 4 neurons fully connected RNN. Teacher signals have an amplitude of 1, and we choose the phase difference vector ϕ^* as $\{0; \pi/3; 2\pi/3; \pi\}$. To generate the torque control we use the output of the neurons 1 and 4 which are in opposite phase, to control each direction of the handlebar angle using the transformation

$$T_c = pos(x_1) - pos(x_4) \tag{12}$$

The use of 3 neurons could have been the minimum acceptable to solve the phase difference of π between the output neurons. But in the scope of recover, with at least 4 of them if one break, we have the opportunity to start the algorithm again and obtain the needed oscillator. The data obtained for the weight matrix $\mathbf{W}$ of the estimated oscillator needed is

$$\mathbf{W} = \begin{bmatrix} 0.493 & -0.18 & -0.673 & -0.493 \\ 0.958 & 0.882 & -0.076 & -0.958 \\ 0.465 & 1.062 & 0.597 & -0.465 \\ -0.493 & 0.18 & 0.673 & 0.493 \end{bmatrix}$$

Convergence is reached in at most 300 time steps, with $\eta = 0.1$.

Will the Free Network Run with an Oscillatory behavior ? As pointed out by [12], it is not guaranteed that once the neural network is released from its forcing teacher signals, it keeps its oscillating. The rate of adaptation parameter η, the numerical approximation, and the local nature of the convergence of such a gradient based method are implied in the uncertainty of the stability of the free system. Three heuristics methods are proposed to enhance the stability. At the time we are writing this document, they methods have not been tested yet.

The first method is to apply some noise to the teacher signals. However this should also add some noise to the parameters, which could result in less stability, particularly where a solution is to be found in a relatively complex parameter space to explore. The second method, which is the one we used in this work, is what we call $weak$ teacher forcing. In this case we use in place of $\mathbf{x}^*$ the mixed vector input

$$\mathbf{d}^* = \mathbf{x} + \alpha(\mathbf{x}^* - \mathbf{x}), \tag{13}$$

with $0 < \alpha < 1$, decreasing the forcing rate α with the error term.

Note that this expression should not be introduced in the sensitivity expression but *only as input* to the network, as changing α would create some undesirable variations of the derivative terms in (9).

During our experiments we noted that the particular maximum value of α was not critical to the learning process. Interestingly, it even improved the convergence speed. We have chosen $\alpha_{max} = 0.6$. The phase relationship between $\mathbf{d}^*$ and $\mathbf{x}$ was the informative criterion which mattered the most to the learning process.

When the learning process has found a solution, we reduce α to remove a part of the residual error which was obtained without weak forcing. However $alpha$ cannot be decreased to 0. For $\eta = 0.1$, we have been able to reduce the teacher signals to $\alpha = 0.2$, i.e. 20% forcing. Even if the residual error is very low, if the teachers are removed the learning process has to be stopped or it will start to be unstable, trying to reduce the error without signals to reset the network to the right position. This means that though the weak forcing signals can help to make the free oscillator more stable, it can not be suppressed and thus this algorithm could not be thought as an *on-line* adaptive process, while the network is controlling the vehicle.

The last heuristic method proposed by [12], is to train the network with teacher forcing for some time, and then to finish the training without forcing. Our experiments of the weak forcing method suggest that this last proposal cannot be applied to temporal signals such as oscillations.

3.2 Feedback Design of the System

The torque amplitude of the oscillatory controller, considering its frequency, may be too large for the needs of the Roller Racer. Beside, the starting energy for a vehicle is generally different from when it is at full speed. So is the Roller Racer with the torque control. It needs a transient oscillatory input which might depend on friction biases and inertia. Thus we create an angle feedback from the Roller Racer to the oscillatory controller, which purpose is to constrain the angle within some limits. We control the vehicle in the forward direction which means the average angle of the handlebar should be π (see [10]).

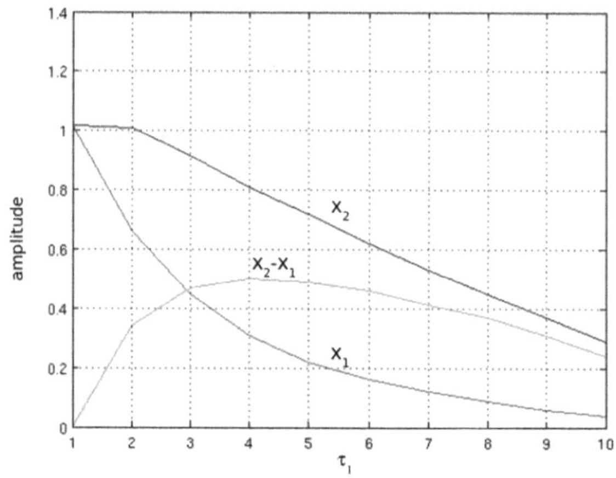

Fig. 1. Effect of the modification of the time constant τ_1 on the state x_1 in the oscillatory neural network. τ_2 is kept fixed($\tau_2 = 1$).

To apply a correction to the torque generated, we use the feedback to modify the time constants τ_i of the oscillatory controller's output neurons. In Figure 1 we illustrate the effect on the amplitude of the output neurons state when changing only one time constant (τ_1 on the figure). As might be expected, the state of each neuron output decreases. However the correction applies more to x_1, and the amplitude difference is maximum at $\tau \approx 4$. It is not really really desirable to have one output which becomes zero as it prevents the Roller Racer to get energy. Therefore we should not have τ_i being set too high.

We now define the transfer function, that constrains the angle within the boundaries $[\pi - 1; \pi + 1]$ and with a correction applied when $\tau_i < 5$, considering the effect on the amplitude reduced above this limit. Note that the frequency is relatively not modified in this range (a frequency modification would take place if all τ_i where equally changed). The transfer function **g** for the feedback is defined as

$$\mathbf{g}(u) = \begin{bmatrix} g_1(u) \\ g_4(u) \end{bmatrix}, \tag{14}$$

with

$$g_1(u) = \frac{6}{1 + e^{-(6u-4.5)}}, \tag{15}$$

g_1 being the transfer function for the feedback to the neuron 1, and $g_4(u) = -g_1(u)$ to the neuron 4.

The formalization of these transfer functions has been chosen so that they can be used as "feedback" neurons, with the same neuron model as in (3) , replacing f_i by g_i.

The signal $u \in \mathbb{R}$ is the actual feedback signal which is the difference between the angle of the handlebar θ_c (in radians) and the desired average angle control $v \in \mathbb{R}$. For the forward direction u is actually set as

$$u = \theta_c - v = \theta_c - (\pi + \delta), \tag{16}$$

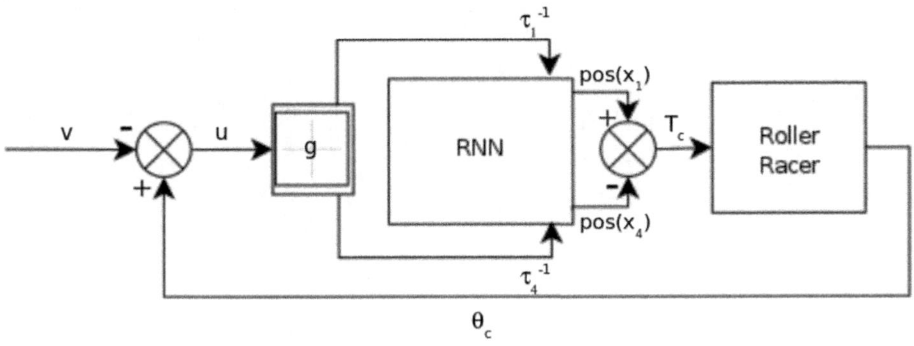

Fig. 2. Control architecture of the Roller Racer. T_c is the torque control applied to the vehicle. v is the angular input which is obtained from the user direction control δ.

where $\delta \in \mathbb{R}$ is the user direction input, which is actually the continuous component control of the RNN.

The feedback information thus obtained is used to modify the time constants τ_1 and τ_4

$$\tau_1^{-1} = \frac{1}{1+g_1(u)} \text{ and } \tau_4^{-1} = \frac{1}{1+g_4(u)} \tag{17}$$

The whole architecture is summarized in Figure 2.

3.3 Results

We present here the results of two different trials. In both of them the purpose is to have a straight trajectory along the x axis of a physical space reference. The architecture is of course able to freely control the vehicle in all directions but the simulations are not shown.

In the first trial we start the RNN with a little energy given to a neuron. The neuron 3 in the following simulation has the initial condition $x_3 = 0.1$.

In Figure 3 left, is plotted the angle of the handlebar θ_c which continuous component is π, for the forward direction. It clearly shows the transient time when the system is extracting itself from reaction forces, until it reaches its permanent speed at around 40 time steps. The transient period has an amplifying oscillation around π because the RNN is also in its transient state with little energy.

This transient activity of the RNN, is shown by the very weak correction applied from the feedback $g(u)$ in Figure 3 right, and the low speed along the x axis in Figure 3 bottom. One can clearly see on this graphics of the right side of the figure, a little deviation during this time. This is only the drawback of the transient state of the RNN which does not provide a symmetric gain, even if the angle is within the boundaries $[\pi - 1; \pi + 1]$.

The correction applied once the vehicle is in its permanent state is only due to the fact that the function g is smooth. Low strength of its output should be ignored. Beside if this correction was too high it would show that the RNN's own oscillation is not optimal and that a relearning could fix this. The trajectory however is quite straight.

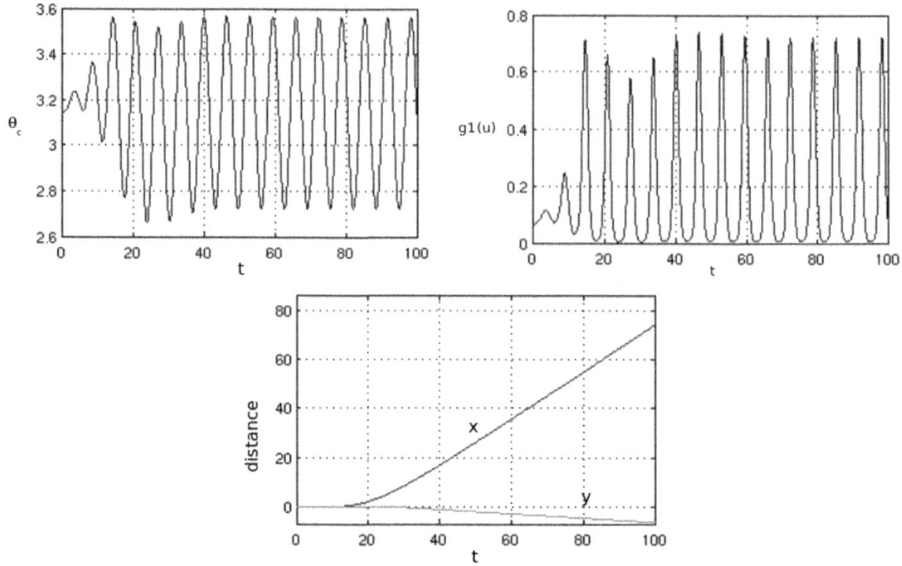

Fig. 3. Simulation results when the RNN is initialized with a weak energy. Left: angle of the handlebar θ_c. Right: feedback correction $g_1(u)$. Bottom: trajectory evolution. During and after the transient times, little correction is applied.

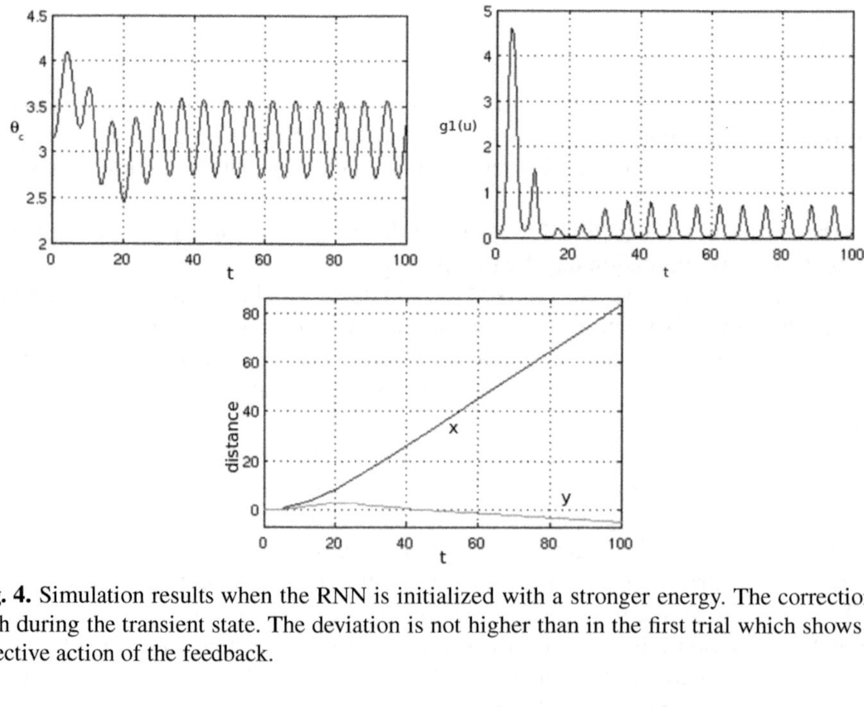

Fig. 4. Simulation results when the RNN is initialized with a stronger energy. The correction is high during the transient state. The deviation is not higher than in the first trial which shows the effective action of the feedback.

In the second trial we initialize the RNN with a strong gain to see how the feedback behaves during the transient period. We set $x_3 = 1$.

In Figure 4 left we can see that the amplitude of the transient state of the RNN has been pushed too high. The angle of the handlebar θ_c does not show anymore an amplifying oscillatory behavior, and reach the boundaries we have specified. The correction from the feedback apply a high gain correction (see Figure 4 right).

After $t \approx 40$, as in the first trial, the symmetric oscillations are recovered, and the correction reduces to the steady-state time in Figure 3 right. Interestingly the correction has constrained the deviation well, which is not higher than in the first trial, except in the transient period. The vehicle also gets speed earlier and the trajectory is also straight (Figure 4 bottom).

4 Conclusions and Discussion

Nature is a great source of inspiration for engineers who deal with autonomous robots. Evolution has found optimally-designed solutions for robustness and adaptability in a changing environment, which are exciting to discover. However, most of the research in the control aspects of robots with neural networks tackle the question of complex mechanics with many degrees of freedom, and massive neural architectures, which appear as "black boxes" designed by genetics algorithms. This hides the dynamics of the neural system and correlatively the opportunity to constitute adaptive strategies.

In this work, we present a generalized version of the teacher forcing learning algorithm, to build up an estimated oscillatory controller for a vehicle with one degree of freedom, the Roller Racer. We discuss the use of a $weak$ teacher forcing heuristic principle, to lower the neural network oscillation dependence on the teacher signals, to improve the stability of the controller once these signals are removed. Actually in our experiments we have seen that this weak forcing cannot vanish. An on-line learning process for a neural network controller is still to be discovered.

We also create an angular feedback such that the degree of freedom is constrained within some boundaries. The purpose is to prevent the vehicle to go out of control during its transient state when it starts moving, as a consequence of the oscillator being not adapted to this particular moment. Our simulation results show that the feedback makes the vehicle to behave relatively well during transient state, when the oscillator is initialized with a weak energy or even a strong one. The deviation also stays little. When the steady-state period is reached, the vehicle moves in a straight line as expected.

From a design point of view, the correction applied by the feedback system to the RNN, never completely vanishes. This shows that a more optimal oscillatory behavior can be obtained, though the "default" one does not critically affect the system with the help of a *mutual entrainment* between the vehicle dynamics and the controller, as described first by [5].

We are currently studying how an adaptive process or observer, can modify permanently the RNN when the average correction is too high. The outputs of the network, with the help of the feedback, could be the desired targets for a second network which could thus be trained in parallel, with a partial teacher forcing. However this is highly computationally expensive, and not necessarily biologically viable.

The teacher forcing principle is made such that an oscillatory behavior can be obtained with a gradient descent algorithm. However, forcing the outputs of the network means disconnecting it, and thus loosing the interesting desired target obtained with feedback. Algorithms without direct gradient descent evaluation techniques may be more appropriate (for e.g. [13]).

Beside, the *constraint method* we used in this article can have some interest when studying the coupling of oscillatory neural networks, for e.g. to synchronize different joints. When we attach an oscillatory RNN to another one which has a constraining feedback, we can find coupling parameters which do not yield an increase of the correction in the feedback. This constraint method thus helps to reduce the space to search for suitable coupling parameters, and to better match the desired phase relationship.

References

1. Buono, P., Golubitsky, M.: Models of central pattern generators for Quadruped Locomotion. Journal of Mathematical Biology 42, 291–326 (2001)
2. Ghigliazza, R.M., Holmes, P.: A Minimal Model of a central pattern generator and Motoneurons for Insects Locomotion. SIAM Journal on Applied Dynamical Systems 3(4), 671–700 (2004)
3. Ishiguro, A., Otsu, K., Fujii, A., Uchikawa, Y., Aoki, T., Eggenberger, P.: Evolving and Adaptive Controller for a Legged-robot with Dynamically-rearranging Neural Networks. In: Proceedings of the Sixth International Conference on Simulation of Adaptive Behavior. MIT Press, Cambridge (2000)
4. Kamimura, A., Kurokawa, H., Yoshida, E., Tomita, K., Murata, S., Kokaji, S.: Automatic Locomotion pattern generation for Modular Robots. In: Proceedings of IEEE International Conference on Robotics and Automation, pp. 714–720 (2003)
5. Taga, G.: Emergence of Bipedal Locomotion through Entrainment among the Neuromusculo-skeletal System and the environment. Physica D 75, 190–208 (1994)
6. Ijspeert, A.: A connectionnist Central pattern generator for the aquatic and Terrestrial gaits of a Simulated Salamander. Biological Cybernetics 84, 331–348 (2001)
7. Mori, T., Nakamura, Y., Sato, M.A., Ishii, S.: Reinforcement Learning for Cpg-driven biped Robot. In: Nineteenth National Conference on Artificial Intelligence, pp. 623–630 (2004)
8. Tsung, F.S., Cottrell, G.W.: Phase-space Learning for recurrent Networks. Technical Report CS93-285, Dept. Computer Science and Engineering, University of California, San Diego (1993)
9. Weiss, M.: Learning oscillations using Adaptive Control. In: Inbternational Conference on Artificial Neural Networks, pp. 331–336 (1997)
10. Jouffroy, G.: Evaluating Adaptive oscillatory Neural Network controllers using a simple Vehicle Model. In: IEEE International Conference on Robotics and Biomimetics (ROBIO), pp. 1670–1675 (December 2006)
11. Krishnaprasad, P., Tsakriris, D.: Oscillations, se(2)-snakes and Motion Control, New Orleans, Lousiana (December 1995)
12. Doya, K.: Recurrent Networks: Learning Algorithms. In: Handbook of Brain Theory and Neural Networks, 2nd edn. MIT Press, Cambridge (2002)
13. Dembo, A., Kailath, T.: Model-free Distributed Learning. IEEE Trans Neural Networks 1(1), 58–70 (1990)

Concept and Design of the Intellwheels Platform for Developing Intelligent Wheelchairs

Rodrigo A.M. Braga[1,2], Marcelo Petry[2], Antonio Paulo Moreira[2], and Luis Paulo Reis[1,2]

[1] LIACC - Artificial Intelligence and Computer Science Lab
[2] FEUP-DEI - Faculty of Engineering of the University of Porto
Informatics Engineering Department, Rua Dr. Roberto Frias, s/n 4200-465, Porto, Portugal
rodrigo.braga@fe.up.pt, marcelo.petry@gmail.com, amoreira@fe.up.pt, lpreis@fe.up.pt

Abstract. Many people with severe disabilities find it difficult or even impossible to use traditional powered wheelchairs independently by manually controlling these electrical devices. Intelligent wheelchairs are a very good solution to assist severely handicapped people who are unable to operate classical electrical wheelchair by themselves in their daily activities. This paper describes a development platform for intelligent wheelchairs called IntellWheels. The intelligent system developed may be added to commercial powered wheelchairs with minimal modifications in a very straightforward manner. The paper describes the concept and design of the platform, including the hardware and software, multimodal input interface and the intelligent wheelchair prototype developed to validate the approach. Preliminary results concerning automatic movement of the IntellWheels prototype are also described showing the autonomous movement capabilities of the prototype.

Keywords: Intelligent Wheelchair, Intelligent Robotics, Human-Robot Interfaces.

1 Introduction

Wheelchairs are important locomotion devices for handicapped and senior people. With the increase in the number of senior citizens and the increment of people bearing physical deficiencies in the social activities, there is a growing demand for safer and more comfortable Intelligent Wheelchairs (IW) for practical applications. The main functions of IWs are [9][3][4]:

- Interaction with the user, including hand based control (such as, joystick, keyboard, mouse, touch screen); voice based control; vision based control and other sensor based control (such as pressure sensors);
- Autonomous navigation (with safety, flexibility and robust obstacle avoidance);
- Communication with other devices (like automatic doors and other Wheelchairs).

This paper discusses the concept and the design of a development platform for intelligent wheelchairs. The project, called IntellWheels, is composed of a control

J.A. Cetto et al. (eds.): Informatics in Control, Automation and Robotics, LNEE 37, pp. 191–203.
springerlink.com

software, simulator/supervisor and a real prototype of an intelligent wheelchair. In the study, shared control and high-level planning algorithms applied in an IW operating in a hospital environment were developed and tested. A shared control algorithm was tested, allowing the IW to automatically avoid dangerous situations. Also, typical algorithms used in most intelligent robotics applications were applied in the control of the IW and simulated in a hospital scenario. Blended with the control, a motion planner was developed capable of generating the behavior/path commands according to an a-priori created map of the world. This motion planner is capable of instructing the low-level motion controller module to achieve the high-level commands desired by the user [12].

The rest of the paper is subdivided as following different sections: Section 2 presents some related work; section 3 explains the hardware design of our development platform; section 4 and 5 contain a complete description of the software design and control system; section 6 provides experimental tests and result discussion and section 7 presents the final conclusions and points out some future research topics.

2 Related Work

This section presents a brief to the state of the art about Intelligent Wheelchairs with focus on the most recent research work on the development of IW prototypes.

In recent years, many intelligent wheelchairs have been developed and a large number of scientific projects were started in the area [19]. Only in the year of 2006 more than 30 publications in IEEE, about IW, may be found.

One of the first concept projects of an autonomous wheelchair for handicapped physicists was proposed by Madarasz [13]. He presented a wheelchair equipped with a microcomputer, a digital camera, and ultrasound scanner. His objective was to develop a vehicle capable to operate without human intervention in populated environments with little or no collision with the objects or people contained in it.

Hoyer and Holper [7] presented an architecture of a modular control for a omni-directional wheelchair. NavChair is described in [20], [2], [11] and has some interesting characteristic such as wall following, automatic obstacle avoidance, and doorways passing capabilities.

Miller and Slak [14], [15] projected the Tin Man I system, initially with three ways of operation: human guided with obstacle avoidance, move forward along a heading, move to a point (x, y). Afterwards, the project Tin Man I evolved, resulting in several new functions, extended in the Tin Man II, by including new capabilities such as: backup, backtracking, wall following, doorway passing, and docking.

Wellman [22] proposed a hybrid wheelchair which is equipped with two legs additionally to the four wheels, which enable the wheelchair to climb over steps and move through rough terrain.

Some projects presented solutions for people with tetraplegia, by using the recognition of facial expressions as the main input to guide the wheelchair [8], [16], [1]. Another method of control is by using the user "thoughts". This technology typically uses sensors that measure the electromagnetic waves of the brain [10], [17].

ACCoMo [6] is a prototype of an IW that allows for safe movement in indoor environments for the handicapped physicists. ACCoMo is an agent based prototype with simple autonomous, cooperative and collaborative behaviors.

Although several prototypes have been developed and different approaches have been proposed for IWs, there is not, at the moment, a proposal for a platform enabling easy development of Intelligent Wheelchairs using common electric powered wheelchairs with minor modifications.

3 Hardware Design

The hardware architecture of IntellWheels prototype is shown in Figure 1. The IntellWheels_chair1 is based on a commercial electrical wheelchair model Powertec, manufactured by Sunrise in England [21]. The Powertec wheelchair has following features: Two differentially driven rear wheels; Two passive castor front; Two 12V batteries (45Ah); Traditional Joystick; Power Module.

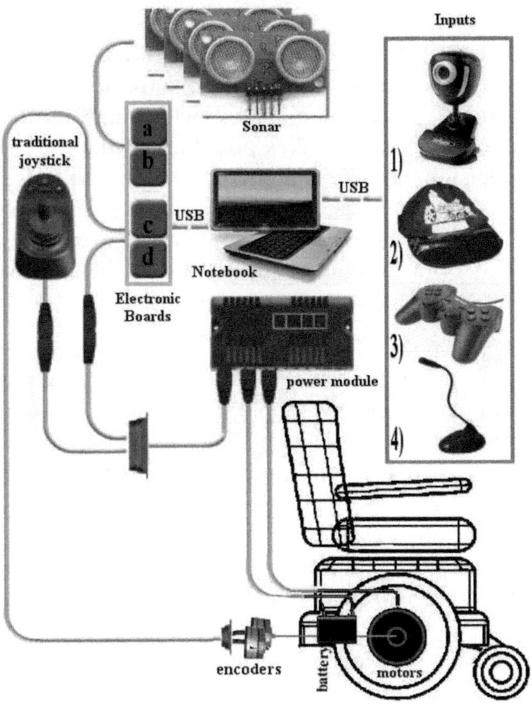

Fig. 1. Hardware Architecture of IntellWheels

The IntellWheels hardware parts are divided in three functional blocks: user inputs, IW sensor, hardware devices. These blocks are depicted in Figure 2.

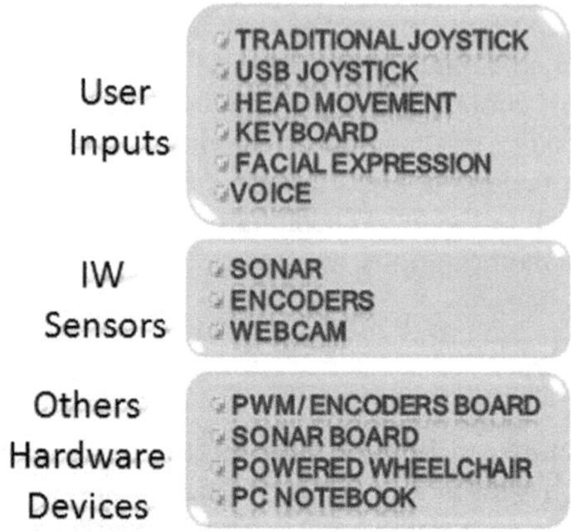

Fig. 2. Hardware Functional Blocks

3.1 User Inputs

To enable people with different kinds of disabilities to drive the intelligent wheelchair, this project has incorporated several sorts of user inputs. The idea is to give options to the patients, and let them choose what control is more comfortable and safer. Besides that, multiple inputs makes possible the IntellWheels integration with an intelligent input decision control, which is responsible to cancel inputs in the case its perception recognizes conflicts, noise and danger. For example, in a noisy environment, voice recognition would have low rate or even would be canceled. The inputs implemented go from traditional joysticks until head movement control, and are explained bellow:

- Traditional Joysticks. These inputs present in ordinary wheelchairs are a robust way to drive a wheelchair. However they may not be accessible to paraplegic or cerebral palsy people. They are present in the prototype due to its simplicity;
- USB Joystick. The USB joysticks are a little bit more sophisticated than traditional joysticks. This game joystick has many configurable buttons that makes the navigation easier;
- Head Movement. This input device is mounted in a cap making it possible to drive the wheelchair just with head movement (Figure 3);
- Keyboard. This device enables that the wheelchair control can be made just pushing some keys in the keyboard;
- Facial Expressions. By using a simple webcam, present in most of the notebooks, this software recognize simple facial expressions of the patient, using them as inputs to execute since basic commands (like: go forward, right and left) to high level command (like: go to nursery, go to bedroom);
- Voice. Using commercial software of voice recognition we developed the necessary conditions and applications to command the wheelchair using voice as an input.

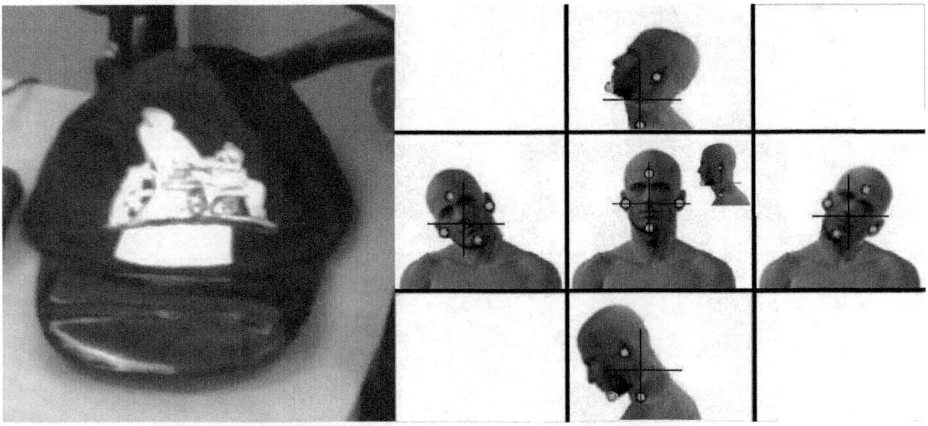

Fig. 3. Sensor for head movement detect

The use of a vast set of input options enables the prototype to be easily controlled by patients suffering from distinct disabilities. All the inputs are integrated in a very flexible multimodal interface [4].

3.2 Sensors

The purpose of this project is to develop a really intelligent wheelchair capable of performing semi-autonomous actions and having a high-level interaction with the user. However, the distinction between an IW and a robotic wheelchair is not just semantic. It means that we want to keep the wheelchair appearance, reducing the visual impact that the sensors mounted on the device produce but, at the same time, increasing the wheelchair regular functionalities. This statement limits the number and the kind of sensors we are able to use due to size, appearance and assembly constraints. To compose the intelligent wheelchair ten sonar sensors were mounted (giving the ability to avoid obstacles, follow walls and perceive unevenness in the ground), two encoders were assembled on the wheels (providing the tools to measure distance, speed, position) and two webcams were placed (one directed to the ground, capable to read ground marks and refine the odometry and the other directed to the patient head to detect facial expressions).

3.3 Hardware Devices

The hardware devices block is composed of (Figure 4):

- 2 sonar board (Electronic board 'a' and 'b' illustrated in the Figure 1 and Figure 2), the function 'a' and 'b' boards are receiving information of the ten sonars and sending it to the PC;
- 2 PWM/Encoders board (Electronic board 'c' and 'd'), these boards have speed control function, as well as sending to the PC the displacement information to enable the odometry;
- Commercial electrical wheelchair and commercial notebook.

Fig. 4. Devices installed in IW

The core of IntellWheels prototype is a PC notebook (HP Pavilion tx1270EP,AMD Turion 64 X2 TI60), although other notebooks could be used without any loss of capabilities.

4 Software Design

The software in this project is composed by a modular system, where each group of tasks interacts with the main application. Figure 5 shows the specific functions and the software responsible to convert data from hardware to be used in the control level.

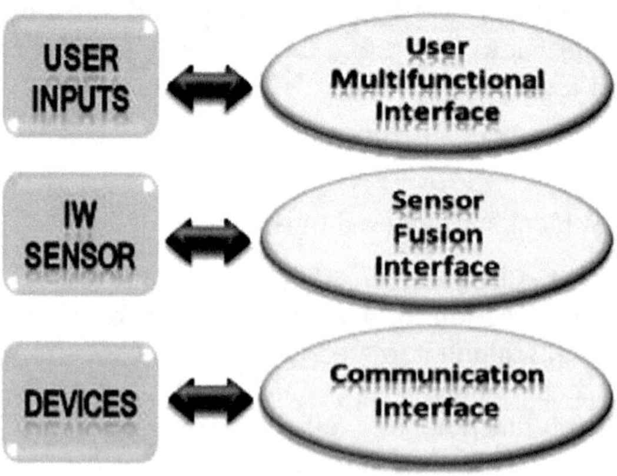

Fig. 5. Intelligent Wheelchair Software

The main application has the function to gather all information from the system, communicate with the control module and set the appropriate output. Moreover, it has other different functions depending on what is the control interface mode set:

- Real. If this option is selected, all data comes from the real world. The main application collects real information from sensors through the Sensor User Interface, calculate the output through the Control module and send these parameters to the PWM Boards;
- Simulated. In this mode, the system works just with virtual information and is used for two purposes: generate the same behavior of a real wheelchair and to test control routines. The main application collects virtual information directly from the simulator, calculates the output through the Control module and send back these parameters to the simulator;
- Augmented Reality. This mode creates an interaction between real and virtual objects, changing a real wheelchair's path to avoid collision with a virtual wall (or with a virtual wheelchair) for example. The objective of the augmented reality is to test the interaction between wheelchairs and environment, reducing its costs once major agents can be simulated. The main application collects real information from the sensors through the Sensor User Interface and mix it with virtual information collected from the simulator, calculate the output through the Control module and send the parameters to the simulator and the PWM Boards.

Due to its capabilities to interact with real, simulated and augmented reality worlds we call this whole system: IW Platform (Figure 6). In other words, the Platform is the fusion of the Simulator, Software modules, Real Wheelchair and Hardware Devices, to test, preview, understand and simulate the behaviour of Intelligent Wheelchairs.

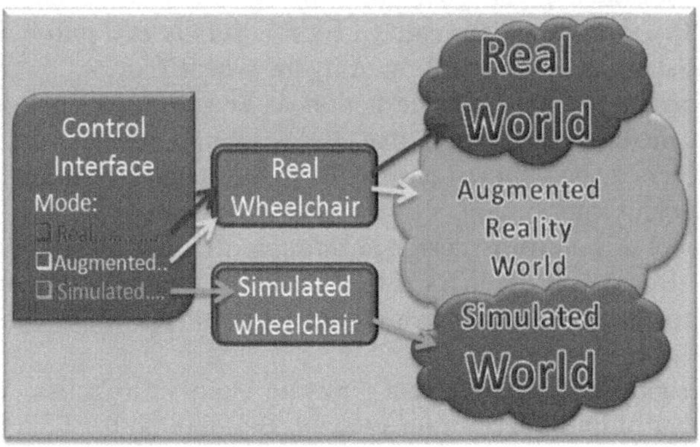

Fig. 6. Intelligent Wheelchair Platform Architecture

5 Control Architecture

IntellWheels has a multi-level control architecture subdivided in three levels: a basic control level, a tactical level and strategy level, as shown in Figure 7.

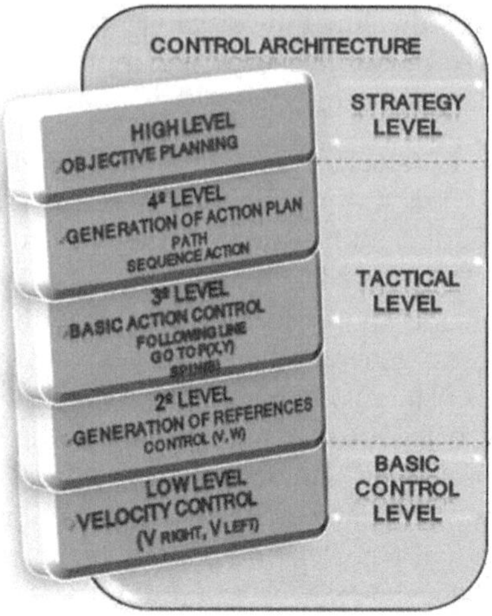

Fig. 7. Control architecture

As focus is primarily testing the Platform, classic algorithms were chosen to validate the system, some other issues are left outside the scope of this paper, such as analyses of its limitations and its performance. High Level strategy planning is responsible to create a sequence of high level actions needed to achieve the global goal. Actually the algorithm implemented to fulfill this task is based on the STRIPS planning algorithm [5].

In the Generation of Action Plans, we order the system to generate a sequence of basic actions aiming to satisfy the objectives proposed previously. To find a path from a given initial node to a given goal node the system has a simple A* Algorithm implemented [18]. Once the path is calculated, it is subdivided into basic forms (lines, circles, points) that are afterwards followed by the wheelchair.

Each basic form represents a basic action to be executed on time in the Basic Action Control module. Following that the system executes the Generation of References, which is responsible to estimate the wheelchair linear and angular speeds, putting the wheelchair into motion.

The lowest level of control is designed through a Digital PID implemented in the PWM/Encoders Boards. References from Basic Action are transferred by serial communication to the boards and then contrasted with real time data to control the speed.

Probabilistic robotics is a new and growing area in robotics, concerned with perception and control in the face of uncertainty. Building on the field of mathematical statistics, probabilistic robotics endows robots with a new level of robustness in real-world situations [23]. In this work was utilized probabilistic odometry motion model for an active localization system. When the uncertainty (variance) in the IW position, given by the odometry, grows up, the path planning changes the trajectory in a way that the IW goes near a landmark and resets its localization. This probabilistic model is also useful to decide how many landmarks are needed and where to put them.

6 Experiments and Results

This section presents the prototype implementation and a simple experiment and the results achieved. Figure 8 shows the mechanical structure and the hardware implementation. As we can see, the prototype is a commercial electric powered wheelchair with minimal modifications.

Fig. 8. IntellWheels_Chair1 Prototype

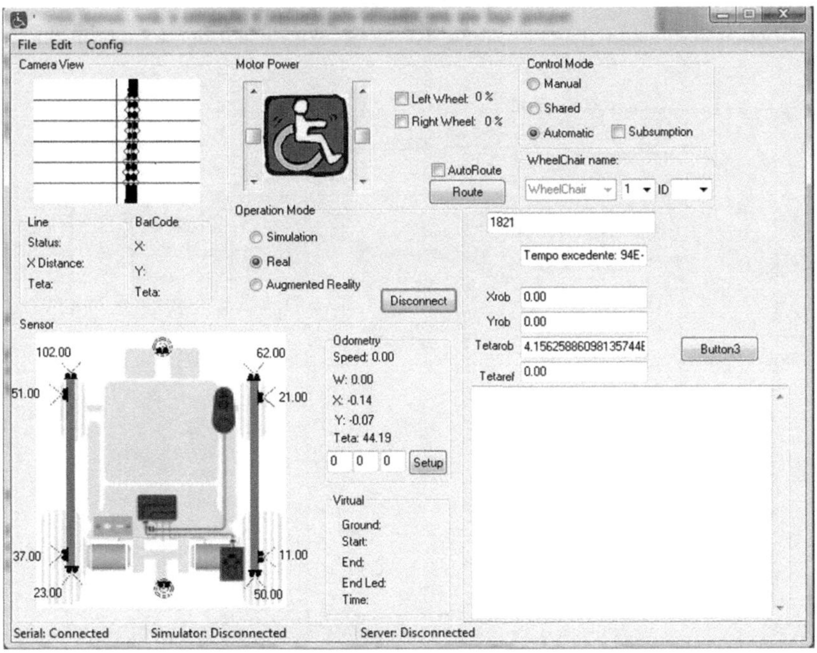

Fig. 9. Interface Control of IntellWheels

User interface software was designed to be as friendly as possible and it is shown in Figure 9. In the main window it contains a webcam window, which is used to recognize landmarks, and the results of its localization. A schematic figure with the position of the sonar mounted in the wheelchair easily shows the distance to nearby objects. By the side, we have a panel with the information provided by the odometry. The application still displays a scrollbar indicating the speed of each wheel as well as buttons to choose the operation mode.

Some basic tests were performed to validate odometry, and consisted in moving the wheelchair around a rectangle path (Figure 10), starting and stopping in the same point. In Figure 11 the results of an automatic test (red line) and a manual test (blue line) are presented. In the manual test the user had all control of the wheelchair.

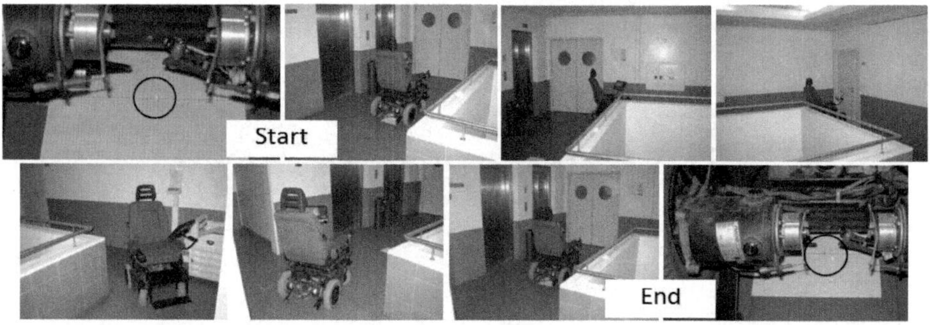

Fig. 10. Rectangle path being followed by the Wheelchair in the experiments

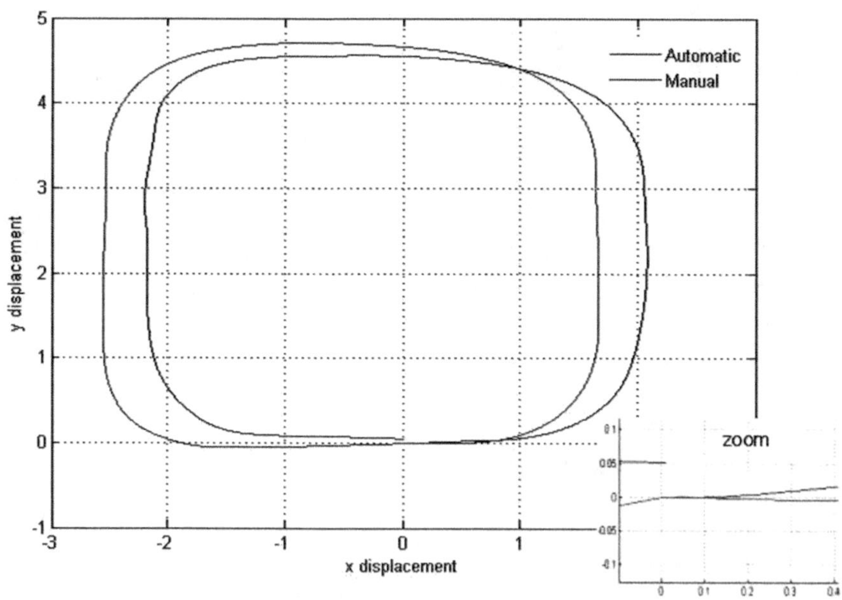

Fig. 11. Path Followed by Wheelchair in the Tests

In the automatic mode the final point of the path of the wheelchair presented a little displacement regarding its start point (physical mark point on the ground). As this error is not displayed in the odometry graph, a Manual test was executed to evaluate the results. In this test, the user drove the wheelchair following the same path, but at this time stopping in the same physical mark point on the ground. Results of this test presented the error displayed in the odometry graph.

The error presented is admissible since it is just 5 cm in a total amount of 1500 cm of displacement, and can be explained by the integration of the odometry systematic error.

Figure 12 and Figure 14 shows the results of distinct tests of displacement in straight line. The objective was evaluate odometry dispersion error for different distances. In this test was valued displacement of 5, 10 and 15 meters, with approximately 20 samples for each path.

Odometry Motion Model presented in [23] is illustrated in Figure 13. The computing of the uncertainty was used to decrease odometry errors through path planning that cross artificial landmarks.

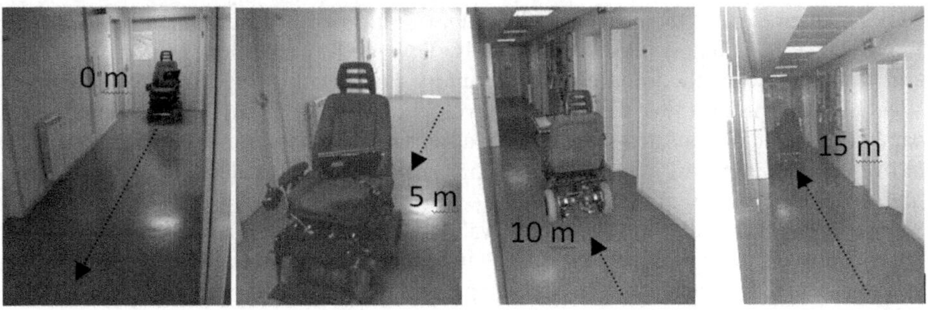

Fig. 12. Test of displacement in straight line

Algorithm **motion_model**(u, x):

$u = \langle \delta_{rot1}, \delta_{rot2}, \delta_{trans} \rangle, x = \langle x, y, \theta \rangle$

$\hat{\delta}_{rot1} = \delta_{rot1} + \text{sample}(\alpha_1 |\delta_{rot1}| + \alpha_2 \delta_{trans})$

$\hat{\delta}_{trans} = \delta_{trans} + \text{sample}(\alpha_3 \delta_{trans} + \alpha_4 (|\delta_{rot1}| + |\delta_{rot2}|))$

$\hat{\delta}_{rot2} = \delta_{rot2} + \text{sample}(\alpha_1 |\delta_{rot2}| + \alpha_2 \delta_{trans})$

$x' = x + \hat{\delta}_{trans} \cos(\theta + \hat{\delta}_{rot1})$

$y' = y + \hat{\delta}_{trans} \sin(\theta + \hat{\delta}_{rot1})$

$\theta' = \theta + \hat{\delta}_{rot1} + \hat{\delta}_{rot2}$

Return $\langle x', y', \theta' \rangle$

normal_distribution

```
Parameters:

sample = 500;
δs =[0.00001 0.1 0.0005
0.0000001];
```

Fig. 13. Probabilistic motion model odometry algorithm

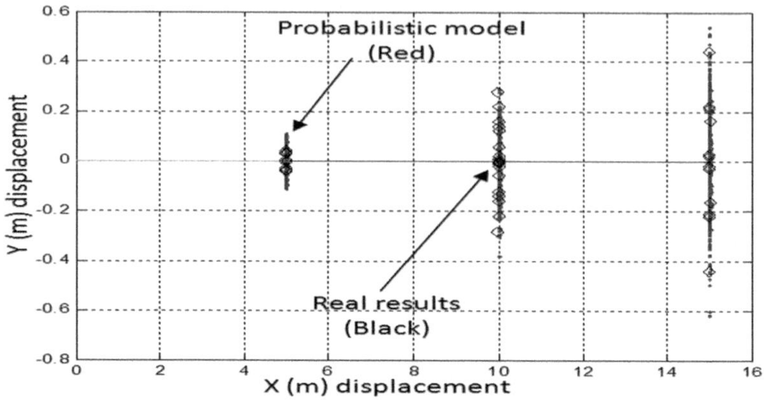

Fig. 14. Odometry dispersion error for 5m, 10m and 15m of displacement in straight line

7 Discussion and Conclusions

This paper presented the design and implementation of a development platform for Intelligent Wheelchairs called IntellWheels. This platform facilitates the development and test of new methodologies and techniques concerning Intelligent Wheelchairs. We believe that the new techniques developed may bring to the wheelchairs real capacities of intelligent action planning, autonomous navigation, and mechanisms to allow the execution, in a semi-autonomous way, of the user desires, expressed in a high-level language of command.

Future research will aim the test and validation for the other sensors mounted in the wheelchair prototype. Moreover, like in many other systems, cooperative and collaborative behaviours are desired to be present in the IW and need to be incorporated in the platform. Another important improvement to be pursued includes a comparative study of the classic implemented algorithms and the new proposal to solve these issues.

The platform will allow that real and virtual IWs interact with each other. These interactions make possible high complexity tests with a substantial number of devices and wheelchairs, representing a reduction in the project costs, once there is not the necessity to build a large number of real IW.

Acknowledgements. The first author would like to thank for CAPES for doctoral course financing. This work was partially supported by the project ACORD – Adaptative Coordination of Robotic Teams - FCT/PTDC/EIA/70695/2006.

References

1. Adachi, Y., Kuno, Y., Shimada, N., Shirai, Y.: Intelligent wheelchair using visual information on human faces. In: International Conference in Intelligent Robots and Systems, vol. 1, pp. 354–359 (1998)
2. Bell, D.A., Borenstein, J., Levine, S.P., Koren, Y., Jaros, L.: An assistive navigation system for wheelchairs based upon mobile robot obstacle avoidance. In: IEEE Conf. on Robotics and Automation (1994)

3. Faria, P.M., Braga, R.A.M., Valgôde, E., Reis, L.P.: Platform to Drive an Intelligent Wheelchair using Facial Expressions. In: Proceedings of the 9th International Conference on Enterprise Information Systems - Human-Computer Interaction (ICEIS 2007), Funchal Madeira, Portugal, June 12–16, pp. 164–169 (2007) ISBN: 978-972-8865-92-4
4. Faria, P.M., Braga, R.A.M., Valgôde, E., Reis, L.P.: Interface Framework to Drive an Intelligent Wheelchair Using Facial Expressions. In: IEEE International Symposium on Industrial Electronics, June 4-7, pp. 1791–1796 (2007) ISBN: 1-4244-0755-9
5. Fikes, R., Nilsson, N.J.: STRIPS: A New Approach to the Application of Theorem Proving to Problem Solving. In: IJCAI 1971. pp. 608–620 (1971)
6. Hamagami, T., Hirata, H.: Development of intelligent wheelchair acquiring autonomous, cooperative, and collaborative behavior. IEEE International Conference on Systems, Man and Cybernetics 4, 3525–3530 (2004)
7. Hoyer, H., Hölper, R.: Open control architecture for an intelligent omnidirectional wheelchair. In: Proc.1st TIDE Congress, Brussels, pp. 93–97 (1993)
8. Jia, P., Hu, H.: Head Gesture based Control of an Intelligent Wheelchair. In: CACSUK - 11th Annual Conference of the Chinese Automation and Computing Society in the UK, Sheffield, UK (September 10, 2005)
9. Jia, P., Hu, H., Lu, T., Yuan, K.: Head Gesture Recognition for Hands-free Control of an Intelligent Wheelchair. Journal of Industrial Robot (2006)
10. Lakany, H.: Steering a wheelchair by thought. In: The IEEE International Workshop on Intelligent Environments, pp. 199–202 (June 2005)
11. Levine, S.P., Bell, D.A., Jaros, L.A., Simpson, R.C., Koren, Y., Borenstein, J.: The NavChair assistive wheelchair navigation system. In: IEEE Transactions on Rehabilitation Engineering, pp. 443–451 (1997)
12. Luo, R.C., Chen, T.M., Lin, M.H.: Automatic Guided Intelligent Wheelchair System Using Hierarchical Grey-Fuzzy Motion Decision-Making Algorithms. In: Proc. of IEEVRSJ 1999 - International Conference on Intelligent Robots and Systems (1999)
13. Madarasz, R.L., Heiny, L.C., Cromp, R.F., Mazur, N.M.: The design of an autonomous vehicle for the disabled. IEEE J. Robotics and Automat, 117 – 126 (September 1986)
14. Miller, D., Slack, M.: Design and testing of a low-cost robotic wheelchair prototype. Autonomous Robots 2, 77–88 (1995)
15. Miller, D.: Assistive Robotics: An Overview. Assistive Technology and AI, 126–136 (1998)
16. Ng, P.C., DeSilva, L.C.: Head gestures recognition. In: Proceedings of the International Conference on Image Processing, vol. 3, pp. 266–269 (2001) ISBN: 0-7803-6725-1
17. Rebsamen, B., Burdet, E., Guan, C., Zhang, H., Teo, C.L., Zeng, Q., Laugier, C., Ang Jr., M.H.: Controlling a Wheelchair Indoors Using Thought. IEEE Intelligent Systems and Their Applications 22, 18–24 (2007)
18. Shapiro, S.C.: Encyclopedia of Artificial Intelligence. Wiley Interscience, Hoboken (2000)
19. Simpson, R.C.: Smart wheelchairs: A literature review. Journal of Rehabilitation Research & Development 42, 423–436 (2005)
20. Simpson, R., et al.: NavChair: An Assistive Wheelchair Navigation System with Automatic Adaptation. In: Mittal, et al. (eds.) Assistive Technology and AI, pp. 235–255 (1998)
21. Sunrise, Sunrise Medical manufacturers (2007), `http://www.sunrisemedical.co.uk`
22. Wellman, P., Krovi, V., Kumar, V.: An adaptive mobility system for the disabled. In: Proceedings of the IEEE International Conference on Robotics and Automation (1994)
23. Thrun, S., Burgard, W., Fox, D.: Probabilistic Robotics. MIT Press, Cambridge (2005)

Part III

Signal Processing, Systems Modeling and Control

Hybrid Controller for Tracking Systems

Christian Wachten[1], Christoph Ament[2], Lars Friedrich[1], Claas Müller[1], and Holger Reinecke[1]

[1] Department of Microsystems Technology, University of Freiburg
Georges-Koehler-Allee 103, 79110 Freiburg, Germany
{christian.wachten,lars.friedrich,claas.mueller, holger.reinecke}@imtek.de
[2] Institute for Automation and Systems Engineering, TU Ilmenau
Gustav-Kirchhoff-Straße 1, 98693 Ilmenau, Germany
christoph.ament@tu-ilmenau.de

Abstract. This chapter presents a hybrid controller which can be used with tracking systems especially with laser tracker systems. The controller consists of an analog *PI* controller and a microcontroller. Special features like e.g. an absolute distance measurement technique or a search algorithm in case of beam loss is implemented. With the help of the distance measurement the controller gain is adjusted automatically by digital potentiometers. Due to the usage of standard elements it is cost effective and easy to use. The controller is designed for working with an interferometer in a laser tracker system but can also act as an autonomous system because of the integrated absolute distance measurement technique. A distance of about eight meters between the tracker unit and the reflector is easily achieved in experiment without static tracking errors. Furthermore, the lateral offset of the laser beam does not exceed a quarter of the beam diameter and thus guarantees stable interferometer functionality.

Keywords: Tracking unit, absolute distance measurement, hybrid design, μC.

1 Introduction

Laser trackers are devices which are used in position metrology and in calibration tasks due to their capability of doing static as well as dynamic high accuracy measurements [3]. A HeNe laser emits a light beam with slightly different frequencies f_1 and f_2. This beam is divided by an interferometer into a reference beam and a measurement beam. The measurement beam leaves the interferometer and is deflected by the mirrors of an actuator in such a manner that it follows the movement of a retroreflector. The reflected light is analyzed by a position sensitive detector, the analog output signals of which are used to determine the position of the incoming beam. The reflected light beam also interferes with the reference beam in the interferometer. So, by measuring the two mirror angles of the actuator and the relative distance given by the interferometer the position of the reflector is calculated by using an analytical model. Figure 1 shows the operation principle.

An essential part of the laser tracker is the feedback controller in the tracking unit. It provides the input signals for the actuator, so that the laser beam follows the

J.A. Cetto et al. (eds.): Informatics in Control, Automation and Robotics, LNEE 37, pp. 207–220.
springerlink.com

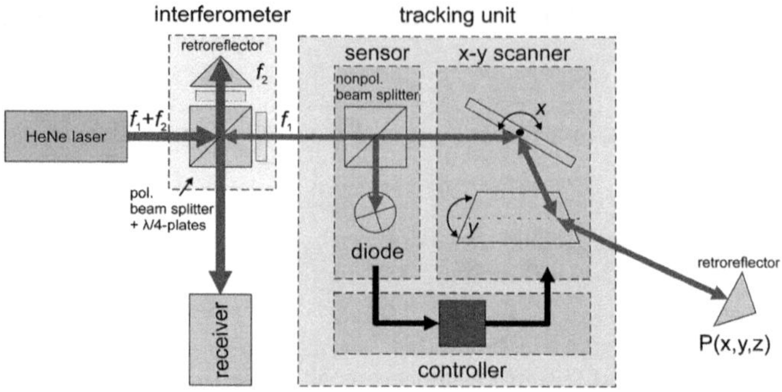

Fig. 1. Operation principle of a laser tracker system. The frequency f_1 represents the measurement beam and f_2 the reference beam.

movement of the retroreflector. The basic task of the controller is to guarantee the proper interferometer function. Dynamic aspects like a high velocity or high acceleration of the retroreflector with a low contouring error are also important.

2 System Components

The tracking unit consists of three components (see figure 1). The first component is the sensor element which is a combination of a nonpolarizing beam splitter and a quadrant diode. The second component is a x-y scanner with two magnetically driven mirrors that deflect the laser beam (galvanometer scanner). The third component is the controller that connects the sensor element with the actuator.

Figure 2 shows a photograph of the complete tracking unit. The laser beam hits a nonpolarizing beam splitter. Afterwards, it is deflected by the scanner and is reflected by the retroreflector. The retroreflector has the unique property that the incoming

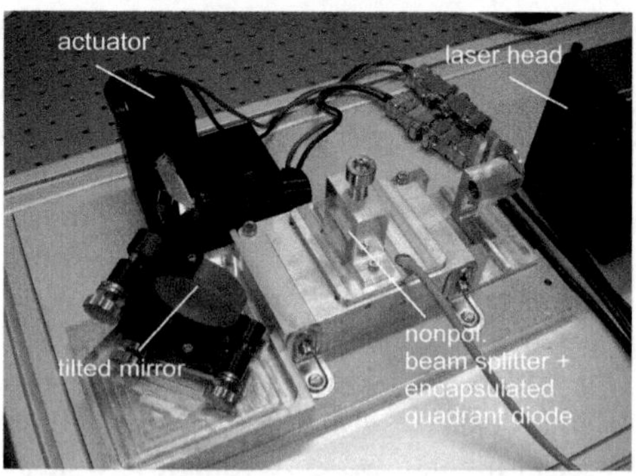

Fig. 2. Photograph of the laser tracker system. The two mirrors of the actuator deflect the incoming laser beam. A tilted mirror is used to adjust the light path.

laser beam is reflected into the same direction where it came from. On its returning path the reflected beam hits the nonpolarizing beam splitter again and a part of the beam is deflected on a quadrant diode. This diode provides the analog input voltages for the feedback controller since it has an integrated transimpedance amplifier. The feedback controller generates the input signals for the scanner. Ideally, the laser beam is centered on the diode and the position signals equal zero volts. By moving the reflector the laser beam leaves the center on the diode and the position signals change. The controller compensates the position change and modifies the input signals of the actuator. The two mirrors rotate and deflect the laser beam in such a way that the offset becomes zero.

2.1 Magnetic Actuator

The magnetic actuator is a galvanometer scanner. Each mirror is magnetically driven and has its own analog *PID* controller to hold the desired position. The transfer factor is 0.83 V/° (mechanical) at the input side of the controller and 0.5 V/° (mechanical) at the output side of the position detector. Integrated sensors allow the measurement of the rotation angle of the mirrors. The short term stability is about 8 μrad. The maximal mechanical rotation angle is ± 12.5° limited by the used assembly.

2.2 Quadrant Diode

A standard quadrant diode is used as active sensor element. Each quadrant is sensitive to light with a sensitivity that is specified to 0.54 A/W at a wavelength of 900 nm. The spacing between the quadrants is 0.2 mm, the quadrant radius is 3.99 mm. The quadrant sensor provides three output signals V_{LR}, V_{TB} and V_{Sum}. The signals V_{LR} and V_{TB} represent the left-right voltage and the top-bottom voltage. The signal V_{Sum} represents the summation voltage of all quadrants and thus is an indicator for the total incident power.

3 Modeling of the Plant

A block diagram of the system is shown in figure 3. The output signal y of the system is the deviation of the light beam from the center of the quadrant diode. This deviation should become zero for each component. The output signal y is a superposition of the movement of the reflector and the compensation part of the actuator. The gain between the mirror angle and the movement of the light beam on the diode depends on the distance between the reflector and the tracking unit. This is modeled in the block *light path*. The symbol z represents the position of the reflector in space and thus is a three-dimensional vector. The signal y represents the beam position on the diode area and thus is a two-dimensional vector.

To design an optimal feedback controller, models are developed and the model parameters are identified for the blocks *actuator, quadrant diode* and *light path*.

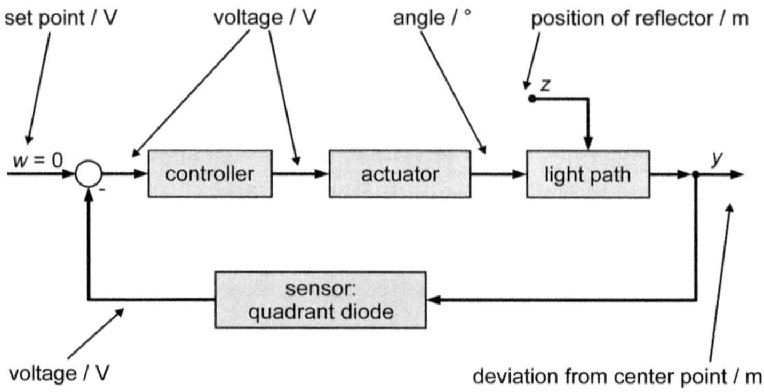

Fig. 3. Block diagram of the controlled plant. The different inputs and outputs with their units are shown.

3.1 Modeling of the Block Actuator

To obtain a transfer function for the magnetic actuator the step response is recorded for each axis monitoring the position output of the integrated angle encoders. A square wave with a peak-peak voltage of 100 mV (corresponding to an angle of about 0.06°) and a frequency of 30 Hz is applied to the inputs of the scanner.

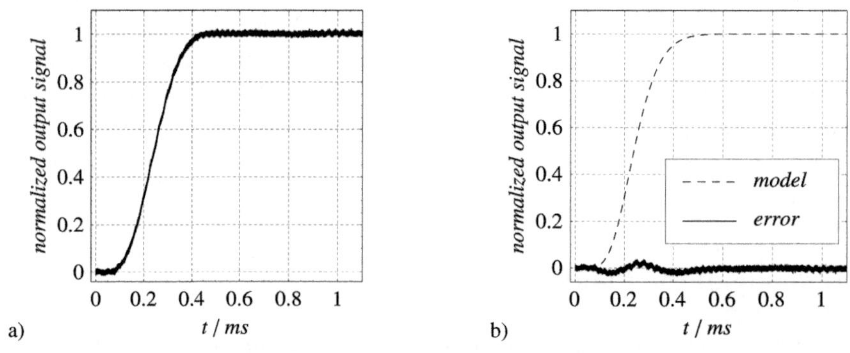

Fig. 4. Normalized and averaged measurement a) of the step response of the y mirror and the model b) of the step response with its error. The step rises at $t = 0$ s.

Figure 4 shows the measured step response and the step response of the model for the y mirror. It is assumed that the scanner has PT_n behavior and thus can be modeled with a PT_n element in (1).

$$H_i(s) = \frac{K_i}{(1 + T_i \cdot s)^{n_i}}, \; i = x,y . \tag{1}$$

The parameter K_i describes the gain and the parameter T_i represents the time constant for axis i. The parameter n_i stands for the order of the element. To obtain the parameters the method of the time percentage values is applied. Since the measurement is not supposed to have ideal PT_n behavior a least-squares fit is done to optimize the parameters so that the error between the model and the measurement becomes minimal. The start parameters for the optimization are the results given by the method of the time percentage values [4]. The final optimization yields in $n_y = 9$ and $T_y = 27.51$ µs for the y mirror. The parameters for the x mirror result in $n_x = 10$ and $T_x = 23.42$ µs. Taking into account that the gain between the output signal and the mechanical deflection is 0.5 V/° the gains are calculated to $K_x = 1.210$ °/V and $K_y = 1.207$ °/V, respectively. The -3 dB frequency is about 1.8 kHz (model).

3.2 Modeling of the Block Diode

The time response of the diode can be modeled in the same way as the time response for the actuator. A red LED is used to generate the step response because it is fast enough and its time response can be neglected. The output signal V_{Sum} is measured. The LED has a power of about 7 µW. Figure 5 shows the measurement and the model of the step response.

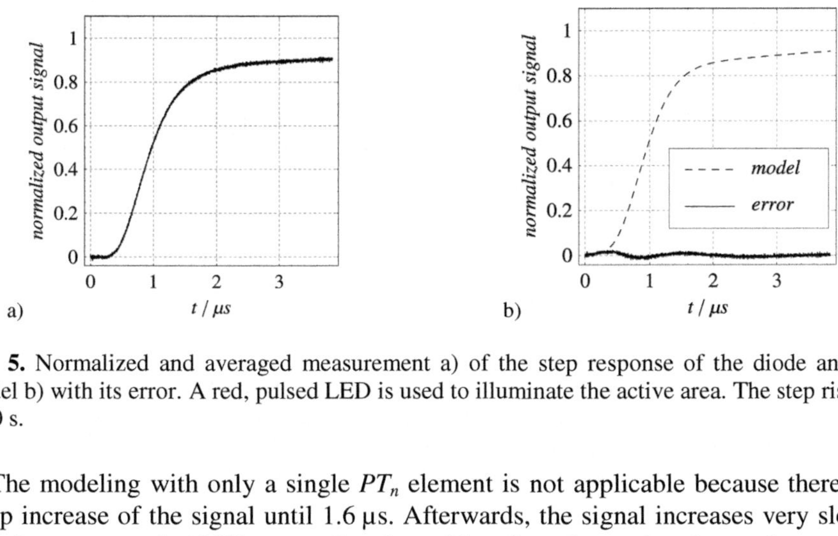

Fig. 5. Normalized and averaged measurement a) of the step response of the diode and the model b) with its error. A red, pulsed LED is used to illuminate the active area. The step rises at $t = 0$ s.

The modeling with only a single PT_n element is not applicable because there is a steep increase of the signal until 1.6 µs. Afterwards, the signal increases very slowly and does not reach 100% even after 4 µs. Therefore, it can be shown that a good approximation is a combination of a PT_n element and a PT_1 element. This is done in (2).

$$H_D(s) = \frac{g}{(1+T_1 \cdot s)^{n_D}} + \frac{1-g}{1+T_2 \cdot s} \quad . \tag{2}$$

The parameters T_1 and T_2 represent time constants of the two elements, the parameter g normalizes the output and the parameter n_D represents the order of the PT_n element. All parameters are obtained using a least-squares fit so that the deviation of the model and the measurement becomes minimal. The optimal parameters are $n_D = 7$, $T_1 = 136$ ns, $T_2 = 4.75$ µs, $g = 0.793$. The model predicts a -3 dB frequency of 250 kHz.

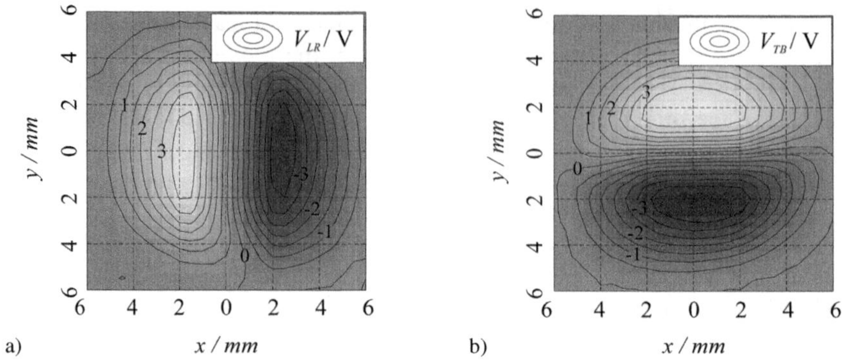

Fig. 6. Local behavior of the diode. There is a nonlinear relation between the output voltages V_{LR} a) and V_{TB} b) and the position of the beam.

Because the diode generates position signals not only the time response is important but also its local behavior. Figure 6 shows the local behavior of the diode. The quadrant diode was put onto an x-y station and a laser diode with a power of 771 μW was installed in front of it. The station moves to 900 defined positions that are placed in an equally spaced square.

In figure 6 it can be seen that there is a nonlinear relation between the position and the output voltages. In the center of the diode the contour lines are nearly parallel to the corresponding axis and so a linearization is possible. It is obvious that the signal V_{TB} only depends on a movement in y direction and V_{LR} only depends on a movement in x direction. As a result, each axis of the scanner can be regarded as independent.

3.3 Modeling of the Block Light Path

The block *light path* depends on the position of the reflector. If the angular errors are neglected the reflector can be approximated as its center point in a plane. The hitting point of the incoming laser beam is point reflected with the center. An offset between the incoming and the reflected laser beam can have different reasons, for example a rotation or a lateral displacement of the reflector (see figure 7).

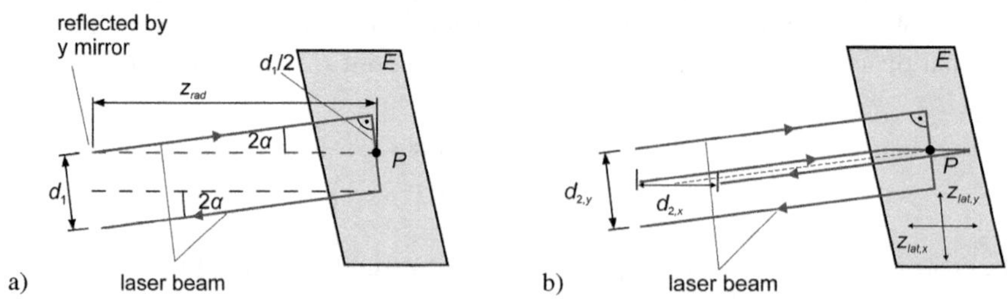

Fig. 7. Modeling of the reflector. An incoming beam is reflected at the center P of the reflector. There is an lateral offset due to a rotation of the mirror a) and a lateral movement z_{lat} of the retroreflector b).

The first effect is shown in figure 7a. The mirror rotation angle is α. It is assumed that the reflector is z_{rad} away from the tracking unit. So the offset d_1 can be written as

$$d_1 = 2 \cdot z_{rad} \cdot \sin(2\alpha) \approx 4 \cdot z_{rad} \cdot \alpha \,. \tag{3}$$

The second effect resulting in a beam offset is the lateral displacement z_{lat} of the reflector. This effect is shown in figure 7b. The indices represent the coordinate frame of the diode. So, this can be expressed as

$$d_{2,i} = 2 \cdot z_{lat,i} \,, \; i = x,y \,. \tag{4}$$

The total deflection on the quadrant diode is a combination of these two effects as shown in (5).

$$y_i = d_1 + d_{2,i} \,, \; i = x,y \,. \tag{5}$$

4 Controller Design

Because the decoupling of the axes is possible (see figure 6) the problem is reduced to the controller design for one axis. Exemplarily, the x-axis is used to demonstrate the design process. It can be derived from the block *light path* that the gain of the plant correlates with the distance z_{rad}. In a first step it is assumed that z_{rad} is constant. During normal operation the controller should be automatically adapted to the distance so the restriction z_{rad} = const. is dropped. This adaptation to the distance is known as gain scheduling.

Several requirements have to be defined for the use in laser tracker systems. First, the offset of the beam should be smaller then a quarter of the beam diameter to guarantee the interferometer function. Second, a high velocity of the reflector is necessary to allow rapid movements of the object. The third requirement is the robustness against vibrations without any disturbance of the measurement accuracy. Of course, a large measurement volume is desirable, too.

To achieve stationary accuracy the controller should possess an integrating part because the plant only consists of proportional blocks. A pure *I* controller is also possible but a proportional part enhances the performance [2]. A *PI* controller is well suited for plants with PT_n behavior.

A standard *PI* controller is proposed due to its simple design and to reduce the analog circuit complexity because the derivative part is missing. The transfer function is well known and given in (6).

$$H_C(s) = K_C \cdot \left(1 + \frac{1}{T_C \cdot s}\right). \tag{6}$$

The parameter K_C describes the gain of the controller and T_C its time constant. First, these parameters are determined by the classic frequency response method as described in [1]. The time constant T_C is set to T_x = 23.42 μs because this is the dominating time constant of the x mirror in the plant. The gain $K_{C,30°}$ is set to 0.1231 to obtain a phase margin of 30° for the gain crossover frequency. It is of interest to examine the disturbance transfer function because the set point ($w(t) = 0$) is constant.

$H_z(s)$ is calculated in (7) and is derived from the block diagram in figure 3.

$$H_z(s) = \frac{2}{1 + H_x(s) \cdot H_C(s) \cdot H_D(s)} \quad (7)$$

To estimate the performance of the classic feedback controller the response to a ramp in $z_{lat,i}$ and the magnification factor is simulated for the disturbance transfer function in (7).

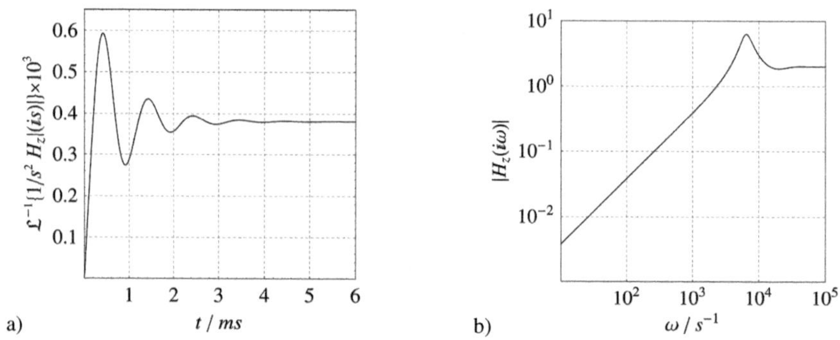

Fig. 8. Simulation of the ramp response a) and the frequency response b) for the disturbance transfer function with a classic *PI* controller

A ramp is chosen because it is the strongest requirement concerning the reflector movement. A step is not applicable because in reality the position of the retroreflector cannot rapidly change.

Figure 8a shows the response on a ramp in $z_{lat}(t)$ at the output y_x of the system. The maximal value of the overshoot at the output y_x is $0.60 \cdot 10^{-3}$ m until a constant contouring error of $0.38 \cdot 10^{-3}$ m is reached. Figure 8b shows the frequency response. Low frequencies are damped due to the integrating part. But there is a magnification factor of 6.3 at a frequency of 1 kHz.

The maximal value is reached at the overshoot, but the contouring error is much lower. Therefore, the maximum of the ramp response has to be minimized so that the overshoot is reduced at the cost of the contouring error. The aim is the adaptation of the overshoot to the contouring error. For $\omega \rightarrow \infty$, $|H_z(i\omega)|$ converges to 2, so an arbitrary factor of 4 is proposed for all frequencies to guarantee robustness against vibrations. To identify the optimal controller parameters K_C and T_C were varied in a range of $K_C = 0.1 \cdot K_{C,30°} \ldots 10 \cdot K_{C,30°}$ and $T_C = 0.1 \cdot T_x \ldots 10 \cdot T_x$. The grid was $\Delta K_C = 0.05 \cdot K_{C,30°}$ and $\Delta T_C = 0.05 \cdot T_x$. Figure 9 shows the ramp response and frequency response for the optimized parameters $K_{C,opt} = 0.542$ and $T_{C,opt} = 135$ µs. The maximal value of the response is about $0.51 \cdot 10^{-3}$ m with a very low overshoot and the magnification factor does not exceed 4. So, the quality of control is much better in comparison with the classic approach.

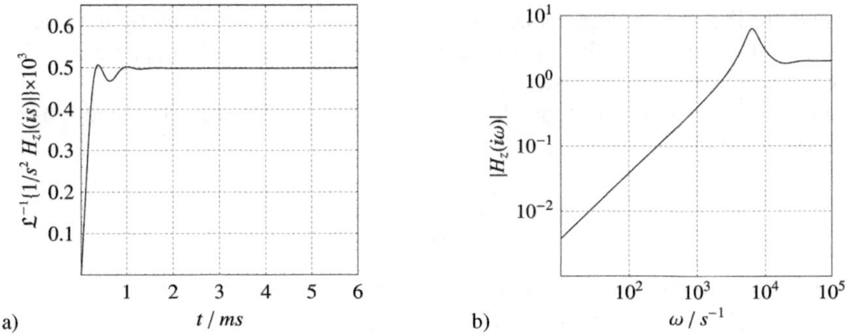

Fig. 9. Simulation of the ramp response a) and the frequency response b) of the disturbance transfer function with optimal controller parameters in regard to contouring error and magnification factor.

It was assumed that the total gain is focused in the parameter $K_{C,opt}$. This is not the case in a real system. The real total gain is a product of the controller gain, the mirror gain, the light path and the sensitivity of the quadrant diode. It can be written as

$$K_{C,opt} = K_C \cdot K_x \cdot 4 \cdot z_{rad} \cdot K_D . \tag{8}$$

$$\Leftrightarrow K_C = \frac{K_{C,opt}}{K_x \cdot 4 \cdot z_{rad} \cdot K_D} . \tag{9}$$

To adapt the controller gain, the parameters K_C and K_D have to be updated during operation (gain scheduling). The parameter K_D is obtained by measuring the voltage V_{Sum} of the quadrant diode. The parameter z_{rad} has to be estimated.

To estimate z_{rad} a known signal $u(t)$ is introduced into the control loop (figure 10). Only the spectral components are considered which are below the cut-off frequency of the diode. So, it is possible to simplify the transfer functions and consider only their proportional parts. But the time response of the controller cannot be neglected

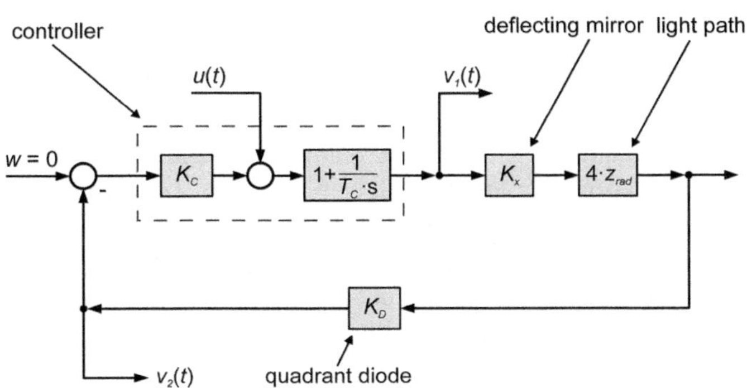

Fig. 10. Control loop with introduced signal $u(t)$ to estimate z_{rad}. The signals $v_1(t)$ and $v_2(t)$ can be measured and depend on z_{rad} .

because there is a strong dependency on low frequencies introduced by the integrating part. The multiplication with z_{rad} is modeled as proportional part because z_{rad} changes slowly in comparison to the system dynamic.

The signals $v_1(t)$ and $v_2(t)$ depend on the distance z_{rad}. The transfer functions can be calculated with (10).

$$H_i(s) = \frac{\mathcal{L}\{v_i(t)\}}{\mathcal{L}\{u(t)\}},\ i = 1,2\,. \tag{10}$$

It can be shown that there is a higher sensitivity of $|H_1(i\omega_0)|$ to a change in z_{rad} if signal $v_1(t)$ is measured. With (10) and figure 10 the transfer function $H_1(s)$ is calculated in (11).

$$H_1(s) = \frac{1 + \frac{1}{T_C \cdot s}}{1 + \left(1 + \frac{1}{T_C \cdot s}\right) \cdot K_C \cdot K_D \cdot 4 \cdot z_{rad} \cdot K_x}\,. \tag{11}$$

The signal $v_1(t)$ is a superposition of the movement of the reflector and the introduced signal $u(t)$. A sinusoidal signal with a frequency ω_0 is proposed. So, a subsequent band-pass filter with the center frequency ω_0 suppresses the disturbance signal introduced by the reflector. To estimate the distance z_{rad} the amplitude of $u(t)$ is compared with the amplitude $v_1(t)$ and so $|H_1(i\omega)|$ is calculated. With the experimentally obtained relationship for K_D in (12), z_{rad} is calculated in (13). The proportional value m has a value of 45.13 $V^{-1}m^{-1}$ for the used system components.

$$K_D = \frac{m \cdot V_{Sum}}{4 \cdot K_x}\,. \tag{12}$$

$$z_{rad} = \frac{\sqrt{(1 + (2 - A^2) \cdot T_C^2 \cdot \omega_0^2 + T_C^4 \cdot \omega_0^4)} - A \cdot T_C^2 \cdot \omega_0^2}{A \cdot K_C \cdot m \cdot V_{Sum} \cdot (1 + T_C^2 \cdot \omega_0^2)}\,. \tag{13}$$

with $A = |H_1(i\omega_0)|$.

To obtain an appropriate value for the amplitude of the signal $u(t)$ the beam deviation $y_x(t)$ from the center of the quadrant diode is analyzed. The deviation should be smaller than a quarter of the beam diameter to guarantee the interferometer function. During distance estimation the deviation $y_x(t)$ is a superposition of the lateral movement of the reflector $z_{lat,x}(t)$ and the introduced signal $u(t)$. Therefore, the amplitude of the signal $u(t)$ is chosen to be only 10% of the maximal deflection so that the maximal velocity v_{lat} is not reduced. With the transfer function $H_2(s)$ and a maximal amplitude $y_{x,max}$ of $y_x(t)$ the amplitude u_0 is calculated in (14).

$$u_0 = y_{x,\max} \cdot \frac{K_D}{|H_2(i\omega_0)|}\,. \tag{14}$$

Because z_{rad} is located in the denominator of (14) its increase leads to a decreasing amplitude u_0. According to the described limit of 10%, $y_{x,max}$ is set to $R/20$ with R being the radius of the beam.

5 Practical Considerations

The feedback controller is implemented in an analog design. To generate the signal $u(t)$, to measure the signal $v_1(t)$ and to adapt the gain of the controller, a microcontroller is used. The microcontroller is an ATmega128 and is programmed in C. The variable gain control is realized by digital potentiometers that are set by the microcontroller.

Figure 11 shows the used circuit for one axis. The digital potentiometer R_2 offers 127 linearly arranged steps. The resistance can be adjusted between 1 kΩ and 50 kΩ. To test the controller a square signal with a peak-peak voltage of 200 mV and a frequency of 100 Hz is applied on the input pin V_{LR}. During the measurement the resistor value is changed from 3.7 kΩ to 1 kΩ at point (2) in figure 11b. The change of K_C results in a step in the output signal. Further tests have shown that the resistor changes of the potentiometers have nearly no influence on system accuracy.

The analog *PI* controller is built with standard components without complex serial or parallel circuits resulting in a time constant of T_C = 132 μs and a adjustable gain between $K_C = 6.00 \cdot 10^{-3} \ldots 300 \cdot 10^{-3}$. The parameter T_C remains constant even if the potentiometers change their value.

Because of the discrete potentiometer positions there is an error between the optimal controller gain and the achieved controller gain. There are only integer positions n_{int} available. To obtain an optimal value for $K_C = K^*$ the theoretical real number $n_{real} = n^*$ for the potentiometer position is calculated. Afterwards, n_{real} is rounded down and up and the lower and the upper controller gains K_u and K_o are calculated. The controller gain with the minimal error in regard to the optimal value is chosen. The procedure can be seen in figure 12.

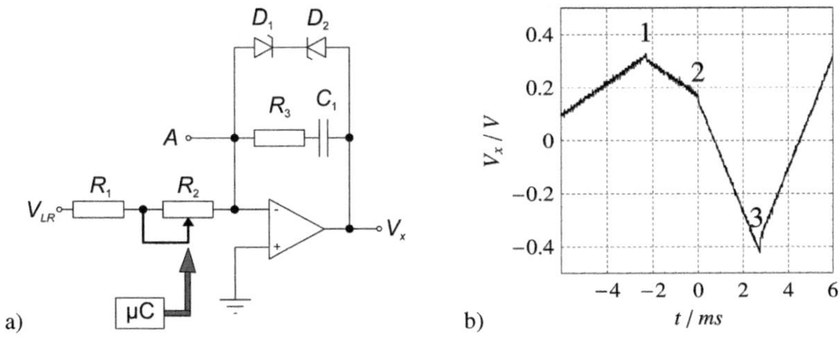

Fig. 11. Microcontroller controlling a digital potentiometer and analog *PI* controller exemplarily shown for the x-axis a) and its response to a square signal b). At t = 0 s (2) the resistor value of R_2 changes from 3.7 kΩ to 1 kΩ which equals seven steps of the potentiometer. The input voltage changes at t = -2.3 ms (1) and t = 2.7 ms (3) due to the rising and falling edge of the square signal.

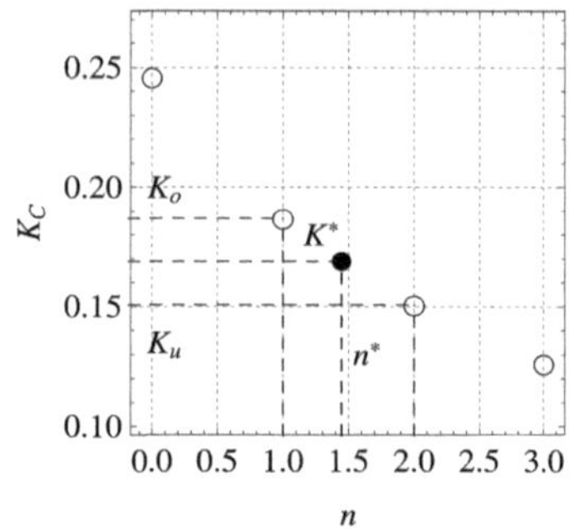

Fig. 12. Calculation of the optimal value n for the digital potentiometers for a desired controller gain K^*

After power up sequence and beam loss during operation, the laser beam searches the reflector in a defined area. This is done by deflecting the mirrors of the actuator without opening the feedback loop. The influence of the analog part is reduced and only the digital part controls the mirror deflection. Figure 13 shows the operation principle. The digital potentiometer is set to its maximal value. So, the influence of the sensor signal to the input of the *PI* controller is weak. This is comparable to an opening of the feedback loop and the signal can be cross talked easily.

The microcontroller introduces a signal at the input of the *PI* controller. At the same time the reduced sensor signal of the diode acts as a disturbance variable. The output of the controller is digitalized and is compared to the set up variable w_α. The microcontroller multiplies the gain K_μ with the deviation e. So, the plant with integrating behavior is controlled via a *P* controller which is a good combination [2]. The DA conversion at the output of the microcontroller is done via a pulse width modulation (PWM) with a frequency of 14.4 kHz and a low-pass filter with a cut-off frequency of 300 Hz (smoothing function).The output of the *PI* controller is digitized by the integrated AD converter of the microcontroller. To reduce alias effects a low-pass filter with a cut-off frequency of 300 Hz is used, too. K_μ represents the total gain of the control loop and is set to $K_\mu = 41$. This is a tenth of the value of the stability limit. So, the safety margin is high enough to avoid instability.

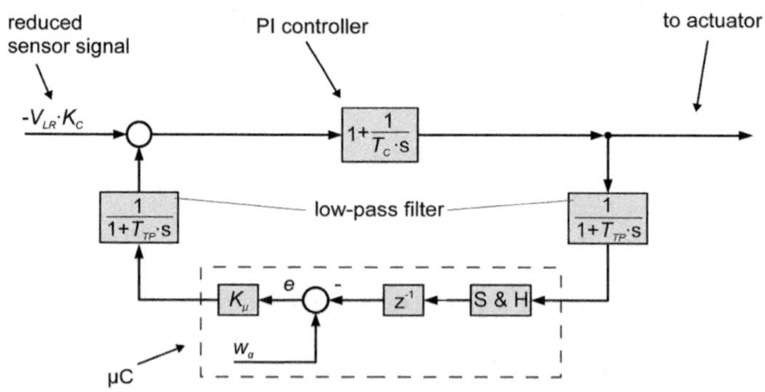

Fig. 13. Operation principle for deflecting the mirrors in a defined way by using a µC

Figure 14 shows the distance estimation for different gains of the controller. To reduce the time effort only distances between 0.5 m and 2.3 m are measured. The estimated value shows a good accordance to the real values. The variance increases with the distance because of the reduced sensitivity of the measured voltage to the distance. As explained before the distance estimation of z_{rad} is done by introducing a signal in the x-axis (figure 10). To reduce the computing and programming complexity a square wave with a frequency of 200 Hz instead of a sinusoidal signal is used. An analog band pass filter with the same frequency is used for signal pre-processing. Only the basic frequency is considered in the signal analysis. Because the movement of the retroreflector has only few spectral components in the pass-band the gain can be increased without leaving the input range of the AD converter. So, a small value of u_0 is sufficient to detect the amplitude of $v_1(t)$ with a DTFT.

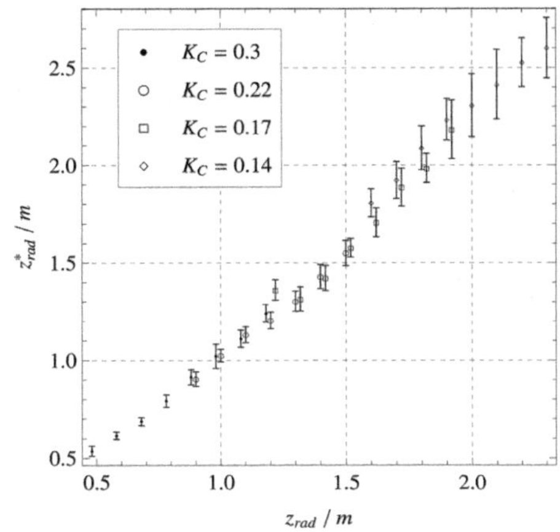

Fig. 14. Estimated z'_{rad} as a function of the real z_{rad}. Each point is averaged 20 times with its two-time standard deviation. To increase readability the measurements for $K_C = 0.3$ show a horizontal offset of -0.02 m and for $K_C = 0.17$ an offset of +0.02 m.

Further experiments have shown that the retroreflector can be moved with a maximal velocity of $v_{max} = 2.5$ m/s at unlimited acceleration and a beam offset of a quarter of the beam diameter. Distances between 0.5 m and 7.8 m (length of laboratory) were tested without tracking problems.

6 Conclusions

An analog *PI* controller with additional features was presented for the use in tracking systems. It is fast to detect a rapid beam movement and shows good control accuracy. Functions as gain scheduling and distance estimation are integrated in this hybrid design consisting of a digital and analog part.

References

1. Föllinger, O.: Regelungstechnik. Hüthig GmbH, Heidelberg (1994)
2. Merz, L., Jaschek, H.: Grundkurs der Regelungstechnik. R. Oldenburg Verlag, München (1996)
3. Riemensperger, M., Gottwald, R.: Kern Smart 310 – Leica's approach to high precision 3d coordinate determination. In: International Workshop on Accelerator Alignment, Hamburg, pp. 182–200 (1990)
4. Schwarze, G.: Bestimmung der regelungstechnischen Kennwerte von P-Gliedern aus der Übertragungsfunktion ohne Wendetangentenkonstruktion. Journal Messen, Steuern, Regeln (1962)

Modeling the Buoyancy System of a Wave Energy Power Plant

Tom Søndergaard Pedersen and Kirsten Mølgaard Nielsen

Department of Automation and Control, Aalborg University
Fr. Bajersvej 7, 9000 Aalborg, Denmark
{tom,kmn}@es.aau.dk

Abstract. A nonlinear dynamic model of the buoyancy system in a wave energy power plant is presented. The plant ("Wave Dragon") is a floating device using the potential energy in overtopping waves to produce power. A water reservoir is placed on top of the WD, and hydro turbines lead the water to the sea producing electrical power. Through air chambers it is possible to control the level of the WD. It is important to control the level in order to maximize the power production in proportion to the wave height, here the amount of overtopping water and the amount of potential energy is conflicting. The pressures in the air chambers may be individually controlled by an air fan through an array of valves. The paper presents a model describing the dynamics from the air inlet to the level. Results from validation of the model against plant data are presented.

Keywords: Dynamic model, wave energy, simulation, buoyancy control, verification, renewable energy.

1 Introduction

Renewable energy is an important issue due to the global warming problem and utilisation of wave power is one of the energy resources to be exploited.

The wave power system "Wave Dragon", on which this paper focuses, was invented by Erik Friis Madsen, Löwenmark and tested at Aalborg University and University of Cork. An EU based European consortium has been involved in the construction and implementation of a 1:4 scaled test site - 57x27 m wide and with a weight of 237 tonnes- which is placed in Nissum Bredning in Denmark. Large numbers of tests have been carried out during a two years operating period. One goal for energy production improvement is a better control of the Wave Dragon buoyancy.

Wave Dragon (WD) is an offshore wave energy converter of the overtopping type, a description is found in [3] and [4]. The main structure consists of a ramp where the waves are overtopping and led to a reservoir (basin). Two reflectors are focusing the waves towards the ramp as seen on figure 1. WD is fastened to an anchor making it possible to turn the ramp towards the dominant wave direction.

The WD use the potential energy of the waves, meaning that for a given wave type there exist an optimal level of the reservoir. As shown on figure 2 the reservoir water is led through a turbine.

J.A. Cetto et al. (eds.): Informatics in Control, Automation and Robotics, LNEE 37, pp. 221–230.
springerlink.com

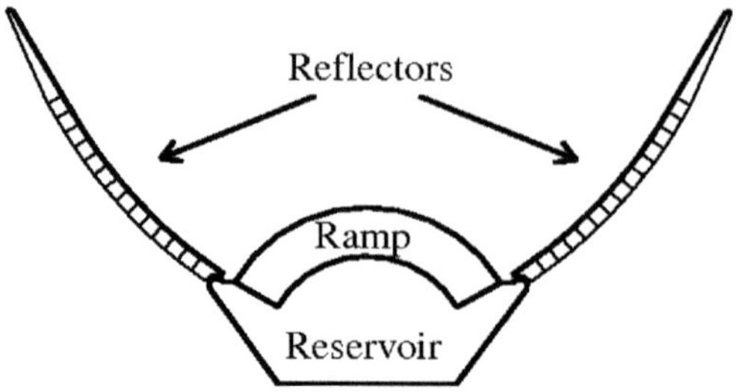

Fig. 1. Main components of the Wave Dragon [3]

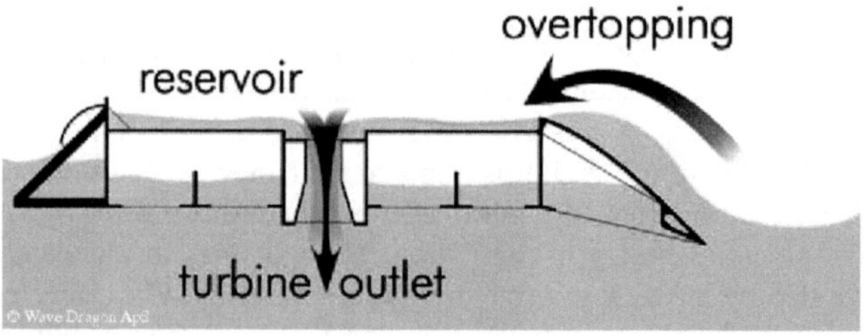

Fig. 2. The basic principle of WD showing waves loading the reservoir via the ramp [3]

The WD floats on open air chambers used to adjust the floating level. Control of the floating level is a part of optimizing the overtopping and a dynamic model for a model-based control system is the topic of this paper. It should be noted that the wave conditions are measured online and may be used as reference to the level control system.

First the wave dragon buoyancy system is presented. A dynamic model of the air supply system controlling the pressure in the air chambers is set up. The model parameters have been adjusted to the test site Wave Dragon. Finally the buoyancy model is verified by comparing simulation results and measurements from the test site Wave Dragon.

2 Dynamic Model of the Buoyancy System

The buoyancy system consists of five air chambers all open to the water surface. The five chambers are shown on figure 3 (21,22,23,24,25,26,27), (15,16,18,19,20), (3,4,5,10,11), (1,2,8,9,14) and (6,7,12,13,17). Furthermore 9 small chambers contain a constant amount of air.

The air pressures in the chambers are controlled by an air supply system using an on/off driven air fan and input/output valves to each chamber, see figure 4. The valves are operated as on/off valves and only one valve is allowed to be active at the time in order to prevent pressure equalizing in the chambers. A PWM scheme is in [1] proposed to handle this problem. The air supply system model consist of two parts, one describing air inlet to the chamber and one describing air flow out of the chamber. In both models tube pressure drops are ignored. The inlet air mass flow, m_{ai}, is given by

$$m_{ai} = -K_{pa}(p_c - p_a) + K_{pb} \tag{1}$$

which is an approximation to the fan characteristic. p_c is the chamber pressure, p_a is the inlet pressure to the fan (atmospheric pressure) and the two constants K_{pa}, K_{pb} are from the fan data sheet.

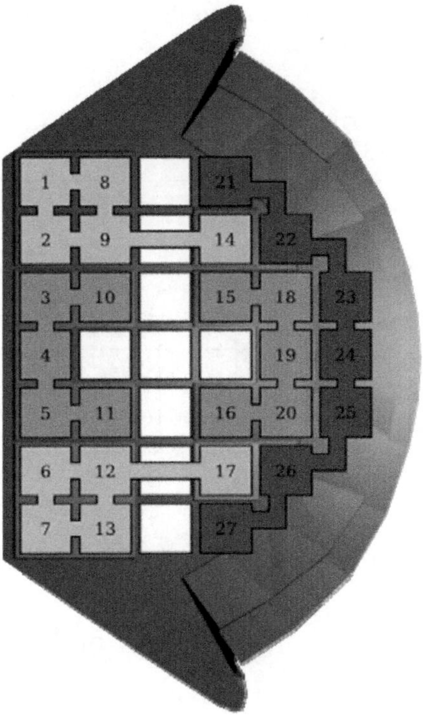

Fig. 3. The air chambers in the Wave Dragon

Air outlet mass flow, m_{ao}, is given by the Bernoulli equation:

$$m_{ao} = K_{vo}\sqrt{p_c - p_a} \tag{2}$$

where the constant K_{vo} depend on the outlet tube dimension and the air density.

Each input/output valve pair is controlled by a signal u, where u=1 allows an airflow into the chamber, u=0 closes both valves and u=-1 open the outlet valve. This gives the air mass flow equation:

$$m_a = m_{ai} g_1(u) - m_{ao} g_2(u) \tag{3}$$

where

$$g_1(u) = \begin{cases} 1 & if \quad u \geq 1 \\ 0 & if \quad u < 1 \end{cases} \tag{4}$$

$$g_2(u) = \begin{cases} 1 & if \quad u \leq -1 \\ 0 & if \quad u > -1 \end{cases}. \tag{5}$$

Eq. (3) describes the air mass flow to the chamber and is valid when only one chamber is operated at a time.

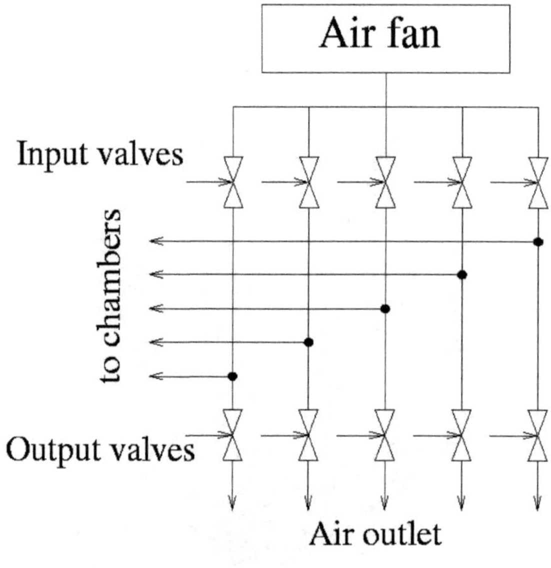

Fig. 4. Fan/valve system

3 One Chamber Dynamic Model

This section outlines the dynamic describing equations for a single chamber. The equations assumes that the air density is constant (variations could be included using the ideal gas equation), the chamber is only moving along a vertical axis perpendicular to the water surface, the cross section area is constant along this axis and there is only one rigid moving body. The model then consists of a mass balance equation describing the air in the chamber and a Newton equation describing the motion of the chamber.

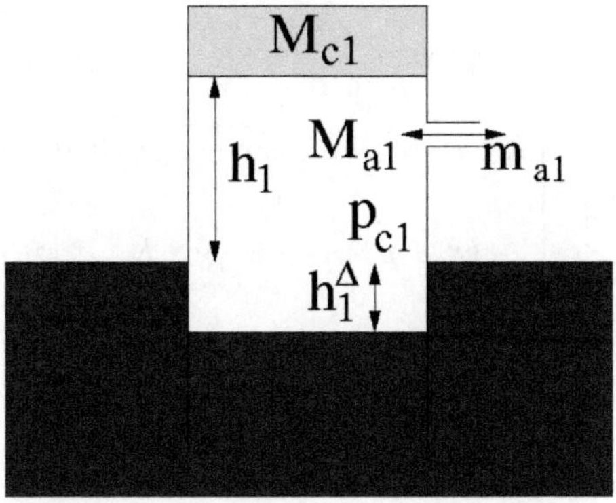

Fig. 5. One chamber model variables

The mass balance equation for the chamber is

$$\frac{dM_{a1}}{dt} = m_{a1} = A_1 \rho_a \frac{d(h_1^{\Delta} + h_1)}{dt} \tag{6}$$

where M_{a1} is the air mass in the chamber, m_{a1} is the mass flow to the chamber, A_1 is the cross section area, ρ_a is the air density, h_1^{Δ} is the distance from the chamber water surface to the ambient water surface and h_1 is the distance from the ambient water surface to the top of the chamber.

The pressure force from the chamber is assumed to equal the buoyancy force. This implies that the acceleration of the water volume in the chamber is small.

$$h_1^{\Delta} A_1 \rho_w g = p_{c1} A_1 \Rightarrow h_1^{\Delta} = \frac{1}{\rho_w g} p_{c1} \tag{7}$$

where ρ_w is the water density. Insertion of Eq. (6) in Eq. (7) gives

$$\frac{dp_{c1}}{dt} = \frac{g\rho_w}{A_1 \rho_a} m_{a1} - \rho_w g \frac{dh_1}{dt} \tag{8}$$

The other main equation for describing the dynamics of the chamber is Newton's 2'th law used on the free body with the mass $M_{c1.}$ Pressure forces, gravity and a friction force proportional with the chamber velocity is assumed to act on the body.

$$M_{c1} \frac{d^2 h_1}{dt^2} = (p_{c1} - p_a) A_1 - M_{c1} g - K_{f1} \frac{dh_1}{dt} \tag{9}$$

In order to use an ODE-solver to simulate the one chamber model, the three states $x_1 = p_{c1}$, $x_2 = h_1$ and $x_3 = \frac{dh_1}{dt}$ may be selected resulting in the differential equation system

$$\begin{bmatrix} \frac{dx_1}{dt} \\ \frac{dx_2}{dt} \\ \frac{dx_3}{dt} \end{bmatrix} = \begin{bmatrix} -\rho_w g x_3 + \frac{g\rho_w}{A_1\rho_a}((-K_{pa}(x_1 - p_a) + \\ K_{pb})g_1(u) - K_{vo}\sqrt{x_1 - p_a}\, g_2(u)) \\ x_3 \\ (x_1 - p_a)A_1 - M_{c1}g - K_{f1}x_3 \end{bmatrix} \tag{10}$$

4 Five-Chamber Dynamic Model

In this section a 5 chamber dynamic model is described. Although the WD has a complex geometry the model assumes that each of the five chambers may be regarded as a control volume with position independent internal variables and that each chambers pressure force actuate the WD bode in a single point. The model consist of 8 differential equations, 5 equations describing the pressures in the five chambers derived from mass balance equations, and 3 equations describing height, trim and heel derived from Newton's law.

The five chambers are placed in a "wave dragon" coordinate system {WD} with the coordinates $(\bar{x}_n, \bar{y}_n, \bar{z}_n)$. An inertial coordinate system {I} is placed in the water level as shown on the figure. The states in the model are the five chamber pressures (p_{c1}, p_{c2}, p_{c3}, p_{c4}, p_{c5}), the trim angle (θ), the heel angle (γ) and the level (h) of the wave dragon.

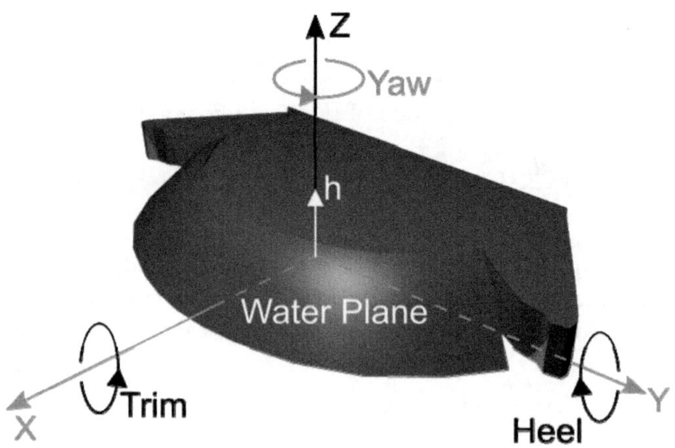

Fig. 6. Wave dragon coordinate systems [1]

The mass balance for the n'th chamber is

$$\frac{dp_{cn}}{dt} = \frac{g\rho_w}{A_n\rho_a} m_{an} - \rho_w g \frac{dh_n}{dt} \tag{11}$$

h_n is the height in the chamber and not a model state. Using a rotation matrix ${}_{WD}^{\ I}R(\theta,\gamma,\alpha)$ (see [2]), h_n and the states are related through

$$h_n = -\bar{x}_n \sin(\gamma) + \bar{y}_n \cos(\gamma)\sin(\theta) + \bar{z}_n \cos(\gamma)\cos(\theta) + h. \tag{12}$$

The time derivative of h_n is a tedious equation but using the approximations $\cos(\theta)= 1$, $\cos(\gamma) = 1$, $\sin(\theta) = \theta$, $\sin(\gamma)= \gamma$ assuming small heal and trim angles (all z_n's are 0) gives the simple relation

$$\frac{dh_n}{dt} = \bar{x}_n \frac{d\gamma}{dt} + \bar{y}_n \frac{d\theta}{dt} + \frac{dh}{dt} \tag{13}$$

which inserted gives

$$\frac{dp_{cn}}{dt} = \frac{g\rho_w}{A_n\rho_a} m_{an} - \rho_w g(\bar{x}_n \frac{d\lambda}{dt} + \bar{y}_n \frac{d\theta}{dt} + \frac{dh}{dt}). \tag{14}$$

Newton's 2'th law for the translational system is

$$\sum_{n=1}^{5} A_n p_{cn} + K_{flh}\frac{1}{h} - M_{wd}\, g - p_w A_w - K_f \frac{dh}{dt} = (M_{wd} + \frac{p_w A_w}{g})\frac{d^2 h}{dt^2} \tag{15}$$

where the sum is chamber pressure forces acting on the wave dragon body. The second term is the forces from the air chambers containing a constant air mass (see figure 3). In the model these are approximated with two chambers. They may be modelled using the mass balance Eq. (4) but in order to keep the model order low they are modelled using a static balance. K_{flh} may be found using the ideal gas law and the geometrics of the chambers. M_{wd} is the total mass of the wave dragon. The term $p_w A_w$ represents the force from the water reservoir on top the Wave Dragon. It may be noted that the pressure p_w is an input to the model.

The rotational trim equation is

$$\sum_{n=1}^{5} A_n p_{cn} \bar{y}_n - \frac{K_{fl\theta}}{h^2}\theta - K_\theta \frac{d\theta}{dt} = J_\theta \frac{d^2\theta}{dt^2} \tag{16}$$

where the sum is the torque from the chambers. The second term is the torque from the constant air mass chambers modelled as two symmetrical chambers. It should be noted that K_{flh} as well as $K_{fl\theta}$ are very dependent on the total air mass in the chambers. The moment of inertia is calculated as a constant although it depends on the water level in a very complex manner. In the simulation a situation with low water level has been used.

The heel equation is

$$\sum_{n=1}^{5} A_n p_{cn} \bar{y}_n - K_{fl\gamma}\gamma - K_\gamma \frac{d\gamma}{dt} = J_\gamma \frac{d^2\gamma}{dt^2} \tag{17}$$

This linear second order equation captures the gross behaviour of the heel dynamics.

The total model now consist of eleven equations, equation (14) used 5 times for the five chambers giving the five states $x_1 = p_{c1}$, $x_2 = p_{c2}$, $x_3 = p_{c3}$, $x_4 = p_{c4}$, $x_5 = p_{c5}$, equation (15) with the states $x_6 = h$, $x_7 = \frac{dh}{dt}$ and the equation $\frac{dx_6}{dt} = x_7$ gives 2 equations, equation (16) with the states $x_8 = \theta$, $x_9 = \frac{d\theta}{dt}$ and the equation $\frac{dx_8}{dt} = x_9$ gives 2 equations, and finally equation (17) with the states $x_{10} = \gamma$, $x_{11} = \frac{d\gamma}{dt}$ and the equation $\frac{dx_{10}}{dt} = x_{11}$ gives 2 equations. Inserting the states $(x_1, x_2, ..., x_{11})$ gives eleven nonlinear first order equations. The equations are solved using an ODE-solver.

5 Simulation

The model is tested using measured data from the wave dragon. The inputs to the model are the chamber pressures as well as the water level in the dragon represented by the pressure measurement p_w. The reason for not using the valve signals is that these were not recorded in the data acquisition equipment. As seen there is a good

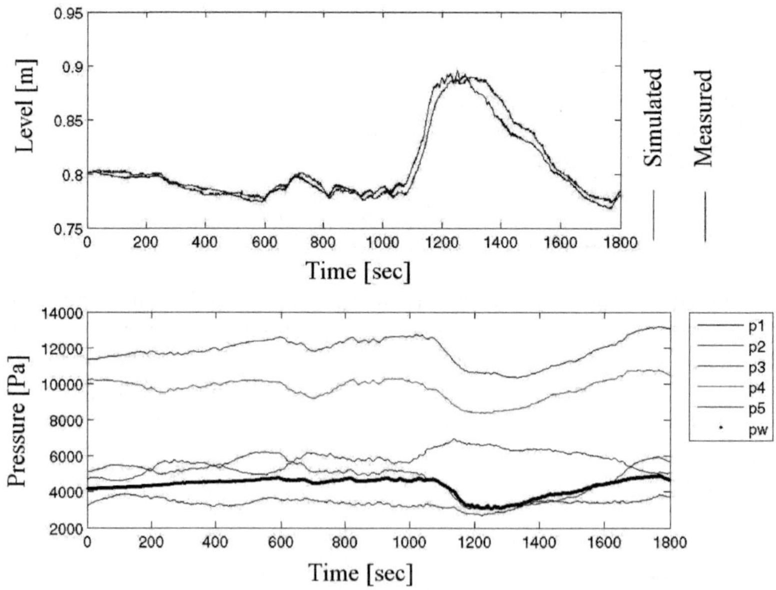

Fig. 7. Level simulated and measured

agreement between level in the model and the experimental data. (The agreements are found on the measurements from the same day. Measurements on different days are based on different initial pressures causing identification of slightly different model parameters) (This is also found on data measured on the same day.) Because the pressures in the constant air chambers are not measured these have been estimated from steady state observations.

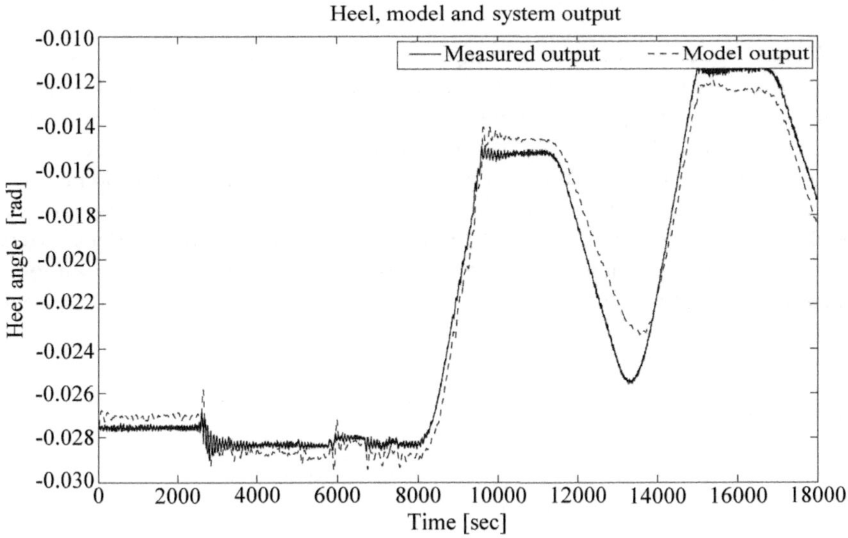

Fig. 8. Heel simulated and measured [1]

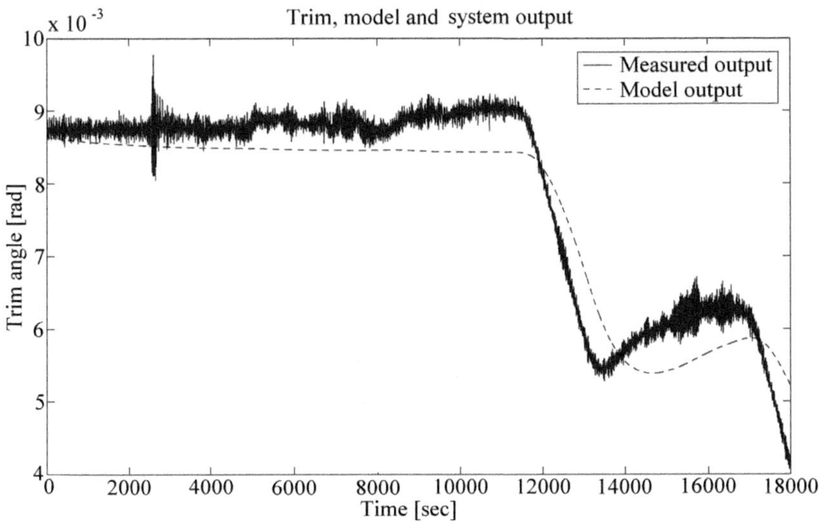

Fig. 9. Trim simulated and measured [1]

The variations on the heel and trim angles are small. As seen on the figures the behaviour is captures by the model. All the data were recording prior to this project and not prepared for this study, unfortunately the WD run severely aground during the project meaning that controlled input signal could not be tested. Regardless of this the model performed well on the recorded data.

6 Conclusions

A nonlinear physical model with a complexity that is suitable for model based control has been presented. The model is partly based on physical parameters for the Wave Dragon and may be scaled to a future larger version. The model has four main equations, one describing the state of the air in a chamber, and three accounting for the level, trim and heel motion of the WD. The model has been validated against measured WD data, where it captures the gross behaviour of the Wave Dragon. In particular it describes the response of the level very well. The model does, however, have one serious deficiency because it does not capture the distribution and movement of water in the water reservoir. A comprehensive study of this is outside the scope of this paper.

References

1. Andersen, J., Hundsdahl, M.Y., Jensen, P.K., Vilbergsson, K.S., Vidarsson, O., Skagestad, R.: Control of Wave Dragon Buoyancy, master thesis, Aalborg University (2007)
2. Craig, J.J.: Introduction to Robotics, Mechanics and Control, 2nd edn. Addison Wesley, Reading (1989)
3. Kofoed, J.P., Frigaard, P., Friis-Madsen, E., Sørensen, H.C.: Prototype testing of the wave energy converter wave dragon. Renewable Energy 31, 181–189 (2006)
4. Aps, W.D.: Wave Dragon – principles, http://www.wavedragon.net

Outbreaks Detection at the Beginning of Monitoring Process: The CUSUM Test Modification

Julia Bondarenko

Helmut-Schmidt-University Hamburg
Holstenhofweg 85, 22045 Hamburg, Germany
julia.bondarenko@hsu-hh.de

Abstract. Detection and reporting structural breaks arising in system during monitoring, is one of the main task of statistical process control. An ignorance of such breaks can cause an associated regression model misspecification. In practice, early detection and response to outbreaks, causing the changes in a process, is highly important. The famous CUSUM test of Brown, Durbin & Evans, has a poor power in detecting the structural breaks in parameters occurring early (and also late) in the sample.

Keywords. CUSUM test for structural breaks, recursive residuals, power of test, mixture distribution, Monte Carlo simulation.

1 Introduction

Structural breaks in time series can occur as consequences of complex processes, arising in system during a monitoring [1]. An ignorance of such changes can cause an associated regression model misspecification. The evidence of the parameters instability in linear models can be detected by the number of corresponding diagnostic tests. Particularly, the famous and important of them include fluctuation tests, with the CUSUM and CUSUMQ tests of Brown, Durbin & Evans (BDE tests) [2], standing first on the list. These tests are easy for implementing and based on the calculating of the cumulative sums of recursive residuals (CUSUM) and the cumulative sums of the squares of recursive residuals (CUSUMQ) for regression. The CUSUM test is generally used to detect the systematic movements of parameters, whereas CUSUMQ test tends to capture the sudden, haphazard movements.

A number of extensions of the standard BDE CUSUM tests have been developed, like the CUSUM and CUSUM of squares tests using OLS (ordinary least squares) residuals instead of recursive residuals [11], the CUSUM tests with lagged dependent variables in regression [9], the CUSUM tests with non-stationary regressors [5], the MOSUM and MOSUM of squares tests putting emphasis on moving averages rather than cumulative sums [3], the moving-estimates test [4], etc.

It is a well-known fact, that, CUSUM tests, both with recursive and OLS residuals, have a poor power in detecting the structural breaks in parameters occurring early and late in the sample, as well as the ones orthogonal to the mean regressor (see [6]; [7], pp. 50-51; [10]; [11]). A modification of recursive residuals CUSUM test, which is robust

J.A. Cetto et al. (eds.): Informatics in Control, Automation and Robotics, LNEE 37, pp. 231–239.
springerlink.com

to the later problem, was proposed in [8]. The reason for earlier problem is that CUSUM has no chance to cumulate for the such kind of breaks. In particular, some improving, due to alternative test boundaries developing, was suggested, in [13] for OLS CUSUM test.

In practice, early detection and response to outbreaks, causing the changes in a process, is highly important. In this paper, we propose a CUSUM-similar test which, due to the transformation of recursive residuals forces the detection of temporal dependence structure in linear regression model and has a larger power for the early structural breaks. Here our interest centres on the detection of single breaks occurring in parameters of the linear model. Distribution and other probabilistic characteristics of the transformed residuals are also provided, the boundaries for the new test are derived. The new test can be considered then as a complement to the standard CUSUM test. The paper is organized as follows. Section 2 presents the BDE's original formulation of the CUSUM test, discusses its main features. Section 3 offers our modification of the CUSUM testing procedure. In the Section 4 we conduct comparison of two tests via Monte Carlo simulation study. Section 5 summarizes and concludes.

2 Standard BDE CUSUM Test

Consider the linear regression model

$$y_t = x_t'\beta_t + \epsilon_t, t = 1, ..., T, \tag{1}$$

where y_t is the tth observation of the dependent variable, x_t is the $k \times 1$ vector of covariates at time t, β_t are unknown parameters, and ϵ_t are independent and normally distributed with zero mean and variance σ^2. We are interested in testing for a discrete jump in at least one of the β_t at the unknown time point. The null hypothesis can be formulated as

$$H_0 : \beta_t = \beta, t = 1, ..., T, \tag{2}$$

with alternative H_1 that at time T^* at least one of the β_t changes its value:

$$H_1 : \beta_t = \beta, t = 1, ..., T^* < T; \beta_t \neq \beta, t = T^* + 1, ..., T. \tag{3}$$

The CUSUM test suggested by Brown, Durbin and Evans (BDE) is a standard and commonly used diagnostic test for this kind of situations. BDE's CUSUM test uses the standardized residuals defined as

$$w_t = \frac{y_t - x_t'\widehat{\beta}_{t-1}}{\sqrt{1 + x_t\left(X_{t-1}'X_{t-1}\right)^{-1}x_t'}} \; for\; t = k+1, k+2, ..., T, \tag{4}$$

where $\widehat{\beta}_{t-1}$ is the OLS-estimator of β based on the first $t-1$ observations, $\widehat{\beta}_{t-1} = \left(X_{t-1}'X_{t-1}\right)^{-1} X_{t-1}Y_{t-1}$, and X_{t-1} and Y_{t-1} are the $(t-1) \times k$ and $(t-1) \times 1$ matrices that obtain by stacking x_s and y_s, respectively, for $s = 1, 2, ..., t-1$.

The advantage of working with w_ts defined by (4): it can be shown, that under H_0 (2), they are independent normally distributed with zero mean and variance σ^2. BDE CUSUM test is based on the cumulated sums of standardized residuals:

$$C_t = \frac{1}{\widehat{\sigma}} \sum_{s=k+1}^{t} w_s, \tag{5}$$

where $\widehat{\sigma}$ is OLS-estimate of σ, $\widehat{\sigma^2} = \frac{1}{T-k} \sum\limits_{s=k+1}^{t} w_s^2$. We cannot obtain the explicit distribution of C_t. But, under null hypothesis of parameter stability, the expected value and variance of the statistic C_t should be equal to zero and the number of normalized residuals being summed. Hence, the continuous Brownian motion process Z_t can be considered as a good approximation of the discrete path of C_t. Under H_0 the sequence C_t is thus a sequence of the approximately normal variables, where $E(C_t) = 0$, $Var(C_t) = t - k$, $Cov(C_t, C_r) = \min(t, r) - k$. Confidence bounds for the cumulated sums C_t are then obtained by plotting the two straight lines connecting the points $k \pm a\sqrt{T-k}$ and $T \pm 3a\sqrt{T-k}$, where a is a parameter depending on the α - significance level chosen, and is calculated on the basis of results for the Gaussian process. Namely, $a = 1.143$ for $\alpha = 0.01$; $a = 0.948$ for $\alpha = 0.05$; $a = 0.850$ for $\alpha = 0.1$ (see [2]).

Unfortunately, the CUSUM test suffers from low power, which decreases dramatically if the change point is close to the early beginning or to the end of the sample. In the next section we consider one modification of the CUSUM test based on the sconced cumulative sums.

3 Modified CUSUM Test

3.1 "Penalized" CUSUM Test (PCUSUM)

Let's take into account the temporal structure of the residuals w_t, and construct a new sequence of random variables u_t, $t = k+1, ..., T$:

$$u_t = w_t + c\, w_t\, I\{w_t\, w_{t-1} > 0\}, \quad w_k = 0\,. \tag{6}$$

where, as before, $w_t \sim N(0, \sigma^2)$ and independent, $I\{\cdot\}$ is indicator function, and c is some constant, $c > 0$, which can be considered as a penalty magnitude for upward or downward trends in CUSUM values. We will obtain now a distribution of the variables u_t.

Theorem 1. *Let w_t are independent identically distributed random variables, having a common symmetric about zero distribution (not only normal!), where $F_w(x)$ is a probability function, and $c > 0$. Then*

$$F_{u_t}(x) = \frac{1}{2}\left[F_{w_t}(x) + F_{w_t}\left(\frac{x}{1+c}\right)\right]. \tag{7}$$

Proof. We can perform $F_{u_t}(x)$ as a sum of the following probabilities:

$$\begin{aligned} F_{u_t}(x) = P[u_t < x] = P[w_t + c\, w_t\, I\{w_t\, w_{t-1} > 0\} < x] = \\ P[w_t + c\, w_t < x\,;\, w_t\, w_{t-1} > 0] + P[w_t < x\,;\, w_t\, w_{t-1} \le 0]\,. \end{aligned} \tag{8}$$

For simplicity let's consider two cases:
1) $x \le 0$. *Then (8) can be written as:*
$P\left[w_t + c\,w_t \ < x\,;\ w_{t-1} < 0\right] + P\left[w_t \le x\,;\ w_{t-1} \ge 0\right] = P\left[w_t \ < \frac{x}{1+c}\,;\ w_{t-1} < 0\right] +$
$P\left[w_t \le x\,;\ w_{t-1} \ge 0\right] = F_{w_t}\left(\frac{x}{1+c}\right) F_{w_t}(0) + F_{w_t}(x)\left(1 - F_{w_t}(0)\right) =$
$F_{w_t}\left(\frac{x}{1+c}\right) F_{w_t}(0) + F_{w_t}(x) - F_{w_t}(x)\,F_{w_t}(0) = \frac{1}{2}\left[F_{w_t}\left(\frac{x}{1+c}\right) + F_{w_t}(x)\right].$

2) $x > 0$. *(8) has a form:*
$P\left[w_t + c\,w_t \ < 0\,; w_t\,w_{t-1} > 0\right] + P\left[0 \le w_t + c\,w_t < x\,;\ w_t\,w_{t-1} > 0\right] +$
$P\left[w_t \ < x\,; w_t\,w_{t-1} \le 0\right] = P\left[w_t < 0\,;\ w_{t-1} < 0\right] + P\left[0 \le w_t < \frac{x}{1+c}\,;\ w_{t-1} > 0\right] +$
$P\left[w_t \ < x\,; w_t\,w_{t-1} \le 0\right] = F_{w_t}(0)\,F_{w_t}(0) + \left(F_{w_t}\left(\frac{x}{1+c}\right) - F_{w_t}(0)\right) F_{w_t}(0) +$
$P\left[w_t \ < 0\,; w_{t-1} \ge 0\right] + P\left[0 \le w_t < x\,;\ w_{t-1} < 0\right] = F_w(0)\,F_w(0) +$
$\left(F_{w_t}\left(\frac{x}{1+c}\right) - F_{w_t}(0)\right) F_{w_t}(0) + F_{w_t}(0)\left(1 - F_{w_t}(0)\right) + \left(F_{w_t}(x) - F_{w_t}(0)\right) F_{w_t}(0) \ =$
$\frac{1}{4} + \frac{1}{2}\left(F_{w_t}\left(\frac{x}{1+c}\right) - \frac{1}{2}\right) + \frac{1}{4} + \frac{1}{2}\left(F_{w_t}(x) - \frac{1}{2}\right) = \frac{1}{2}\left[F_{w_t}\left(\frac{x}{1+c}\right) + F_{w_t}(x)\right].$

Thus, the CDF (PDF) of the random variables u_t is a mixture of two normal distributions. It follows from the Theorem that u_ts have zero expectation, $E(u_t) = 0$, and variance $Var(u_t) \ = \ \frac{\sigma^2}{2}\left(1 + (1+c)^2\right)$. Covariance, $Cov(u_t, u_{t+1}) \ = E\left[(u_t - E(u_t))(u_{t+1} - E(u_{t+1}))\right] = E(u_t\,u_{t+1})$, equals to zero. In addition, variables u_ts have zero skewness $s(u_t) = 0$, and kurtosis $\kappa(u_t) = \frac{6\left[1+(1+c)^4\right]}{\left(1+(1+c)^2\right)^2}$, revealing fatter tails. The joint CDF function obtained by analogy with (8), is

$$F_{u_t,u_{t+1}}(x,y) = \begin{cases} \frac{1}{2}F_{w_{t+1}}\left(\frac{y}{1+c}\right)\left[F_{w_t}\left(\frac{x}{1+c}\right) + F_{w_t}(x)\right] + \frac{1}{2}\left[F_{w_{t+1}}(y) - F_{w_{t+1}}\left(\frac{y}{1+c}\right)\right], \\ \text{if } x > 0,\ y > 0; \\ \frac{1}{2}F_{w_{t+1}}(y)\left[F_{w_t}\left(\frac{x}{1+c}\right) + F_{w_t}(x)\right] + \frac{1}{2}\left[F_{w_{t+1}}\left(\frac{y}{1+c}\right) - F_{w_{t+1}}(y)\right], \\ \text{if } x > 0,\ y \le 0; \\ \frac{1}{2}F_{w_{t+1}}(y)\left[F_w\left(\frac{x}{1+c}\right) + F_w(x)\right], \\ \text{if } x \le 0,\ y > 0; \\ \frac{1}{2}F_{w_{t+1}}\left(\frac{y}{1+c}\right)\left[F_{w_t}\left(\frac{x}{1+c}\right) + F_{w_t}(x)\right], \\ \text{if } x \le 0,\ y \le 0. \end{cases} \tag{9}$$

The new statistic, which we will call "penalized" CUSUM (PCUSUM), can be written as following:

$$C_t^p = \frac{1}{\widehat{\sigma_p}} \sum_{s=k+1}^{t} u_s = \frac{1}{\widehat{\sigma^p}}\left(\sum_{s=k+1}^{t} w_s + c\sum_{s=k+1}^{t} w_s I\left\{w_s\,w_{s-1} > 0\right\}\right), \tag{10}$$

where $\widehat{\sigma_p}^2 = \frac{1}{T-k}\sum_{s=k+1}^{t} u_s^2$. It is follows from the properties of u_t, that $E(C_t^p) = 0$ but $\widehat{\sigma_p} \ge \widehat{\sigma}$, thus, we expect that the PCUSUM test has a chance to cumulate well and to detect structural shifts only in very beginning of the sample.

3.2 PCUSUM Test Boundaries

Like the original CUSUM test (5), the modified test (9) is only an asymptotic test and has no any certain distribution. Obviously, under $c \to 0$, the sequence $C^p_{k+1}, ..., C^p_t$ may be approximated by the Brownian motion process mentioned in Section 1, and the bounds for cumulated sums C^p_t are calculated then as the ones for C_t. But for larger values c this can not be applied. Hence, we have here a subproblem of simulation of the presented test boundaries.

We will partially implement the techniques used in [12] for confidence intervals calculation. The simulation algorithm may be carried out as follows. Let us generate L replicates. In each replicate i, $i = 1, 2, ..., L$, we simulate $T - K$ random variables $u_{i,t}$, $t = k + 1, k + 2, ..., T$, pairwise dependent and uncorrelated, drawn from the mixture of two bivariate normal distributions (7). Then the cumulated sums, $s^p_{i,t} = \frac{1}{\sigma^p} \sum_{s=k+1}^{t} u_{i,s}$, are calculated. At the significance level α we have for statistic (10) that $P[L^p_{k+1} < C^p_{k+1} < U^p_{k+1}, ..., L^p_T < C^p_T < U^p_T] = 1 - \alpha$, where L^p_t and U^p_t are correspondingly the lower and upper bounds for the value C^p_t, and

$$P\left[L^p_{k+1} < C^p_{k+1} < U^p_{k+1}, ..., L^p_T < C^p_T < U^p_T\right] \neq \left(P\left[L^p_t < C^p_t < U^p_t\right]\right)^{T-k}$$

for any $t = k+1, ..., T$, since C^p_t are not independent. Let assume that $P\left[L^p_t < C^p_t < U^p_t\right] = 1 - \alpha_p$, and denote $\alpha = f(\alpha_p)$, where f is some unknown function (in the case of independence one has $f(x) = 1 - x^{T-k}$), which we will obtain by simulation. Applying the Newton-Raphson algorithm, we calculate our α_p following the scheme:

$$\alpha_p^{(j)} = \alpha_p^{(j-1)} + d^{(j)} \left(\alpha - f\left(\alpha_p^{(j-1)}\right)\right),$$

where j is iteration number, $d^{(j)} = \delta d^{(j-1)}$ with $\delta = 0.5$ and $d^{(0)} = 1$, $\alpha_p^{(0)} = \alpha$. The convergence criterion is $\left|\alpha_p^{(j)} - \alpha_p^{(j-1)}\right| < 0.0001$. Repeat, that the function $f\left(\alpha_p^{(j)}\right)$ is derived at jth iteration as following: $f\left(\alpha_p^{(j)}\right)$ equals to the number of sequences $\left\{ s^p_{i,k+1}, ..., s^p_{i,T} \right\}_{i=1}^{L}$ within intervals $\left(L^p_{i,k+1}, U^p_{i,k+1}\right)^{(j)}$,..., $\left(L^p_{i,T}, U^p_{i,T}\right)^{(j)}$ divided by L, where intervals $\left(L^p_{i,t}, U^p_{i,t}\right)^{(j)}$ are obtained for the value $\alpha_p^{(j)}$. Note, that unlike the standard CUSUM statistic, which has symmetric lower and upper boundaries, we don't claim here $L^p_t = -U^p_t$ for PCUSUM statistic. To be fair to the standard CUSUM, we use a linear regression to obtain the linear boundaries from the curved ones: $\widetilde{L}^p_t = a_{1L} + a_{2L}\, t$ and $\widetilde{U}^p_t = a_{1U} + a_{2U}\, t$, where a_{1L}, a_{1U} are intercepts, a_{2L}, a_{2U} are slopes, $\widetilde{L}^p_t$, $\widetilde{U}^p_t$ are the best fitting lines.

If we still ignore a dependence between variables u_t (for example, for small c it's allowed), the distribution of the variable C^p_t (9) can be derived. Indeed, the probability distribution of a random variable $S_{(t)}$ which is itself the sum of independent random variables $S_{(t)} = X_1 + X_2 + ... + X_t$, with a common distribution function $G(x)$, is given by the t-fold convolution of $G(x)$:

$$G_t(x) = P\left[S_{(t)} < x\right] = G^1(x) * G^2(x) * ... * G^t(x) = \int_{-\infty}^{x} G(y)\, G_{t-1}(x - y)\, dy,$$

where $G^i(x) = P[X_i < x]$, $i = 1, ..., t$. In our case, when

$$G(x) = \frac{1}{2}\left(F_{w_t}\left(\frac{x\,\widehat{\sigma^p}}{1+c}\right) + F_{w_t}\left(x\,\widehat{\sigma^p}\right)\right),$$

we get that

$$C_t^p \sim \left(\frac{1}{2}\right)^t \sum_{k=0}^{t} \binom{t}{k} \left((F_{w_t})_k\left(\frac{x\,\widehat{\sigma^p}}{1+c}\right)\right) * \left((F_{w_t})_{t-k}\left(x\,\widehat{\sigma^p}\right)\right) =$$

$$\left(\frac{1}{2}\right)^t \sum_{k=0}^{t} \binom{t}{k} N\left(0,\ k\ (1+c)^2\left(\frac{\sigma}{\widehat{\sigma^p}}\right)^2 + (t-k)\ \left(\frac{\sigma}{\widehat{\sigma^p}}\right)^2\right),$$

where $(F_{w_t})_k\left(\frac{x\,\widehat{\sigma^p}}{1+c}\right)$ and $(F_{w_t})_{t-k}\left(x\,\widehat{\sigma^p}\right)$ are k-fold and $(t-k)$-fold convolutions of distribution functions $F_{w_t}\left(\frac{x\,\widehat{\sigma^p}}{1+c}\right)$ and $F_{w_t}\left(x\,\widehat{\sigma^p}\right)$ correspondingly, $\binom{t}{k} = \frac{t!}{k!\,(t-k)!}$ is a binomial coefficient. Thus, we have the mixture of normal distributions, with zero means and variances $k\ (1+c)^2\left(\frac{\sigma}{\widehat{\sigma^p}}\right)^2 + (t-k)\ \left(\frac{\sigma}{\widehat{\sigma^p}}\right)^2$, and binomial weights $\left(\frac{1}{2}\right)^t \binom{t}{k}$, $k = 0, ..., t$.

4 Monte Carlo Study

In this section we present a Monte Carlo study of the new CUSUM test in comparison with the classical one. In order to see the performance of C_t^p, we consider the model (1), where the matrix of independent variables x_t has the following design:

$$X^{MC} = \left\{\left[1, (-1)^t\right]'\right\}_{t=1}^{T},$$

which was also used in simulation study in [5] and [7]. Values ϵ_t, as before, are independent normal, generated with parameters 0 and 1. We simulate our responses y_t for three sample sizes, $T = 20$ (small sample), $T = 50$ (medium sample) and $T = 500$ (large sample), with $\beta = [10, 2]'$ under the null hypothesis of parameters constancy. The significance level $\alpha = 0.05$, the computed empirical levels α_p for samples under $c = \{0.2;\ 0.5;\ 1.5\}$ are correspondingly $\alpha_p = \{0.0075;\ 0.0049;\ 0.0054\}$ $(T = 20)$, $\alpha_p = \{0.0055;\ 0.0036;\ 0.0029\}$ $(T = 50)$, $\alpha_p = \{0.0011;\ 0.0013;\ 0.0016\}$ $(T = 500)$.

Empirical (actual) test sizes, based on the simulated data, at the nominal size of 5%, and $c = \{0.2; 0.5; 1.5\}$ in PCUSUM test, are presented in the Table 1. Generally estimated sizes, calculated as rejection rates under null hypothesis, with Monte-Carlo replications number $N = 5000$, are either below the nominal size for CUSUM test,

Table 1. Empirical Sizes of the Tests, $\alpha = 0.05$

Test	$T = 20$	$T = 50$	$T = 500$
CUSUM	0.0154	0.0276	0.0458
PCUSUM, $c = 0.2$	0.0338	0.0453	0.0476
PCUSUM, $c = 0.5$	0.0512	0.0510	0.0444
PCUSUM, $c = 1.5$	0.0686	0.0586	0.0646

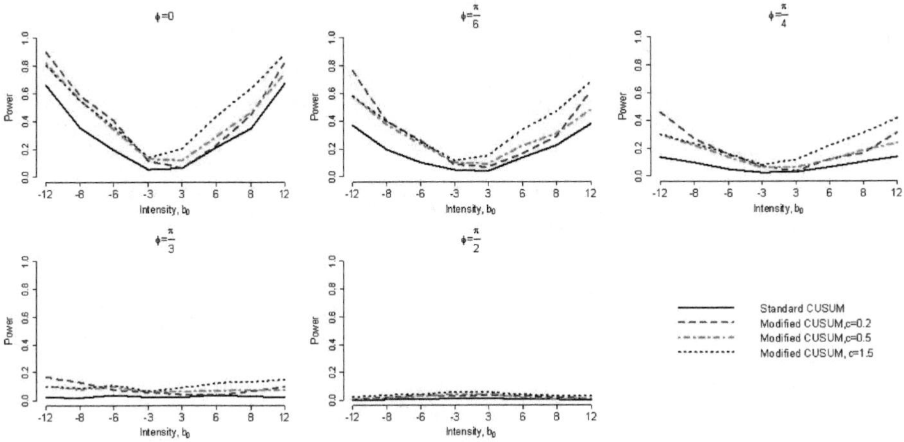

Fig. 1. Power plots under $T = 20$, $\lambda = 0.3$

resulting in more "liberal" test, or almost equal to the nominal size. However, the empirical sizes for PCUSUM test under $c = 1.5$ are larger than 0.05, declaring the more "sensitive" test.

Now, we introduce at time $T^* = [\lambda T]$, where λ can take any values between 0 and 1, some structural shift. Let's consider a single structural shift in parameters β is given by $\Delta\beta = \frac{b_0}{\sqrt{T}}\,[\cos\phi,\ \sin\phi]'$, where ϕ is the angle between $\Delta\beta$ and mean regressor $r = \left[\frac{1}{T}\sum_{t=1}^{T} x_{1t}, \frac{1}{T}\sum_{t=1}^{T} x_{2t}\right]' = [1, 0]'$, b_0 is a constant determining the intensity of the shift , $\|\Delta\beta\| = \frac{|b_0|}{\sqrt{T}}$. We will take a number of different values of b_0 and ϕ, namely, $b_0 = \{-12;\ -8;\ -6;\ -3;\ 3;\ 6;\ 8;\ 12\}$ (positive and negative values, as we don't claim the boundaries of PCUSUM to be symmetric) and $\phi = \{0°;\ 30°;\ 45°;\ 60°;\ 90°\}$. In

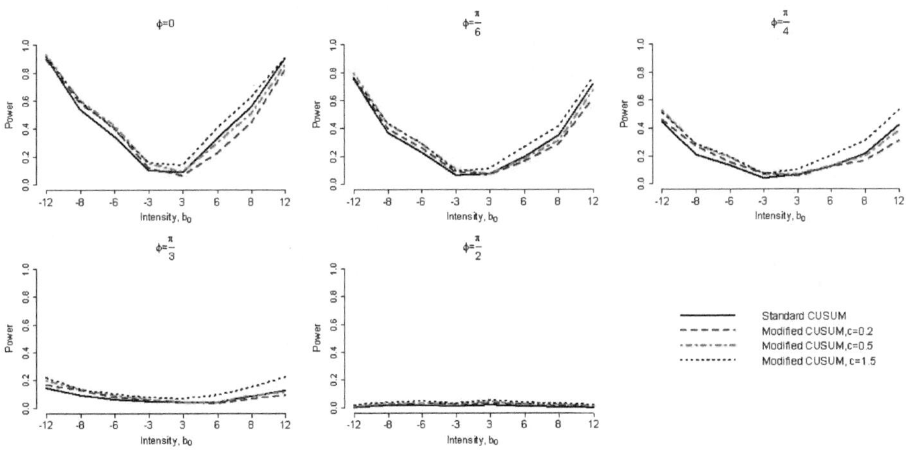

Fig. 2. Power plots under $T = 50$, $\lambda = 0.3$

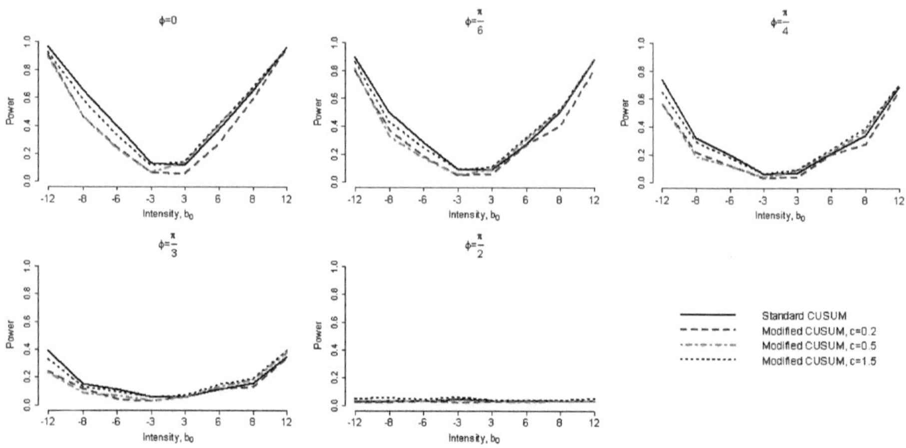

Fig. 3. Power plots under $T = 500$, $\lambda = 0.3$

other words, we are testing two hypotheses: H_0 : parameters are constant for $1 \leq t \leq T$ against H_1 : parameters have two different constant values, for $1 \leq t < T^*$ and $T^* \leq t \leq T$.

It is possible to show that, as we have expected, the linear boundaries for PCUSUM test become narrower than CUSUM test boundaries only at the beginning of the sample. Hence, it makes sense to choose λ corresponding to the structural changes at the beginning of the sample, $\lambda = 0.3$, for example. Empirical power was calculated as a probability that the test statistic under the alternative hypothesis exceeded the significance threshold calculated from the distribution under the null hypothesis (a frequency of the null hypothesis rejection under the alternative hypothesis). The obtained power plots, under $N = 1000$, $\lambda = 0.3$, $\alpha = 5\%$, $c \in \{0.2; 0.5; 1.5\}$ for $T = 20$, $T = 50$ and $T = 500$ are presented on the Figures 1, 2 and 3, correspondingly.

The obtained simulation results for the small sample with $T = 20$ reveal that the PCUSUM outperforms the CUSUM. For small values of shift intensity b_0 and for $\phi = 90°$ this superiority is quite insufficient, for all samples. For the medium sample $T = 50$, PCUSUM with $c = 1.5$ has higher power everywhere, but PCUSUM with $c = 0.5$ and $c = 0.2$ - only for negative values of b_0. In big sample with $T = 500$ an advantage is exhibited only by PCUSUM with $c = 1.5$, for positive b_0s. By virtue of the boundaries non-symmetry, the results are different for positive and negative values of b_0 also among PCUSUM tests: for example, for $T = 20$ PCUSUM with larger parameter c outperforms PCUSUM with smaller one for positive b_0s, but for negative b_0s PCUSUM with smaller value of c remains more effective.

5 Conclusions

In this paper we proposed a modified, based on the penalized residuals, version of the standard BDE CUSUM test for single structural breaks in parameters of linear regression. The new test, PCUSUM, is recommended as a complement to the standard

CUSUM test for better detecting the structural shifts occurring early in the samples. Simulation results have shown, that the modified CUSUM test has the better chance to cumulate parameter breaks, occurred at the beginning of the sample.
The subjects of eventual future research are:

- closer examination of properties of the PCUSUM test boundaries, their comparison with derived curved boundaries of the standard CUSUM test;
- adoption of the modified residuals (6) by CUSUM of squares test (CUSUMQ) for testing structural changes in variance (serial correlation and heteroscedasticity);
- appropriate transformation of the proposed new test for both early and late structural breaks (inclusion of the more lagged residuals in (6), etc.).
- Acknowledgment

Acknowledgements. I would like to acknowledge Professor Wilfried Seidel for his beneficial comments.

References

1. Basseville, M., Nikoforov, I.: Detection of Abrupt Changes - Theory and Application. Prentice Hall, Englewood Cliffs (1993)
2. Brown, R.L., Durbin, J., Evans, J.M.: Techniques for Testing the Constancy of Regression Relationships over Time. Journal of Royal Statistical Society B 27, 149–163 (1975)
3. Chu, C.-S.J., Hornik, K., Kuan, C.-M.: MOSUM Tests for Parameter Constancy. Biometrika 82, 603–617 (1995a)
4. Chu, C.-S.J., Hornik, K., Kuan, C.-M.: The Moving-Estimates Test for Parameter Stability. Econometric Theory 11, 669–720 (1995b)
5. Inder, B., Hao, K.: A New Test for Structural Change. Empirical Economics 21, 475–482 (1996)
6. Krämer, W., Proberger, W., Alt, R.: Testing for Structural Change in Dynamic Models. Econometrica 56, 1355–1369 (1988)
7. Krämer, W., Sonnberg, H.: The Linear Regression Model under Test. Physica-Verlag, Heidelberg (1986)
8. Luger, R.: A Modified CUSUM Test for Orthogonal Structural Changes. Economic Letters 73, 301–306 (2001)
9. Proberger, W., Krämer, W., Alt, R.: A Modification of the CUSUM Test in the Linear Regression Model with Lagged Dependent Variables. Empirical Economics 14, 65–75 (1989)
10. Proberger, W., Krämer, W.: The Local Power of the CUSUM and CUSUM of Squares Tests. Econometric Theory 6, 335–347 (1990)
11. Proberger, W., Krämer, W.: The CUSUM Test with OLS Residuals. Econometrica 60, 271–285 (1992)
12. Tanizaki, H.: Asymptotically Exact Confidence Intervals of CUSUM and CUSUMQ Tests: a Numerical DErivations Using Simulation Technique. Communications in Statistics, Simulation and Computation 24, 1019–1036 (1995)
13. Zeileis, A.: *P* Values and Alternative Boundaries for CUSUM Tests. Working Paper 78, SFB Adaptive Information Systems and Modelling in Economics and Management Science (2000)

A Novel Wearable Instrumentation System for Bomb Disposal Suits

John Kemp, Elena I. Gaura, and James Brusey

Cogent Computing Applied Research Centre
Coventry University, Priory St, Coventry, CV1 5FB, U.K

Abstract. Bomb disposal suits contain a large amount of padding and armour to protect the wearer's vital organs in the case of explosion. The combination of the heavy (roughly 40kg) suit, physical exertion, and the environment in which these suits are worn can cause the wearer's temperature to rise to uncomfortable and potentially dangerous levels during missions. This paper reports on the development of a wearable wireless sensing system suitable for deployment in such manned bomb disposal missions. In its final form, the system will be capable of making in-network autonomous decisions related to the actuation of cooling within the suit, in order to increase the comfort of the wearer. In addition, it will allow an external observer to remotely monitor the health and comfort of the operative. Laboratory experiments with the instrumented suit show how skin temperature varies differently for different skin sites, motivating the need for multiple, distributed sensing. The need for timely application of in-suit cooling is also shown, as well as the importance of monitoring the overall health of the wearer of the suit.

Keywords: Body sensor networks, first responders, actuation.

1 Introduction

The monitoring of hazardous environments, along with the people working within them, is an area which lends itself to the use of wireless and body sensor networks (WSNs and BSNs). The field is rich with potential WSN applications in detecting hazards, providing feedback to remote observers and other critical tasks that can increase the safety and benefit the overall working conditions of people operating in these environments. This paper reports on the work towards the development of wireless body sensor network instrumentation for protective suits worn in bomb disposal missions.

A typical bomb disposal mission will initially involve investigating the site using a remote controlled robot, and if possible, disarming the bomb remotely. Sometimes, however, it is necessary for a human bomb disposal expert to disarm the device. For this, the expert will put on a protective suit and helmet (as shown in figure 1), pick up a tool box of equipment, and walk the 100 or so metres to the site. To reach the bomb's location, it may be necessary to climb stairs, crawl through passageways, or even lie down.

The environment where the suit is most commonly used, such as the hot climate of the Middle East, plays an important role in the design of the protective EOD suit. One of the UK manufacturers of such suits has identified the problem of the suit wearer becoming uncomfortably hot and, in the worst case, suffering heat exhaustion. They have attempted to address this by installing an in-suit cooling system based on a dry-ice pack and a fan that cycles air through the pack and blows cooled air onto the wearer's

J.A. Cetto et al. (eds.): Informatics in Control, Automation and Robotics, LNEE 37, pp. 241–253.
springerlink.com

Fig. 1. Explosive Ordinance Disposal (EOD) Suit

back and into the helmet. The cooling system has a variable control thus both allowing the airflow to be adjusted for comfort and also allowing the life of the batteries that power the fan to be extended, as they would only provide sufficient power for part of the mission otherwise. The problem with this cooling approach, though, is that the bomb disposal expert has other critical concerns during the mission and either does not bother to put the fan on or tends to set it to maximum airflow from the beginning of the mission.

To address the above problems, this work proposes embedding into the suit a body sensor network that primarily aims to:

- sense the temperature of the skin of various parts of the body, in order to assess overall thermal comfort, and
- actuate the cooling dynamically to both remove the need for human intervention, and also to prolong battery life.

The prolonging of battery life is intended to provide cooling over the whole mission duration (compared to the partial coverage provided currently) rather than increasing the mission duration itself.

A secondary goal of this work is to help the manufacturer better understand how the suit material and design choices are affecting the wearer's thermal comfort during use. Finally, the prototype presented here has been designed such as to allow easy integration of additional sensors: accelerometers to monitor posture, heart rate monitors, and CO_2 sensing within the helmet.

The paper is organised as follows: Section 2 examines related work, focusing in particular on body sensor networks and research relating to instrumenting first responders (such as police, fire services etc). Section 3 describes the system design and architecture of the prototype system produced to date. Section 4 contains an evaluation of the prototype. Finally the paper concludes with some observations based on the work so far and outlines future work.

2 Related Work

The work reported in this paper is most closely aligned with respect to the instrumentation design and implementation with the field of Body Sensor Networks (BSNs). This is a sub-area of Wireless Sensor Networks that makes use of a combination of wireless and miniaturised sensor technologies to monitor the human body. The scope of present BSN approaches is patient care. Such systems are either designed to focus on capturing the evolution of a particular physiological parameter and ensuring that alarms are generated when parameters stray outside a safe range [1], or aimed to provide general monitoring solutions for patient status within a hospital or similar environment [2]. In comparison, the work presented here is concerned with increased safety and comfort of human subjects in constrained environments through integrating sensing, actuation, and autonomous decision making. In this context, wireless sensor technology is used as an enabler for the necessary detailed measurement of physiological parameters. This work shares some of the design space of BSN in terms of the type of physiological parameters sensed and the wearability requirements of the implemented system. On the other hand, given that the application is within the safety critical domain, the work here also shares some common characteristics with the area of instrumenting and monitoring first responders. In this section, samples of BSN platforms are reviewed together with commercial instances of first responder monitoring and prior, motivating, physiological findings about the EOD suit.

2.1 Body Sensor Networks—Platforms

BSN systems are often more constrained than ordinary embedded systems. These constraints are mainly in terms of power, size and weight. Power is restricted because mains AC power is not available. Furthermore, wearability requirements imposing size and weight restrictions limit the battery supplies that can be used. Size and weight must be limited as large and heavy devices would be cumbersome, uncomfortable, and in applications such as the one described here, an unnecessary distraction.

In response to the above, some of the BSN systems designed and implemented by research groups integrate within the nodes an appropriate central processing unit, memory and radio transceiver as a single custom chip. An example here is the MITes platform (for monitoring movement of human subjects) developed by Tapia *et al.* [3], which is based around the Nordic VLSI Semiconductors nRF24E1 chip. This chip integrates a radio transceiver and an Intel 8051 based processor core that runs at 16MHz and provides a nine channel 12-bit ADC and various other interfaces, such as SPI (serial peripheral interface) and GPIO (general purpose I/O). This approach is efficient in terms of size and weight due to the integration of several functions into one chip, but has limited generality as it can not be easily adapted for new applications.

Another, more popular design option is to use off-the-shelf components. There is a trade off made between processing and storage capabilities and the size and power consumption of the devices. This means that the devices selected would likely be considered severely computationally under-powered in other systems (often including 16- or even 8-bit processors) and have small amounts of memory (in the order of tens or hundreds of kilobytes). For instance, the Texas Instruments MSP430F149 micro-controller has been used for several systems including those developed by Lo and Yang [4] and Jovanov *et al.* [5]. This is a 16-bit processor running at 8MHz incorporating 60KB of flash memory and 2KB of RAM and provides interfacing opportunities via 48 GPIO lines and a 12-bit ADC. The system developed by Lo and Yang used ECG sensors, accelerometers, and a temperature sensor to monitor patient health. The system developed by Jovanov *et al.*, was used for monitoring the elderly and those undergoing physiotherapy.

Other systems expand upon commercial devices such as the Mica2 and MicaZ motes developed at the University of California, Berkeley, or Intel's Imote platform. This approach often has a disadvantage in that the basic platform is generic, and may not directly provide the facilities required for the specific BSN project. Such commercial platforms are also often larger and heavier than custom developed platforms as they are required to be general purpose in order to achieve any commercial success. The MicaZ mote uses the Atmega128L, an 8-bit processor running at 8MHz and featuring 128KB of flash memory to which an additional 512KB is added externally on the mote itself. A 10-bit ADC, UART and I2C bus are also available. Gao *et al.* [6] developed a system based around this mote, adding various sensors and supporting devices to allow patient tagging and monitoring in an emergency response environment. Walker *et al.* [7] present a blood pressure monitoring system based on the MicaZ platform. In that work, a commercial blood pressure monitoring device is connected to the MicaZ via a serial interface.

2.2 Instrumenting First Responders

The best fit example of a commercial product designed for the purpose of monitoring personnel carrying out missions in dangerous environments is the VivoResponder by VivoMetrics [8]. VivoResponder is based upon an earlier product called the LifeShirt and is aimed at personnel engaged in firefighting and hazardous materials training or emergency response, industrial clean-ups using protective gear, and biohazard-related occupational work. The VivoResponder is supplied in three parts: a lightweight, machine washable chest strap with embedded sensors; a data receiver; and, VivoCommand software for monitoring and data analysis. The sensors embedded in the chest strap monitor the subject's breathing rate, heart rate, activity level, posture, and single point skin temperature.

Monitoring of the subject's breathing is performed using a method called inductive plethysmography, where breathing patterns are monitored by passing a low voltage electrical current through a series of contact points around the subject's ribcage and abdomen. Monitoring of the subject's heart rate is performed via an ECG.

The VivoCommand software, provided with the device, displays the gathered data from the chest strap in real-time on a remote PC. The parameters are updated every second along with 30-second average trends. The parameters are displayed with colour coding intended to allow quick assessment of the status of up to 25 monitored personnel simultaneously. Baseline readings can be set individually per monitored person.

The work developed here differs in intent: the aim here is to provide a detailed thermal assessment based on sensors integrated into the protective suit and deliver remotely abstracted *thermal comfort information*.

2.3 Other Work on EOD Suits

Working from a physiological perspective, Thake and Price [9] have investigated the thermal strain of a subject when wearing EOD suits in hot environments. The work looks at quantifying the level of strain by assessing how hot and tired the suit wearer feels whilst wearing various combinations of suit components. An "activity" regime was developed for the assessment based on the types of activities that a bomb disposal technician would undergo during a mission and included walking on a treadmill, unloading and loading weights from a rucksack, crawling and searching activity, arm cranking and cognitive tests. Aspects of hand-eye coordination and psychological performance were also assessed. The investigations have demonstrated a large increase in physiological strain when wearing the EOD suit, though benefits have been shown when the ambient air is cooled for the suit ventilation purpose and lighter-weight trousers are worn.

It is indeed these types of studies, together with the user requests, that prompted the development of the detailed physiological monitoring system presented in this paper. The activity regimes described by Thake and Price were used in the experiments presented in this paper to allow validation of findings.

3 System Design and Architecture

The main part of the prototype system is designed following a sense-model-decide-act architecture as shown in figure 2. The environment within the suit is sensed in terms of temperature; sensed data is integrated into a model representing the thermal state of the wearer; a decision is made by the system about how to adjust the cooling system based on the thermal state; finally, the determined action is transmitted as an actuation signal to the fan speed controller. In addition to this basic architecture, the system also transmits inferred state values for the purpose of remote, on-line, visualisation of the thermal state of the wearer. From this visualisation, the mission control operator can assess how different parts of the mission, or different actions being taken by the suit wearer are affecting their thermal state and hence assess the wearer's fitness for the mission. (It is expected that such information, collected during field trials and real missions, might lead to changes to future mission planning or to changes in the design of the suit.) In summary, the prototype system can be seen as being composed of two control loops: one giving rapid feedback to autonomously adjust cooling; the other, longer term one, providing support for an iterative design process in terms of both the mission use and construction of the suit.

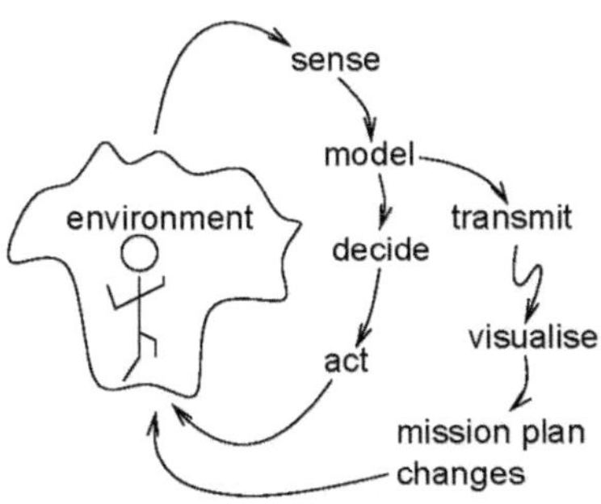

Fig. 2. Conceptual design of prototype system

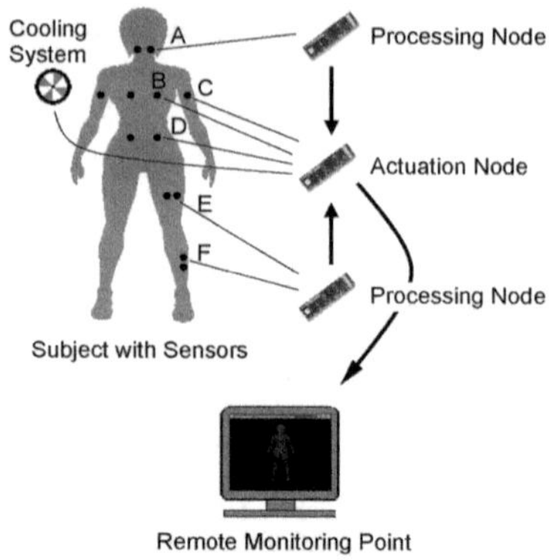

Fig. 3. Prototype system hardware components and sensor positioning (A – neck, B – chest, C – bicep, D – abdomen, E – thigh, F – calf)

The prototype design consists of a number of hardware components, including a remote monitoring station, two processing nodes, one actuation node, 12 temperature sensors, and the cooling system. The connection between these components is shown in figure 3. The processing nodes, actuation nodes, and remote monitoring point form a wireless network. Each processing node is wired to several sensor packages via an I^2C bus. Although it would be possible to integrate all sensor packages used in this prototype into a single processing / actuation node, using separate processing nodes allows the instrumentation for the helmet, jacket, and trousers to be kept separate with no wires running between them. This is essential for ensuring that the product remains easy to use and transparent to the wearer.

The system components and their functionality are described in the remainder of this section.

3.1 Sensor Packages and Sensor Positioning

The prototype system discussed here uses twelve sensor packages based on Analog Devices ADT75A temperature sensor ICs (shown in figure 4). This device has the advantage that it contains the sensor, ADC, and bus interface in a single package. Temperature values are transmitted as 12 bits, which causes rounding to within 1/8°C. The sensor packages are connected to only two nodes in the current version: one actuation node and one processing node. (The actuation node here doubles up as the second processing node.)

The sensor packages were positioned around various parts of the body roughly following the standard positioning used for skin sensors as used by Thake and Price [9], which is a subset of the locations discussed by Shanks [10]. These were: lateral calf muscle, front of thigh (or quadruceps), abdomen, chest, biceps, and neck, as indicated in figure 3. Given that temperatures are known to be symmetrical between left and right sides in healthy people [11], sensors have been placed on a single side during the prototype evaluation experiments. Two sensor packages were used per skin site. This arrangement enables individual data validation.

Fig. 4. Sensor package based on an ADT75A chip

3.2 Processing and Actuation Nodes

Construction. There are a variety of available embedded platforms for sensing and control applications. The hardware choice decisions for the prototype system here were based on the available platforms' processing power, external interfaces, ease of software development, and size.

Gumstix Connex 400xm-bt boards were selected as the main processing platform. Although not as popular as Mica2 motes, they are becoming more prevalent (see Keoh *et al.* [1] for an example). These devices offer more processing power and memory (in terms of both RAM and flash) than many similarly sized platforms. The Connex board includes an Intel XScale PXA255 400MHz processor, 16MB of flash memory, 64MB of RAM, a Bluetooth controller and antenna, and 60-pin and 92-pin connectors for expansion boards. There are no on-board sensors provided. The sensor packages connect to the Connex board via an expansion board, designed in-house.

The prototype system exploits the following capabilities offered by the Gumstix Connex device: Bluetooth communications to transmit data between nodes; I^2C bus interface for the attachment of sensor packages; real-time data modelling and decision-making; and, a small form factor, which enables convenient mounting on or around a subject's body.

Functionality. In the current revision of the prototype, the actuation and processing nodes only transmit data back to the remote monitoring station, upon filtering outlying values. The longer term view is for the processing and actuation nodes to perform in-network modelling of the suit wearer's comfort through collaborative behaviour. Comfort modelling would firstly involve production of a thermal sensation model within the network. This will be followed by integration of supplementary physiological and contextual sensing performed by expanded sensor packages. Work to date in thermal sensation modelling and its integration within the processing and actuation nodes has been reported elsewhere by the authors here [12]. The actuation node will eventually be used to perform decision making on the basis of the wearer's comfort and act by controlling the fan speed.

3.3 Remote Monitoring

The remote monitoring component of this system allows an external observer to monitor both the instrumentation system (to ensure that trustworthy information is being

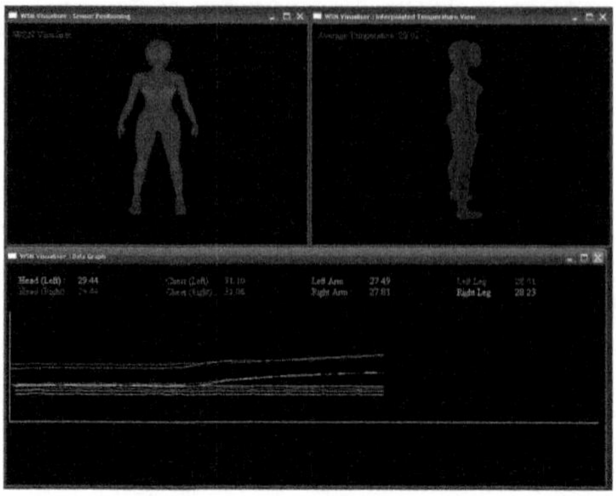

Fig. 5. A snapshot of the remote monitoring component

recorded) and the bomb disposal technician during a mission (which is the main function of the instrumentation). The remote monitoring component displays the health and comfort information and provides alerts to the remote observer if physiological parameters fall outside safe ranges or the wearer is shown to be significantly uncomfortable. A snapshot of the remote monitoring component is shown in figure 5. Currently the remote monitor displays skin site temperature data and a rotating, suggestive, 3-D interpolated model of skin temperatures. Cool to hot zones are displayed dynamically through a range of colours, from blue to red.

4 Prototype Evaluation

4.1 Experimental Setup

The prototype instrumentation system was evaluated through laboratory experiments that attempt to reproduce typical bomb disposal mission situations by having the subject undertake a series of activities and tasks, as discussed in section 2.3. The experiments begin with sensors being attached to the subject, followed by suiting-up. The upper body sensors are integrated into the clothing and thus easier to attach, whilst lower body sensors are attached with PVC tape. The subject wore the outer shell of the bomb

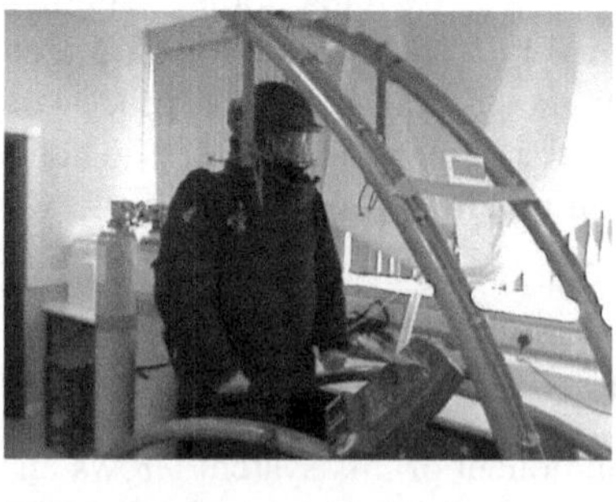

Fig. 6. First activity: walking at 4km/h

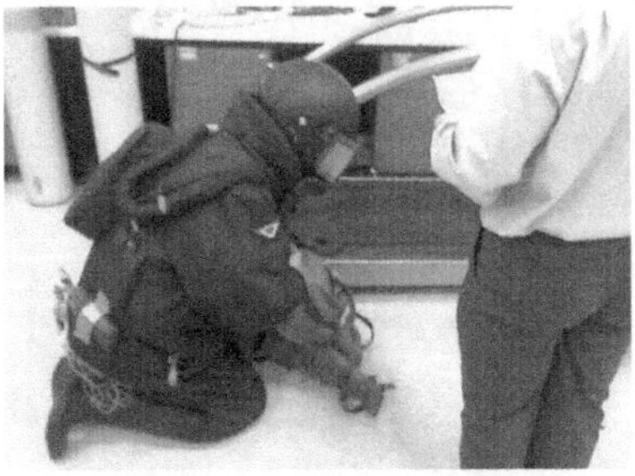

Fig. 7. Second activity: kneeling while removing weights from a sack. The wired-in data logger can be seen taped onto the subject's lower back

disposal suit including the jacket and trouser segments in addition to armour plating and the helmet.

The subject then undertakes an activity regime composed of: (1) walking (3 minutes) (see figure 6); (2) kneeling while putting weights into and out of a rucksack (2 minutes) (see figure 7); (3) crawling (2 minutes); (4) arm exercise (4 minutes); (5) sitting (3 minutes); (6) standing (1 minute). Temperature data is collected both via the prototype wireless system and via a wired-in data logger for validation purposes. Data was gathered during two consecutive runs, both consisting of the same routine and taking place in a 5m x 6m draft free room, with an ambient temperature of 21°C.

4.2 Evaluation Results

The prototype was evaluated according to a number of criteria that follow directly from user requirements. The criteria were: ease of use, data yield, accuracy, robustness, communication range, and information gain.

Ease of use: Instrumentation systems, particularly those used for bomb disposal missions, are expected to have stringent ease of use requirements as they should be transparent to the user and should not interfere with the mission. Ease of use was assessed here subjectively by comparing the ease of application of the sensor packages with sensor mountings for a wired data logger.

As mentioned previously some of the sensor packages have been integrated into clothing, whilst some (neck, thigh, and calf) have been taped to the skin. It is expected that clothing integrated sensors will be less accurate than ones taped to the skin because contact with the skin surface will change when the subject is moving. While avoiding the problem of inconsistent contact, taping on sensors, on the other hand, means that they are less convenient to apply and remove. In comparison to using a standard wired data logger, the wireless system mounting takes considerably less time and has been found to be more comfortable by experimental subjects. Further revisions of the prototype will have all sensor packages mounted on individual elasticated straps, ensuring both firm contact and comfort.

Data Yield is a measure of the proportion of data captured. Wireless sensing systems are inherently prone to low yields due to both transmission errors and sensor faults. For

Fig. 8. Enhanced sensor package, which is comprised of a PIC processor, 3-DOF accelerometer, I2C buffer, and temperature sensor

the system here, during experimentation, no packets were lost in transmission, however 5% of the sensor samples were found to be out of range (95% yield). Most of the out of range values were from particular sensor packages (3.3% from the worst two), with several sensor packages having no out of range values at all. It is likely that the erroneous values were introduced by I^2C bus transmission errors. In comparison, there were no errors apparent in the wired data logger values apart from the chest sensor, which produced incorrect values 61% of the time, leading to 39% yield for this sensor and 88% over all data logger sensors.

Accuracy is a measure of how closely the sensor data obtained corresponds to the underlying physical phenomena being sensed. As the data logger results in figure 9 show, calibration is needed. Discretisation due to the 12 bit resolution causes some information loss that is offset by sampling frequently. The system is currently being calibrated against a newer data logger instrument with higher yield.

Robustness is particularly important for this system as the intended usage scenario involves it functioning in an environment where it may be subjected to large mechanical shocks and radio frequency interference (RFI). The activity regime here reproduces shocks roughly equivalent to normal application usage and the prototype functioned correctly throughout all trials. As yet, no RFI testing has been carried out.

Communication Range is a measure of how far the subject can roam from the monitoring station without losing communications. In line-of-sight tests, a range of 50 metres was achieved, whilst non-line-of-sight range (through several walls) was about 10 metres. Bluetooth communication will be replaced by ZigBee in the next revision of the prototype, with a view of achieving the application imposed 100m range.

Information Gain is a measure of the benefit of the system in terms of providing more (or better) information about the subject. The prototype has demonstrated two advantages. First, by being untethered, it allows data gathering to occur in the field. Second, it provides a means for real-time remote monitoring and actuation as opposed to offline data acquisition, which is the only role fulfilled by the wired data logger.

4.3 Data Analysis

A summary of temperature data obtained from all sensors for a sample run is given in the series of graphs in figure 9. Data was recorded during the experimentation using both the prototype system and a wired-in data logger. These graphs show that the two systems agree in terms of the changes in temperature. It is important to note that the shapes of the graphs, not the exact temperatures measured, are compared here as the sensors in the wireless prototype system had not yet been fully calibrated. The graphs

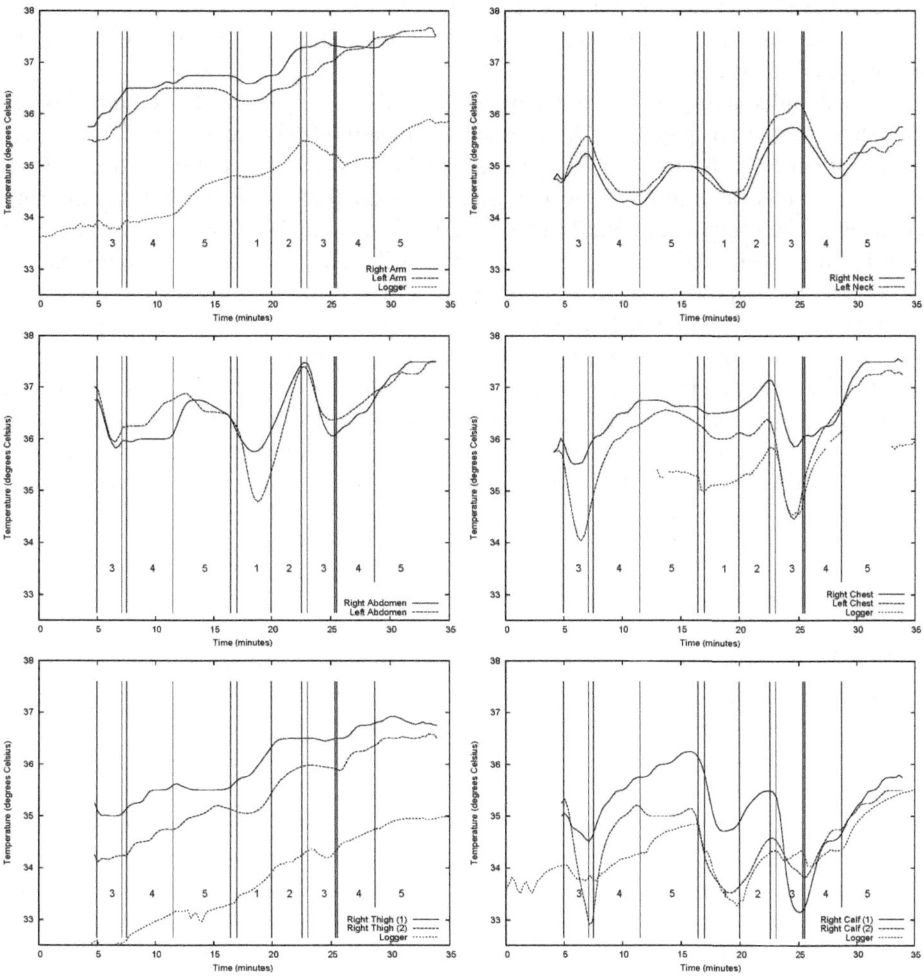

Fig. 9. Skin temperature over time for (a) arm, (b) neck, (c) abdomen, (d) chest, (e) thigh, and (f) calf sites. The two leg sensors (thigh and calf positions) were placed on the right leg only. For several skin sites, temperature values were also obtained using a wired-in data logger (denoted "Logger"). The vertical lines in each graph show the start and end of activities. Each activity is represented by a number.

show sensed temperatures over a period of time starting from the third activity (crawling) through to the fifth activity (sitting) followed by a repeat from the first (walking on a treadmill) through to the last activity again (see section 4.1). In the graphs, the start and end times for each activity are indicated by a vertical bar and the activity number (starting with 3) is given in between each set of bars. Note that there were some rest periods between activities, which are left unmarked.

The aim of the experimentation carried out was two-fold. First, the system under development was compared, in terms of the criteria discussed previously, with a commercially available, wired-in data logger. Second, the data obtained from the two systems was compared to check for consistency. The positioning of the sensors, with several locations having more than one sensor, also meant that the data from the system could be compared internally.

While detailed interpretation of the physiological meaning of the data obtained is beyond the scope of this paper, the data gathered is meaningful in the context of the developed application as follows: 1) Large variations in the skin temperature on some of the sites monitored (maximum 3°C over 30 minutes) indicate the need for both monitoring and accurate cooling actuation; 2) There are uncorrelated skin temperature variations over the sites monitored stressing the need for distributed and detailed measurement (as opposed to single point measurement performed by most developed BSN systems); 3) From the graphs, the relationship between activity and skin temperature at different sites is not an obvious one. (An example here is the sudden dip in temperature which occurs for all chest sensors during crawling (activity 3). Crawling is strenuous with the suit on, so this result is surprising.) This indicates the need for added sensing such as humidity and posture information in order to predict the physiological effects of wearing the suit during such exercise regimes. The next prototype, currently under production contains such enhanced sensor packages.

5 Conclusions and Future Work

WSN technology is clearly an enabler for detailed measurement in domains such as the one discussed in this paper, domains which are currently not sufficiently understood and lack the necessary instrumentation to further scientific investigation.

Experimental results obtained with a detailed, WSN-based temperature monitoring instrument showed that 1) under a set of activities typical to a bomb disposal mission, skin temperatures for different parts of the body (arms, thigh, chest, and so forth) vary differently thus there is value in sampling at many points; 2) skin temperatures exhibit large variations leading potentially to heat exhaustion hence the need for health monitoring of subjects; 3) autonomous feedback control of the in-suit cooling system, based on a detailed map of how temperature is changing over time, is enabled by the prototype developed so far but more work is needed to determine how best to respond to changes in temperature to ensure that the wearer is kept comfortable.

In the next revision of the prototype, it is planned to integrate temperature sensors into a multi-modal sensor board designed in-house, as shown in figure 8. Each board has a temperature sensor and an accelerometer, along with a PIC micro-controller. The two sensors allow the combined monitoring of temperature and acceleration data at any point on a subject's body. The acceleration data will be used for posture identification in further work, which will allow enhanced, activity based remote visualisation of the subject and lead to improved estimates about how the thermal state and thus comfort of the subject is changing. This will hence improve the timeliness and appropriateness of autonomous cooling decisions.

Acknowledgements. The project is kindly sponsored by NP Aerospace LTD, Coventry, UK and the Integrated Electronics Manufacturing Research Centre, Loughborough, UK. The authors wish to thank Doug Thake and his students for the use of and assistance with the Coventry University Health and Life Sciences laboratory and equipment. The authors also wish to thank Professor R.M. Newman for his contribution to hardware development in the early stages of this project.

References

1. Keoh, S.L., Dulay, N., Lupu, E., Twidle, K., Schaeffer-Filho, A.E., Sloman, M., Heeps, S., Strowes, S., Sventek, J.: Self managed cell: A middleware for managing body sensor networks. In: Proceedings of the 4th International Conference on Mobile and Ubiquitous Systems: Computing, Networking and Services (Mobiquitous), Philadelphia, USA (2007)
2. Shnayder, V., Chen, B., Lorincz, K., Fulford-Jones, T.R.F., Welsh, M.: Sensor networks for medical care. Technical report, Division of Engineering and Applied Sciences, Harvard University (2005)
3. Tapia, E.M., Marmasse, N., Intille, S.S., Larson, K.: Mites: Wireless portable sensors for studying behavior. In: Proceedings of Extended Abstracts UbiComp 2004: Ubiquitous Computing (2004)
4. Lo, B., Yang, G.Z.: Architecture for body sensor networks. In: Perspective in Pervasive Computing, pp. 23–28 (2005)
5. Jovanov, E., Raskovic, D., Price, J., Chapman, J., Moore, A., Krishnamurthy, A.: Patient monitoring using personal area networks of wireless intelligent sensors. Biomedical Sciences Instrumentation 37, 373–378 (2001)
6. Gao, T., Greenspan, D., Welsh, M., Juang, R.R., Alm, A.: Vital signs monitoring and patient tracking over a wireless network. In: 27th Annual International Conference of the IEEE EMBS, Shanghai, pp. 102–105 (2005)
7. Walker, W., Polk, T., Hande, A., Bhatia, D.: Remote blood pressure monitoring using a wireless sensor network. In: 6th Annual IEEE Emerging Information Technology Conference, Dallas, Texas (2006)
8. Vivometrics: (Vivometrics: Better results through non-invasive monitoring) (accessed November 16, 2007), `http://www.vivometrics.com`
9. Thake, C.D., Price, M.J.: Reducing uncompensable heat stress in a bomb disposal (EOD) suit: a laboratory based assessment. In: Proceedings of the 12th International Conference on Environmental Ergonomics (ICEE 2007), Piran, Slovenia (2007) ISBN 978-961-90545-1-2
10. Shanks, C.A.: Mean skin temperature during anaesthesia: An assessment of formulae in the supine surgical patient. British Journal of Anaesthesia 47, 871–876 (1975)
11. Silberstein, E., Bahra, G., Kattan, J.: Thermographically measured normal skin temperature asymmetry in the human male. Cancer 36, 1506–1510 (1975)
12. Kemp, J., Gaura, E.I., Brusey, J., Thake, C.D.: Using body sensor networks for increased safety in bomb disposal missions. In: Proceedings of the IEEE International Conference on Sensor Networks, Ubiquitous, and Trustworthy Computing (SUTC2008), Taichung, Taiwan, pp. 81–89. IEEE Computer Society, Los Alamitos (2008)

Hybrid Modeling for Set-Membership State Estimation with Uncertain Nonlinear Continuous-Time Systems

Nacim Meslem[1], Nacim Ramdani[2], and Yves Candau[1]

[1] Université Paris Est, CERTES EA 3481, 61 av. G. de Gaulle, 94010 Créteil, France
[2] INRIA Sophia Antipolis - Méditerranée and LIRMM CNRS Univ. Montpellier 2
161 rue Ada, 34392 Montpellier Cedex 5, France
{meslem, candau}@univ-paris12.fr,
nacim.ramdani@inria.fr

Abstract. This work is about state estimation in the bounded error context for non linear continuous time systems. The main idea is to seek to estimate not an optimal value for the unknown state vector but the set of feasible values, thus to characterize simultaneously the value of the vector and its uncertainty. Our contribution resides in the use of comparison theorems for differential inequalities and the analysis of the monotonicity of the dynamical systems with respect to the uncertain variables. The uncertain dynamical system is then bracketed between two hybrid dynamical systems. We show how to obtain this systems and to use them for state estimation with a *prediction-correction* type observer. An example is given with bioreactors.

Keywords: State estimation, non linear systems, bounded error context, non linear differential equations, interval analysis, guaranteed numerical integration, hybrid automata.

1 Introduction

State estimation with continuous dynamic systems is recognized as problem of great importance in practice. Indeed, to apply advanced methods for the control or the diagnosis of dynamical systems one often needs to compute on-line their internal state. Generally the direct measurement of this state by means of sensors may not be available for various reasons such as physical, practical, economic, ... etc. However, it is possible to carry out this task by software sensors, i.e. observer or estimator which can, under certain conditions of observability [1, 2], provide on-line an estimate of the real state system.

In fact, there are always uncertainties in the mathematical models used for characterizing studied systems. Consequently, the classical approaches for building observers are insufficient [3]. Thus, a new approach was developed recently in a deterministic set-membership context, which aims to reconstruct all the state trajectories which are consistent with both the uncertain models and the uncertain measurements.

This approach can be used easily when measurements are available at discrete time. It is of *prediction-correction* type: (*i*) The prediction phase consists in computing a guaranteed (conservative) over-approximation of the reachable state space generated by the uncertain system, (*ii*) and the correction phase consists in removing from this over-approximation, at each time a measurement is available, all the part which are not consistent with feasible measurements domains.

J.A. Cetto et al. (eds.): Informatics in Control, Automation and Robotics, LNEE 37, pp. 255–266.
springerlink.com

In the literature, several geometrical forms are used to implement this set-membership approach with linear systems. For example, parallelotopes [4], ellipsoidal [5] and zonotopes [6].

With nonlinear systems, a guaranteed numerical integration method for ordinary differential equations (ODE) [7] based on intervals Taylor models [8] was used recently to solve this estimation problem [9]. Generally, the wrapping effect [8, 7] associated with the interval representation of uncertain variables limits considerably their use in order to deal with practical cases. Thereafter, in order to circumvent the wrapping effect, the authors of [10] used the Müller's existence theorem [11, 12] as a tool for deriving a guaranteed bracketing for an uncertain dynamical system between two deterministic dynamical systems. The main difficulty resides in the definition of the bracketing systems. Our contribution resides in the continuation of these work since we show how to build deterministic hybrid dynamical systems as bracketing systems and thus make it possible to use the Müller theorem with a larger class of nonlinear dynamical systems.

This work is organized as follow. In section 2 we present the context and the main ideas of set-membership estimation. Then, we will recall the Müller's theorem in section 3. We introduce the hybrid bracketing approach for uncertain dynamical systems in section 4. Finally we illustrate our method on a model taken from the biotechnology domain in section 5.

2 Problem Statement

2.1 Some Definitions of Interval Analysis

Interval analysis was initially developed to account for the quantification errors introduced by the floating point representation of real numbers with computers and was extended to validated numerics [13]. A real interval $[x] = [\underline{x}, \overline{x}]$ is a connected and closed subset of $\mathbb{R}$. We have $\text{Inf}([x]) = \underline{x}$ and $\text{Sup}(x) = \overline{x}$. The set of all real intervals of $\mathbb{R}$ is denoted by $\mathbb{IR}$. Real arithmetic operations are extended to intervals. Consider an operation $\circ \in \{+, -, *, \div\}$ and two intervals $[x]$ and $[y]$. Then:

$$[x] \circ [y] = [\mathtt{inf}_{u\in[x],v\in[y]}(u \circ v),\ \mathtt{sup}_{u\in[x],v\in[y]}(u \circ v)] \ . \tag{1}$$

An interval vector is the Cartesian product of n intervals. The set of n-dimensional real interval vectors is denoted by $\mathbb{IR}^n$. For a n-dimensional real vector $\mathbf{z} \in \mathbb{R}^n$ and n-dimensional real interval vectors $[\mathbf{x}]$, $[\mathbf{y}] \in \mathbb{IR}^n$, we have $\mathbf{z} \in [\mathbf{x}] \Leftrightarrow \forall i \ z_i \in [x_i]$ and $[\mathbf{x}] \subseteq [\mathbf{y}] \Leftrightarrow \forall i \ [x_i] \subseteq [y_i]$.

Consider $\mathbf{g} : \mathbb{R}^n \mapsto \mathbb{R}^m$; the range of this function over an interval vector $[\mathbf{x}]$ is given by:

$$\mathbf{g}([\mathbf{x}]) = \{\mathbf{g}(\mathbf{x}) \mid \mathbf{x} \in [\mathbf{x}]\} \ . \tag{2}$$

The interval function $[\mathbf{g}] : \mathbb{IR}^n \mapsto \mathbb{IR}^m$ is an inclusion function for $\mathbf{g}$ if

$$\forall [\mathbf{x}] \in \mathbb{IR}^n,\ \mathbf{g}([\mathbf{x}]) \subseteq [\mathbf{g}]([\mathbf{x}]) \ . \tag{3}$$

An inclusion function for $\mathbf{g}$ can obtained by replacing each occurrence of a real variable by the corresponding interval and each standard function by its interval counterpart. The resulting function is called the natural inclusion function. The performances of this inclusion function depend on the formal expression for $\mathbf{g}$.

2.2 Context of Set-Membership State Estimation

Let us consider the uncertain continuous dynamic system (4) where uncertainties are naturally represented by bounded intervals with *a priori* known bounds,

$$\left\{\begin{array}{ll} \dot{\mathbf{x}}(t) & = \mathbf{f}(\mathbf{x}, \mathbf{p}, \mathbf{u}(t)) \\ \mathbf{y}(t) & = \mathbf{g}(\mathbf{x}, \mathbf{p}, \mathbf{u}(t)) \\ \mathbf{x}(t_0) & \in [\mathbf{x}_0] \subset \mathbb{D} \\ \mathbf{p} & \in [\mathbf{p}] \end{array}\right\} \tag{4}$$

where $t \in [t_0, T]$, $\mathbf{f} \in C^{k-1}(\mathbb{D} \times \mathbb{U} \times [\mathbf{p}])$, $\mathbb{D} \times \mathbb{U} \times [\mathbf{p}] \subseteq \mathbb{R}^{n+n_u+n_p}$ is an open set; n, n_u, m and n_p are the dimension of respectively the state vector $\mathbf{x}$, the input vector $\mathbf{u}$, the output vector $\mathbf{y}$ and the parameter vector $\mathbf{p}$. The functions $\mathbf{f} : \mathbb{D} \times \mathbb{U} \times [\mathbf{p}] \rightarrow \mathbb{R}^n$ and $\mathbf{h} : \mathbb{D} \times \mathbb{U} \times [\mathbf{p}] \rightarrow \mathbb{R}^m$ are possibly nonlinear. The initial state $\mathbf{x}_0$ is assumed to belong to a prior known set $[\mathbf{x}_0]$. We assume that measurements $\mathbf{y}_j$ of the output vector are available at sampling times $t_i \in \{t_1, t_2, \ldots, t_n\}$ in $I = [t_0, t_{n_T}]$. Note that the sampling interval needs not be constant. The measurement noise is a discrete time signal assumed additive and bounded with known bounds. Denote by $\mathbb{E}_j$ a feasible domain for output error at time t_j. So the feasible domain for output model at time t_j is then given by

$$\mathbb{Y}_j = \mathbf{y}_j + \mathbb{E}_j \ . \tag{5}$$

Under these considerations, estimating the state vector $\mathbf{x}$ consists in determining an over-approximation of the set $\mathbb{X}(t)$ of all acceptable state trajectories

$$\mathbb{X}(t) = \left\{\begin{array}{c} \mathbf{x}(t) \mid (\forall \mathbf{x}_0 \in [\mathbf{x}_0], \forall t \in I,\ \dot{\mathbf{x}}(t) \in [\mathbf{f}](\mathbf{x}, [\mathbf{p}], \mathbf{u}(t))) \\ \wedge \\ (\forall t_j \in \{t_1, t_2, \ldots, t_{nT}\},\ \mathbf{x}(t_j) \in [\mathbf{g}]^{-1}(\mathbb{Y}(t_j), [\mathbf{p}], \mathbf{u}(t_j))) \end{array}\right\} . \tag{6}$$

2.3 Principle: Prediction-Correction Method

Figure 1 shows the principle of the prediction and correction phases for this set-member-ship approach between two successive time measurement indexes t_j and t_{j+1}. Indeed, by using one of the guaranteed numerical simulation methods for uncertain ODEs, the *prediction* phase computes a guaranteed over-enclosure $[\mathbf{x}_{j+1}]^p$ for the set $\mathcal{X}(t_{j+1}; t_j, [\mathbf{x}_j])$ of all the solutions of (4) at the time t_{j+1} with $(t_j, [\mathbf{x}_j])$ as initial conditions,

$$\mathcal{X}(t_{j+1}; t_j, [\mathbf{x}_j]) \subseteq [\mathbf{x}_{j+1}]^p \ . \tag{7}$$

The *correction* phase uses set inversion, consistency techniques and intervals analysis [13] to characterize the reciprocal image $[\mathbf{x}_{j+1}]^{inv}$ at time t_{j+1} of the admissible measurement domain $\mathbb{Y}_{j+1}$ by the output model $\mathbf{g}$

$$[\mathbf{x}_{j+1}]^{inv} = [\mathbf{g}]^{-1}([\mathbf{y}_{j+1}], [\mathbf{p}]) \tag{8}$$

where $[\mathbf{y}_{j+1}] = \mathbb{Y}_{j+1}$.

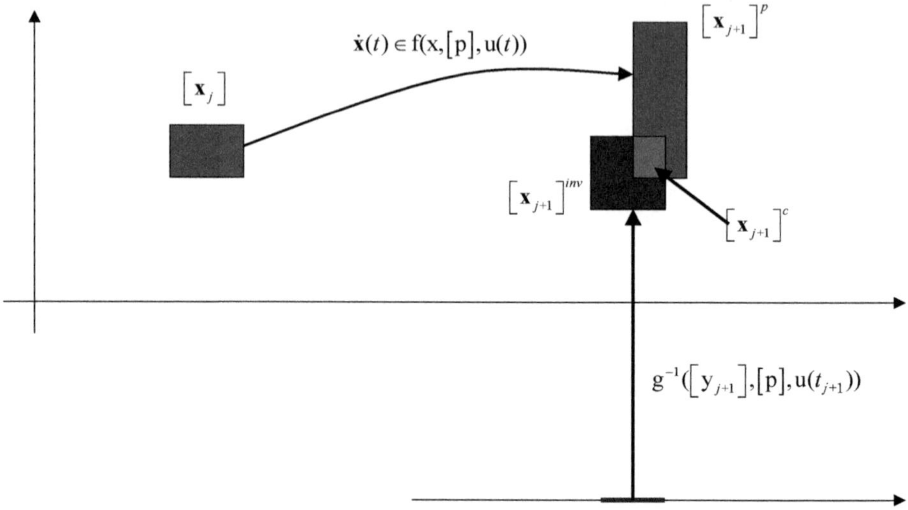

Fig. 1. Prediction and correction phases

Thereafter, it contracts the predicted state intervals vector by comparing it with the reciprocal image $[\mathbf{x}_{j+1}]^{inv}$ and eliminating the inconsistent state vectors

$$[\mathbf{x}_{j+1}]^c = [\mathbf{x}_{j+1}]^{inv} \cap [\mathbf{x}_{j+1}]^p \ . \tag{9}$$

Thus for the next measurement the prediction phase will be initialized by $[\mathbf{x}_{j+1}] = [\mathbf{x}_{j+1}]^c$. In fact, by repeating these two phases each time a new measurement is available, one improves considerably the precision of the guaranteed over approximation of the set $\mathbb{X}(t)$.

The algorithm below shows the process for this set-membership estimation approach.

Algorithm. Prediction-Correction-Estimation.

1. **Input**: $([\mathbf{x}_0], [\mathbf{p}], \mathbf{f}, \mathbf{g}, [\mathbf{y}_1], \ldots, [\mathbf{y}_{nT}])$
2. $t_j = t_0$; $[\mathbf{x}_j] = [\mathbf{x}_0]$;
3. **while** $(t_j < t_{nT})$ **do**
4. $\{t_{j+1}, [\mathbf{x}_{j+1}]^p\}$=**Validated-Integration**$([\mathbf{x}_j], [\mathbf{p}], t_j)$;
5. $[\mathbf{x}_{j+1}]^{inv} = [\mathbf{g}]^{-1}([\mathbf{y}_{j+1}], [\mathbf{p}], \mathbf{u}(t_{j+1}))$;
6. $[\mathbf{x}_{j+1}]^c = [\mathbf{x}_{j+1}]^{inv} \cap [\mathbf{x}_{j+1}]^p$;
7. $[\mathbf{x}_{j+1}] = [\mathbf{x}_{j+1}]^c$
8. $j = j + 1$;
9. **end**
10. **Output**: $\mathbb{X}(t)$.

3 Müller's Theorem

In this section, we introduce an approach for bracketing an uncertain dynamical systems when both the initial state and parameter vectors are defined by boxes (interval vectors). The main idea consists in building a lower and an upper dynamical system which involve no uncertainty and enclose in a guaranteed way, all the state trajectories generated by the original uncertain system. This approach relies on comparison theorems for differential inequalities [14, 15], and in particular the work of Müller [11, 16]. Below, we present the Müller's theorem as given in [10].

Theorem 1. *Consider the dynamical system*

$$\dot{\mathbf{x}}(t) = \mathbf{f}(\mathbf{x}, \mathbf{p}, t), \tag{10}$$

where function $\mathbf{f}$ *is continuous over a domain* $\mathbb{T}$ *defined by*

$$\mathbb{T} : \begin{cases} \omega(t) \leq \mathbf{x}(t) \leq \Omega(t) \\ \underline{\mathbf{p}} \leq \mathbf{p} \leq \overline{\mathbf{p}} \\ t_0 \leq t \leq t_{n_T} \end{cases} \tag{11}$$

Functions $\omega_i(t)$ *and* $\Omega_i(t)$ *are continuous over* $[t_0, t_{n_T}]$ *for all* i *and satisfy the following properties*

1. $\omega(t_0) = \underline{\mathbf{x}}_0$ *and* $\Omega(t_0) = \overline{\mathbf{x}}_0$
2. *the left derivatives* $D^-\omega_i(t)$ *and* $D^-\Omega_i(t)$ *and the right derivatives* $D^+\omega_i(t)$ *and* $D^+\Omega_i(t)$ *of* $\omega_i(t)$ *and* $\Omega_i(t)$ *are such that*

$$\forall i,\ D^{\pm}\omega_i(t) \leq \min_{\underline{\mathbb{T}}(t)} f_i(\mathbf{x}, \mathbf{p}, t) \tag{12}$$

$$\forall i,\ D^{\pm}\Omega_i(t) \geq \max_{\overline{\mathbb{T}}(t)} f_i(\mathbf{x}, \mathbf{p}, t) \tag{13}$$

where $\underline{\mathbb{T}}(t)$ *is the subset of* $\mathbb{T}(t)$ *defined by*

$$\underline{\mathbb{T}}_i : \begin{cases} x_i = \omega_i(t) \\ \omega_j(t) \leq x_j \leq \Omega_j(t),\ j \neq i \\ \underline{\mathbf{p}} \leq \mathbf{p} \leq \overline{\mathbf{p}} \end{cases} \tag{14}$$

and where $\overline{\mathbb{T}}(t)$ *is the subset of* $\mathbb{T}(t)$ *defined by*

$$\overline{\mathbb{T}}_i : \begin{cases} x_i = \Omega_i(t) \\ \omega_j(t) \leq x_j \leq \Omega_j(t),\ j \neq i \\ \underline{\mathbf{p}} \leq \mathbf{p} \leq \overline{\mathbf{p}} \end{cases} \tag{15}$$

Then for all $\mathbf{x}_0 \in [\underline{\mathbf{x}}_0, \overline{\mathbf{x}}_0]$, $\mathbf{p} \in [\underline{\mathbf{p}}, \overline{\mathbf{p}}]$, *system 4 admits a solution* $\mathbf{x}(t)$ *that stays in the domain*

$$\mathbb{X} : \begin{cases} t_0 \leq t \leq t_{n_T} \\ \omega(t) \leq \mathbf{x}(t) \leq \Omega(t) \end{cases} \tag{16}$$

and takes the value $\mathbf{x}_0$ *at* t_0. *If, in addition, for all* $\mathbf{p} \in [\underline{\mathbf{p}}_0, \overline{\mathbf{p}}_0]$, *function* $\mathbf{f}(\mathbf{x}, \mathbf{p}, t)$ *is Lipschitzian with respect to* $\mathbf{x}$ *over* $\mathbb{D}$ *then this solution is unique for any given* $\mathbf{p}$.

Finally, an enclosure for the solution of 10 is given by

$$\forall t \in [t_0, t_{n_T}], \quad [\mathbf{x}](t) = [\omega(t), \Omega(t)] \,. \tag{17}$$

The main difficulty is how to obtain suitable bracketing functions $\omega(t)$ and $\Omega(t)$ in the general case. However, when the components of $\mathbf{f}$ are monotonic with respect to each parameter and each state vector component, it is quite easy to define these systems [17], while avoiding possible divergence that may occur when both upper and lower components of the parameter/state vector appear simultaneously in the same expression of the components of the bracketing systems [18].

Proposition 1. *Use of Monotonicity Property. [17]*
In order to build the upper system, i.e. the one which yields the upper solution $\Omega(t)$, *one can replace in the formal expression of* f_i, x_i *by* Ω_i, x_j $(j \neq i)$ *by* Ω_j *if* $\frac{\partial f_i}{\partial x_j} \geq 0$ *or by* ω_j *if* $\frac{\partial f_i}{\partial x_j} \leq 0$ *and* p_k *by* $\overline{p}_k$ *if* $\frac{\partial f_i}{\partial p_r} \geq 0$ *or by* $\underline{p}_k$ *if* $\frac{\partial f_i}{\partial p_k} \leq 0$. *The components of the lower system, i.e. the one which yields the lower solution* $\omega(t)$ *are derived in the same way by reversing monotonicity conditions.*

Obviously $\omega(t)$ and $\Omega(t)$ are in general, solutions of a system of coupled differential equations, i.e.

$$\begin{cases} \dot{\omega}(t) = \underline{\mathbf{f}}(\omega, \Omega, \underline{\mathbf{p}}, \overline{\mathbf{p}}, t) \\ \dot{\Omega}(t) = \overline{\mathbf{f}}(\omega, \Omega, \underline{\mathbf{p}}, \overline{\mathbf{p}}, t) \\ \omega(t_0) = \underline{\mathbf{x}}_0 \\ \Omega(t_0) = \overline{\mathbf{x}}_0 \end{cases} \tag{18}$$

which involves no *uncertain* quantity. Therefore interval Taylor models such as the one presented in [7] can be used for efficiently solving 18. Indeed when these methods are used for solving differential equations with no uncertainty, they are usually able to curb the pessimism induced by the wrapping effect, even over long integration time.

4 Hybrid Bracketing System

Now, one addresses the case of uncertain dynamical systems 4, for which the signs of the partial derivatives $\partial f_i/\partial p_k$ and $\partial f_i/\partial x_j$ change along the integration time interval $[t_0, t_{n_T}]$. In such a case, the uncertain system (4) admits an enclosure over each time interval where functions f_i are monotonic with respect to variables p_k $(k \in \{1, \ldots, n_p\})$ and x_j $(j \in \{1, \ldots, n\}, j \neq i)$. Therefore both upper and lower bounding systems are defined by piecewise nonlinear ODEs and can thus be regarded as hybrid dynamical systems. Thus, they can be modeled by a *hybrid automaton*[19].

So to compute a guaranteed enclosure of the reachable state space generated by the uncertain system (4), we will build a hybrid system composed of l continuous dynamic modes which satisfies locally the conditions imposed by proposition 1

$$\mathbb{M} = \{M_1, M_2, ..., M_l\} \tag{19}$$

and which is given in state space representation by

$$\begin{cases} \dot{\omega}(t) = \underline{\mathbf{f}}_{M_i}(\omega, \Omega, \underline{\mathbf{p}}, \overline{\mathbf{p}}, t) \\ \dot{\Omega}(t) = \overline{\mathbf{f}}_{M_i}(\omega, \Omega, \underline{\mathbf{p}}, \overline{\mathbf{p}}, t) \end{cases} \quad for \;\; i = 1, ..., l \, . \tag{20}$$

Hence, the evolution of this hybrid system is controlled by the sign changes of the partial derivatives $\partial f_i / \partial p_k$ and $\partial f_i / \partial x_j$ which represents the guard conditions which authorize the transitions between the continuous bracketing modes. Thus, for a given initial conditions the execution of this hybrid automata makes it possible to obtain a guaranteed over approximation of the reachable state space of the continuous time system (4).

Example 1. Let us consider the following system

$$\dot{x}(t) = f(x, [p], t) \, . \tag{21}$$

According to the sign of $\partial f / \partial p$, the system (21) admitted two possible bracketing modes

if $\partial f / \partial p \leq 0$

$$\text{then } M_1 = \begin{cases} \dot{\Omega}(t) = f_{M_1}(\Omega, \underline{p}, t) \\ \dot{\omega}(t) = f_{M_1}(\omega, \overline{p}, t) \end{cases} \tag{22}$$

$$\text{else } M_2 = \begin{cases} \dot{\Omega}(t) = f_{M_2}(\Omega, \overline{p}, t) \\ \dot{\omega}(t) = f_{M_2}(\omega, \underline{p}, t) \end{cases} \tag{23}$$

and its hybrid bracketing automata is represented in the figure 2.

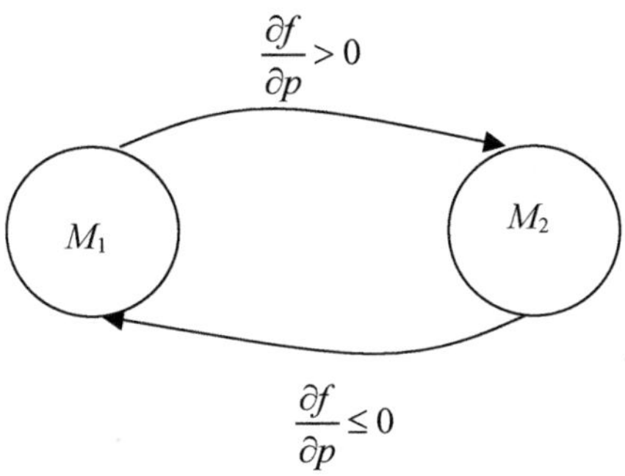

Fig. 2. Hybrid automata

5 Application

5.1 Model

Generally, the mathematical model of biotechnological processes is difficult to establish with accuracy. This is due to the living behavior of the bacteria represented by a complex poorly known function of the bioreactor state. In this section we consider a

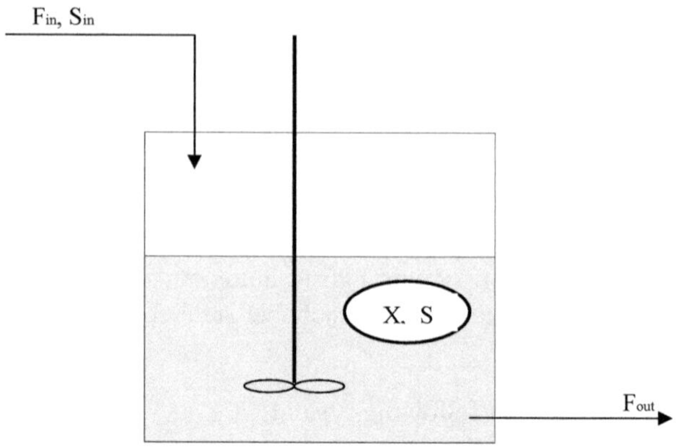

Fig. 3. General representation of a bioreactor

simple model where only one population of bacteria is taken into account. In this context, to describe the state of the bioreactor, two state variables are necessary, the first one represents the bacteria concentration called biomass and denoted X, the second one represents the substrate concentration, denoted S.

Thus the model below shows the evolution of the biomass by consuming the polluting body S

$$\begin{cases} \dot{X}(t) = \mu(S)X - \alpha DX \\ \dot{S}(t) = -k_1\mu(S)X + D(S_{in} - S) \end{cases} \tag{24}$$

where $\mu(s)$ is the growth rate of biomass modeled by the Haldane law:

$$\mu(S) = \mu_0 \frac{S}{S + k_s + S^2/k_i} \tag{25}$$

with the uncertain bounded parameter μ_0

$$\underline{\mu}_0 \leq \mu_0 \leq \overline{\mu}_0 \,.$$

This bioreactor is fed by a solution containing substrate in concentration S_{in} which is not exactly measured

$$\underline{S}_{in}(t) \leq S_{in}(t) \leq \overline{S}_{in}(t)$$

and we suppose that the biomass is accessible to measurement

$$y(t) = X(t) \,.$$

5.2 Building Hybrid Bracketing System

For this application, according to the sign of the derivative of μ with respect to S,

$$\text{sign}(\frac{\partial \mu(S)}{\partial S}) \begin{cases} > 0 \text{ if } \forall S \in [S],\ S < \sqrt{k_s k_i} \\ < 0 \text{ if } \forall S \in [S],\ S > \sqrt{k_s k_i} \\ \text{ambiguous if } \sqrt{k_s k_i} \in [S] \end{cases} \tag{26}$$

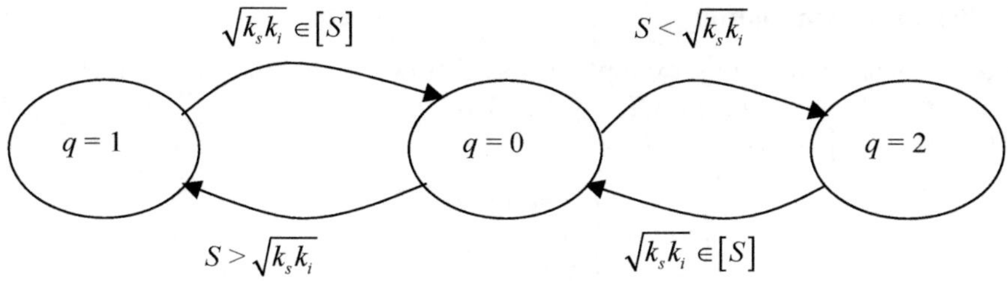

Fig. 4. The hybrid automata used in prediction phase

system (24) allows three possible modes for the bracketing. The first mode, $M_1 = 1$, corresponds to the intervals time when this derivative is negative

$$(M_1 = 1) \begin{cases} \dot{\overline{X}}(t) = \overline{\mu}(\overline{S})\overline{X} - \alpha D\overline{X} \\ \dot{\overline{S}}(t) = -k_1\underline{\mu}(\overline{S})\underline{X} + D(\overline{S}_{in} - \overline{S}) \\ \dot{\underline{X}}(t) = \underline{\mu}(\underline{S})\underline{X} - \alpha D\underline{X} \\ \dot{\underline{S}}(t) = -k_1\overline{\mu}(\underline{S})\overline{X} + D(\underline{S}_{in} - \underline{S}) \end{cases} \tag{27}$$

and the second mode, $M_2 = 2$, is linked to intervals time where this derivative is positive

$$(M_2 = 2) \begin{cases} \dot{\overline{X}}(t) = \overline{\mu}(\underline{S})\overline{X} - \alpha D\overline{X} \\ \dot{\overline{S}}(t) = -k_1\underline{\mu}(\overline{S})\underline{X} + D(\overline{S}_{in} - \overline{S}) \\ \dot{\underline{X}}(t) = \underline{\mu}(\overline{S})\underline{X} - \alpha D\underline{X} \\ \dot{\underline{S}}(t) = -k_1\overline{\mu}(\underline{S})\overline{X} + D(\underline{S}_{in} - \underline{S}) \,. \end{cases} \tag{28}$$

Finally, the third mode $M_3 = 0$ is associated to case when the sign of this derivative is ambiguous. In this case either one uses guaranteed integration methods based on interval Taylor models to bracket 24, or if this is possible, one finds a trivial bracketing for $\mu(s)$. For example,

$$\underline{\mu}_0 \frac{\underline{S}}{\underline{S} + k_s + \underline{S}\overline{S}/k_i} \leq \mu(S) \leq \overline{\mu}_0 \frac{\overline{S}}{\overline{S} + k_s + \underline{S}\overline{S}/k_i} \,.$$

Hence, one proposes the differential equations system below for the third mode $M_3 = 0$

$$(M_3 = 0) \begin{cases} \dot{\overline{X}}(t) = \overline{\mu}_0 \frac{\overline{S}}{\overline{S} + k_s + \underline{S}\overline{S}/k_i}\overline{X} - \alpha D\overline{X} \\ \dot{\overline{S}}(t) = -k_1\underline{\mu}(\overline{S})\underline{X} + D(\overline{S}_{in} - \overline{S}) \\ \dot{\underline{X}}(t) = \underline{\mu}_0 \frac{\underline{S}}{\underline{S} + k_s + \underline{S}\overline{S}/k_i}\underline{X} - \alpha D\underline{X} \\ \dot{\underline{S}}(t) = -k_1\overline{\mu}(\underline{S})\overline{X} + D(\underline{S}_{in} - \underline{S}) \,. \end{cases} \tag{29}$$

Now, function **Validated-Integration** in algorithm **Prediction-Correction-Estimation** selects on-the-fly the local bracketing mode according to the sign of $\frac{d\mu(.)}{dS}$ (see Figure 4) and then uses guaranteed numerical integration methods for ODE based on intervals Taylor models for solving (27), (28) or (29).

5.3 Results of Simulation

The data considered in this example are as follows: $\alpha = 0.5$, $k = 42.14$, $k_s = 9.28 mmol/l$, $k_i = 256 mmol/l$, $\mu_0 \in [0.70, 0.78]$, $X_0 \in [3, 7]$, $S_0 \in [35, 45]$, $S_{in}(t) \in ([63, 67] + 15\cos(1/5t))$,

$$D(t) = \begin{cases} 2 \; si \; 0 \leq t \leq 5 \\ 0.5 \; si \; 5 \leq t \leq 10 \\ 1.14 \; si \; 10 \leq t \leq 20, \end{cases}$$

and the feasible measurement domain

$$\mathbb{Y}(t_j) = [0.98 y_m(t_j), 1.02 y_m(t_j)]$$

with a constant measurements time step $t_{j+1} - t_j = 2h$.

The red continuous lines curves in figures 5(a) and 6(a) show the guaranteed enclosure of all the possible state trajectories of (24) which are compatible with the model and its uncertainties and the acceptable domain of the discrete measurements signal. The discontinuous blue curves represent the real state of (24) which corresponds to the following values of the uncertain parameters and the initial state: $\mu_0 = 0.74$, $X_0 = 5$, $S_0 = 40$ and $S_{in}(t) = 65 + 15\cos(1/5t)$.

So figure 7 shows the actual commutations between the three bracketing modes as obtained during the simulation period. This represents the evolution of the discrete component of the hybrid automata used to bracket the state flow generated by the uncertain system 24.

As a conclusion, guaranteed numerical integration methods for ODE based on the interval Taylor models fail to give non divergent enclosures after few integration step because of the large magnitude in uncertainty in both parameter vector and initial state vector, see red line curves in figures 5(b) and 6(b). In addition, proposition 1 is not applicable over all the simulation period because the sign of the partial derivative $\frac{\partial \dot{X}}{\partial S}$ changes with time. Hence, our hybrid bracketing method is an important alternative to solve set membership estimation problems of this kinds.

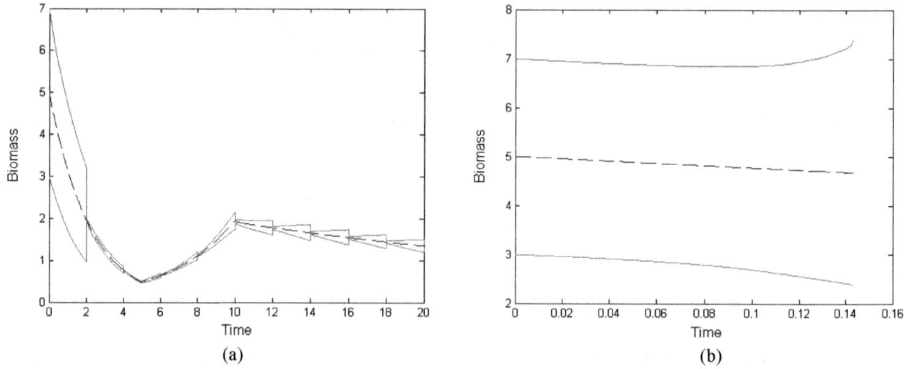

Fig. 5. Both real and estimated biomass evolution. (a) hybrid method, (b) first steps of interval Taylor method.

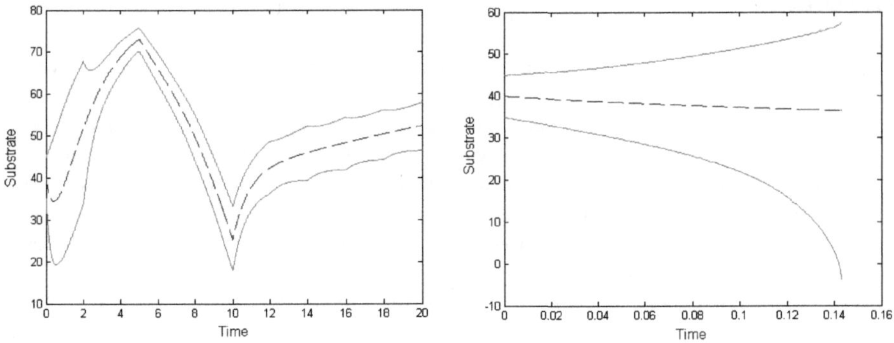

(a) (b)

Fig. 6. Both real and estimated substrate evolution. (a) hybrid method, (b) first steps of interval Taylor method.

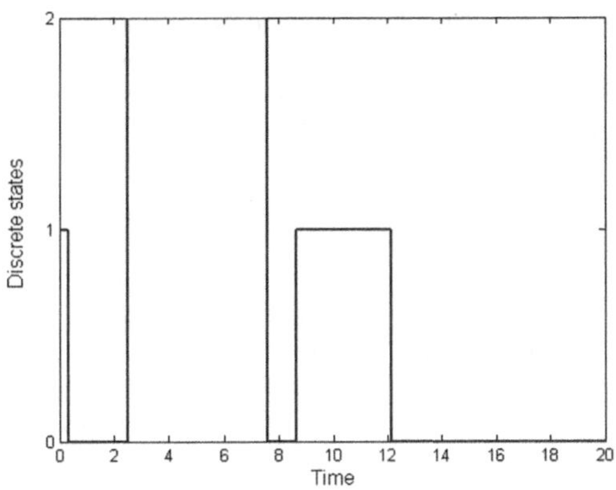

Fig. 7. Indices of the three bracketing modes used over the simulation period

6 Conclusions

In this paper, we wanted to show that by using the Müller's theorem and by analyzing the monotonicity of the uncertain dynamical system with respect to both the uncertain variables and the parameters, one is able to solve the set-membership state estimation problem for a large class of uncertain nonlinear continuous-time dynamical systems. Indeed, the method presented makes it possible to circumvent the propagation of the pessimism due to the wrapping effect which generally causes the divergence of guaranteed numerical integration methods based on interval Taylor models when used with uncertain ODEs. In the future, we wish to extend this approach for systems with higher dimension and also to hybrid dynamical systems.

References

1. Hermann, R., Krener, A.J.: Nonlinear controllability and observability. Transactions on Automatic Control AC-22, 728–740 (1977)
2. Hermann, R.: On the accessibility problem in control theory. In: Int. Symp. on nonlinear differential equations and nonliear mechanics, pp. 325–332. Academic Press, New York (1963)
3. Dochain, D.: State and parameter estimation in chemical and biochemical processes: a tutorial. Journal of process control 13, 801–818 (2003)
4. Chisci, L., Garulli, A., Zappa, G.: Recursive state bounding by parallelotopes. Automatica 32, 1049–1055 (1996)
5. Chernousko, F.L.: Ellipsoidal state estimation for dynamical systems. Nonlinear analysis (2005)
6. Combastel, C.: A state bounding observer for uncertain nonlinear continuous-time systems based on zonotopes. In: 44th IEEE Conference on decision and control and European control conference ECC 2005 (2005)
7. Nedialkov, N.: Computing rigorous bounds on the solution of an initial value problem for an ordinary differential equation. PhD University of Toronto (1999)
8. Moore, R.: Interval analysis. Prentice-Hall, Englewood Cliffs (1966)
9. Raïssi, T., Ramdani, N., Candau, Y.: Set membership state and parameter estimation for systems described by nonlinear differential equations. Automatica 40(10), 1771–1777 (2004)
10. Kieffer, M., Walter, E.: Guaranteed nonlinear state estimation for continuous-time dynamical models from discrete-time measurements. In: Proceedings 6th IFAC Symposium on Robust Control, Toulouse (2006)
11. Müller, M.: Uber das fundamentaltheorem in der theorie der gewöhnlichen differentialgleichungen. Math. Z. 26, 619–645 (1926)
12. Walter, W.: Differential inequalities and maximum principles: Theory, new methods and applications. Nonlinear Analysis, Theory, Methods & Applications 30(8), 4695–4711 (1997)
13. Jaulin, L., Kieffer, M., Didrit, O., Walter, E.: Applied interval analysis: with examples in parameter and state estimation, robust control and robotics. Springer, London (2001)
14. Smith, H.: Monotone dynamical systems: An introduction to the theory of competitive and cooperative systems. Ams. providence, ri edn. (1995)
15. Hirsch, M., Smith, H.: Monotone dynamical systems. In: Canada, A., Drabek, P., Fonda, A. (eds.) Handbook of Differential Equations, Ordinary Differential Equations, vol. 2. Elsevier, Amsterdam (2005)
16. Marcelli, C., Rubbioni, P.: A new extension of classical Müller theorem. Nonlinear Analysis, Theory, Methods & Applications 28(11), 1759–1767 (1997)
17. Kieffer, M., Walter, E., Simeonov, I.: Guaranteed nonlinear parameter estimation for continuous-time dynamical models. In: Proceedings 14th IFAC Symposium on System Identification, Newcastle, Aus., pp. 843–848 (2006)
18. Ramdani, N., Meslem, N., Raïssi, T., Candau, Y.: Set-membership identification of continuous-time systems. In: Proceedings 14th IFAC Symposium on System Identification, Newcastle, Aus., pp. 446–451 (2006)
19. Alur, R., Courcoubetis, C., Halbwachs, N., Henzinger, T., Ho, P.H., Nicollin, X., Olivero, A., Sifakis, J., Yovine, S.: The algorithmic analysis of hybrid systems. Theoretical Computer Science 138, 3–34 (1995)

An Embedded Attitude and Heading Reference System Based on a Nonlinear Filter

Philippe Martin and Erwan Salaün

École des Mines de Paris, Centre Automatique et Systèmes
60 boulevard Saint-Michel, 75272 Paris Cedex 06, France
{philippe.martin,okerwan.salaun}@ensmp.fr

Abstract. In this paper we propose a nonlinear filter for estimating the orientation of a flying rigid body, using measurements from low-cost inertial and magnetic sensors. It has by design a nice geometrical structure appealing from an engineering viewpoint; it is easy to tune, computationally very economic, and with guaranteed (at least local) convergence around every trajectory. Moreover it behaves sensibly in the presence of acceleration and magnetic disturbances. We illustrate its good performance on experimental comparisons with a commercial system, and demonstrate its simplicity by implementing it on an 8-bit microcontroller.

1 Introduction

Aircraft, especially Unmanned Aerial Vehicles (UAV), commonly need to know their orientation to be operated, whether manually or with computer assistance. When cost or weight is an issue, using very accurate inertial sensors for "true" (i.e. based on the Schuler effect due to a non-flat rotating Earth) inertial navigation is excluded. Instead, low-cost systems –often called Attitude Heading Reference Systems (AHRS)– rely on light and cheap "strapdown" gyroscopes, accelerometers and magnetometers. The various measurements are "merged" according to the motion equations of the aircraft assuming a flat non-rotating Earth, usually with some kind of Extended Kalman Filter (EKF). For more details about avionics, various inertial navigation systems and sensor fusion, see for instance [2, 3, 5] and the references therein.

While the EKF is a general method capable of good performance when properly tuned, it suffers several drawbacks: it is not easy to choose the numerous parameters; it is computationally expensive, which is a problem in low-cost embedded systems; it is usually difficult to prove the convergence, even at first-order, and the designer has to rely on extensive simulations.

An alternative route is to use an ad hoc nonlinear observer as proposed in [4, 6, 7, 8, 10]. In the absence of a general method the main difficulty is of course to find such an observer. In this paper we use the rich geometric structure of the attitude-heading problem to derive an observer by the method developed in [1], building up on the preliminary work [8]. It has by design a nice geometrical structure appealing from an engineering viewpoint; it is easy to tune, computationally very economic, and with guaranteed (at least local) convergence around every trajectory. Moreover it behaves sensibly in the presence of acceleration and magnetic disturbances. We illustrate its good performance

J.A. Cetto et al. (eds.): Informatics in Control, Automation and Robotics, LNEE 37, pp. 267–281.
springerlink.com

on experimental comparisons with the commercial Microbotics MIDG II system, and demonstrate its simplicity by implementing it on an 8-bit Atmel AVR microcontroller.

2 The Physical System

2.1 Motion Equations

The motion of a flying rigid body (assuming the Earth is flat and defines an inertial frame) is described by

$$\dot{q} = \frac{1}{2} q * \omega$$
$$\dot{V} = A + q * a * q^{-1},$$

where:

- q is the quaternion representing the orientation of the body with respect to the Earth-fixed frame
- ω is the instantaneous angular velocity vector
- V is the velocity vector of the center of mass with respect to the Earth-fixed frame
- $A = ge_3$ is the (constant) gravity vector in North-East-Down coordinates (the unit vectors e_1, e_2, e_3 point respectively North, East, Down)
- a is the specific acceleration vector, in this case the aerodynamic forces divided by the body mass.

The first equation describes the kinematics of the body, the second is Newton's force law. It is customary to use quaternions instead of Euler angles since they provide a global parametrization of the body orientation, and are well-suited for calculations and computer simulations. For more details see [9] or any other good textbook on aircraft modeling, and section 8 for useful formulas used in this paper.

2.2 Measurements

We use three triaxial sensors providing nine scalar measurements: 3 gyros measure $\omega_m = \omega + \omega_b$, where ω_b is a constant vector bias; 3 magnetometers measure $y_B = b_s q^{-1} * B * q$, where $B = B_1 e_1 + B_3 e_3$ is the Earth magnetic field in NED coordinates and $b_s > 0$ is a constant scaling factor; 3 accelerometers measure $a_m = a_s a$, where $a_s > 0$ is a constant scaling factor. As customary in low-cost unaided attitude heading systems we assume the linear acceleration $\dot{V}$ small, hence approximate the accelerometers measurements by $y_A := a_s q^{-1} * A * q$ (the sign is reversed for convenience). All the nine measurements are of course also corrupted by noise.

There is some freedom in modeling the sensors imperfections. A simple first-order observability analysis reveals that up to six unknown constants can be estimated: besides the gyro bias ω_b, it is possible to estimate two imperfections on y_B and one on a_m, or one on y_B and two on a_m. Nevertheless it is impossible to model three imperfections on a_m: in particular if we write $a_m = a + a_b$, with a_b a constant vector bias, only two components of a_b are observable; moreover, only one imperfection on a_m can be estimated without relying on the possibly disturbed magnetic measurements. On the other hand it is also impossible to estimate the three components of the magnetic field B, but only the North and Down components.

In an AHRS it is usually desirable to use the magnetic measurements to estimate only the heading, so that a magnetic disturbance does not affect the estimated attitude. We will see this can be achieved by considering $y_C := y_A \times y_B = c_s q^{-1} * C * q$, where $c_s := a_s b_s > 0$ and $C := A \times B$, rather than the direct measurement y_B. Notice that $\langle y_A, y_C \rangle = \langle A, C \rangle = 0$, so that we are left with 8 independent measurements; as a consequence only five unknown constants can now be estimated. This is not a drawback and is even beneficial since the observer will then not depend on the latitude-varying B_3.

2.3 The Model

To design our observer we thus consider the system

$$\dot{q} = \frac{1}{2} q * (\omega_m - \omega_b) \tag{1}$$

$$\dot{\omega}_b = 0 \tag{2}$$

$$\dot{a}_s = 0 \tag{3}$$

$$\dot{c}_s = 0 \tag{4}$$

with the output

$$\begin{pmatrix} y_A \\ y_C \end{pmatrix} = \begin{pmatrix} a_s q^{-1} * A * q \\ c_s q^{-1} * C * q \end{pmatrix}. \tag{5}$$

This system is observable since all the state variables can be recovered from the known quantities ω_m, y_A, y_C and their derivatives: from (5), $a_s = \frac{1}{g} \|y_A\|$ and $c_s = \frac{1}{B_1 g} \|y_C\|$; hence we know the action of q on the two independent vectors A and C, which completely defines q in function of y_A, y_C, a_s, c_s. Finally (1) yields $\omega_b = \omega_m - 2q^{-1}\dot{q}$.

3 The Nonlinear Observer

3.1 Invariance of the System Equations

There is no general method for designing a nonlinear observer for a given system. When the system has a rich geometric structure the recent paper [1] provides a constructive method to this problem. In short, when a system of state dimension n is invariant by a transformation group of dimension $r \leq n$, the method yields the general invariant preobserver. More importantly the error between the estimated and actual states is described in suitable invariant coordinates by a system of dimension $2n - r$; in particular the error system has dimension n when the group has dimension n. This result, which is reminiscent of the linear case, greatly simplifies the convergence analysis.

In our case the transformation generated by constant rotations and translations in the body-fixed frame and output scaling

$$\varphi_{(q_0,\omega_0,a_0,c_0)}\begin{pmatrix} q \\ \omega_b \\ a_s \\ c_s \end{pmatrix} = \begin{pmatrix} q * q_0 \\ q_0^{-1} * \omega_b * q_0 + \omega_0 \\ a_0 a_s \\ c_0 c_s \end{pmatrix}$$

$$\psi_{(q_0,\omega_0,a_0,c_0)}(\omega_m) = q_0^{-1} * \omega_m * q_0 + \omega_0$$

$$\rho_{(q_0,\omega_0,a_0,c_0)}\begin{pmatrix} y_A \\ y_C \end{pmatrix} = \begin{pmatrix} a_0 q_0^{-1} * y_A * q_0 \\ c_0 q_0^{-1} * y_C * q_0 \end{pmatrix},$$

where q_0 is a unit quaternion, ω_0 a vector in $\mathbb{R}^3$ and $a_0, c_0 > 0$, is easily seen to be a transformation group with the same dimension as the system (1)–(4).

The system (1)–(4) is indeed invariant by the transformation group since

$$\overbrace{q * q_0}^{\cdot} = \dot{q} * q_0 = \frac{1}{2}(q * q_0) * \big((q_0^{-1} * \omega_m * q_0 + \omega_0) - (q_0^{-1} * \omega_b * q_0 + \omega_0)\big)$$

$$\overbrace{q_0^{-1} * \omega_b * q_0 + \omega_0}^{\cdot} = q_0^{-1} * \dot{\omega}_b * q_0 = 0$$

$$\overbrace{a_0 a_s}^{\cdot} = a_0 \dot{a}_s = 0$$

$$\overbrace{c_0 c_s}^{\cdot} = c_0 \dot{c}_s = 0.$$

Notice also that from a physical and engineering viewpoint, it is perfectly sensible for an observer using measurements expressed in the body-fixed frame not to be affected by the actual choice of coordinates, i.e. by a constant rotation in the body-fixed frame. Similarly, a translation of the gyro bias by a vector constant in the body-fixed frame and output scalings should not affect the observer. This is precisely what the method of [1] achieves.

3.2 The General Invariant Observer

Following the theory developed in [1], see also [8] for details, the general invariant observer writes

$$\dot{\hat{q}} = \frac{1}{2}\hat{q} * (\omega_m - \hat{\omega}_b) + (LE) * \hat{q} \tag{6}$$

$$\dot{\hat{\omega}}_b = \hat{q}^{-1} * (ME) * \hat{q} \tag{7}$$

$$\dot{\hat{a}}_s = \hat{a}_s NE \tag{8}$$

$$\dot{\hat{c}}_s = \hat{c}_s OE. \tag{9}$$

Here the invariant output error E is the 5×1 vector

$$E := \Big(\langle E_A, e_1\rangle, \langle E_A, e_2\rangle, \langle E_A, e_3\rangle, \langle E_C, e_1\rangle, \langle E_C, e_2\rangle\Big)^T$$

made up of the projections of the vectors

$$E_A := A - \frac{1}{\hat{a}_s}\hat{q} * y_A * \hat{q}^{-1} \quad \text{and} \quad E_C := C - \frac{1}{\hat{c}_s}\hat{q} * y_C * \hat{q}^{-1}.$$

Only 5 of the 6 possible projections are independent since $\langle A, C\rangle = 0$ and $\langle y_A, y_C\rangle = 0$ imply

$$\langle E_A, E_C\rangle = \langle A, E_C\rangle + \langle E_A, C\rangle;$$

L, M are 3×5 matrices and N, O are 1×5 matrices with entries possibly depending on the components of E and of the complete invariant I defined by

$$I := \hat{q} * (\omega_m - \hat{\omega}_b) * \hat{q}^{-1}.$$

It is easy to check this observer is invariant. Notice also the two built-in desirable geometric features: $\|\hat{q}(t)\| = \|\hat{q}(0)\| = 1$ since LE is a vector of $\mathbb{R}^3$ (see section 8); $\hat{a}_s(t), \hat{c}_s(t) > 0$ provided $\hat{a}_s(0), \hat{c}_s(0) > 0$.

3.3 The Invariant Error System

Following [1], an invariant state error is given by

$$\begin{pmatrix} \eta \\ \beta \\ \alpha \\ \gamma \end{pmatrix} = \begin{pmatrix} \hat{q} * q^{-1} \\ q * (\hat{\omega}_b - \omega_b) * q^{-1} \\ \frac{a_s}{\hat{a}_s} \\ \frac{c_s}{\hat{c}_s} \end{pmatrix}.$$

Therefore,

$$\begin{aligned}
\dot{\eta} &= \dot{\hat{q}} * q^{-1} - \hat{q} * (q^{-1} * \dot{q} * q^{-1}) = (LE) * \eta - \frac{1}{2}\eta * \beta \\
\dot{\beta} &= q * (\dot{\hat{\omega}}_b - \dot{\omega}_b) * q^{-1} + \dot{q} * (\hat{\omega}_b - \omega_b) * q^{-1} - q * (\hat{\omega}_b - \omega_b) * q^{-1} * \dot{q} * q^{-1} \\
&= (\eta^{-1} * I * \eta) \times \beta + \eta^{-1} * (ME) * \eta \\
\dot{\alpha} &= -\frac{a_s \dot{\hat{a}}_s}{\hat{a}_s^2} = -\alpha NE \\
\dot{\gamma} &= -\frac{c_s \dot{\hat{c}}_s}{\hat{c}_s^2} = -\gamma OE.
\end{aligned}$$

Since E is obtained from

$$\begin{aligned}
E_A &= A - \frac{a_s}{\hat{a}_s}\hat{q} * (q^{-1} * A * q) * \hat{q}^{-1} = A - \alpha\eta * A * \eta^{-1} \\
E_C &= C - \gamma\eta * C * \eta^{-1}
\end{aligned}$$

we find as expected that the error system

$$\dot{\eta} = (LE) * \eta - \frac{1}{2}\eta * \beta \tag{10}$$

$$\dot{\beta} = (\eta^{-1} * I * \eta) \times \beta + \eta^{-1} * (ME) * \eta \tag{11}$$

$$\dot{\alpha} = -\alpha NE \tag{12}$$

$$\dot{\gamma} = -\gamma OE \tag{13}$$

depends only on the invariant state error $(\eta, \beta, \alpha, \gamma)$ and the "free" known invariant I, but not on the trajectory of the observed system (1)–(4). This property greatly simplifies the convergence analysis of the observer.

The linearized error system around the no-error equilibrium point $(\overline{\eta}, \overline{\beta}, \overline{\alpha}, \overline{\gamma}) = (1, 0, 1, 1)$ then reads

$$\delta\dot{\eta} = L\delta E - \frac{1}{2}\delta\beta \tag{14}$$

$$\delta\dot{\beta} = I \times \delta\beta + M\delta E \tag{15}$$

$$\delta\dot{\alpha} = -N\delta E \tag{16}$$

$$\delta\dot{\gamma} = -O\delta E, \tag{17}$$

where δE is the 5×1 vector

$$\Big(\langle\delta E_A, e_1\rangle, \langle\delta E_A, e_2\rangle, \langle\delta E_A, e_3\rangle, \langle\delta E_C, e_1\rangle, \langle\delta E_C, e_2\rangle\Big)^T = g\big(-2\delta\eta_2, 2\delta\eta_1, -\delta\alpha, 2B_1\delta\eta_3, -B_1\delta\gamma\big)^T$$

made up from the projections of the vectors

$$\delta E_A = A * \delta\eta - \delta\eta * A - \delta\alpha A = 2A \times \delta\eta - \delta\alpha A$$
$$\delta E_C = 2C \times \delta\eta - \delta\gamma C.$$

4 Design of L, M, N, O

Up to now, we have only investigated the structure of the observer. We now must choose the gain matrices L, M, N, O to meet the following requirements:

- the error must converge to zero, at least locally
- the local error behavior should be easily tunable, if possible with a clear physical interpretation
- the magnetic measurements should not affect the attitude estimate, but only the heading
- the behavior in the face of acceleration and/or magnetic disturbances should be sensible and understandable.

4.1 Local Design

It turns out that the previous requirements can easily be met *locally around every trajectory* by taking

$$L := \frac{1}{2g}\begin{pmatrix} 0 & -l_1 & 0 & 0 & 0 \\ l_2 & 0 & 0 & 0 & 0 \\ 0 & 0 & 0 & -\frac{1}{B_1}l_3 & 0 \end{pmatrix} \qquad M := \frac{1}{2g}\begin{pmatrix} 0 & m_1 & 0 & 0 & 0 \\ -m_2 & 0 & 0 & 0 & 0 \\ 0 & 0 & 0 & \frac{1}{B_1}m_3 & 0 \end{pmatrix}$$

$$N := \frac{1}{g}\begin{pmatrix} 0 & 0 & -n & 0 & 0 \end{pmatrix} \qquad O := \frac{1}{B_1 g}\begin{pmatrix} 0 & 0 & 0 & 0 & -o \end{pmatrix}$$

for all constant $l_1, l_2, l_3, m_1, m_2, m_3, n, o > 0$. This will follow from the very simple form of the linearized error system (14)–(17). We insist that it is not usually obvious to come up with a similar convergence result for an EKF.

Indeed, (14)–(17) now reads

$$\delta\dot{\eta} = \mathrm{D_l}\delta\eta - \frac{1}{2}\delta\beta \tag{18}$$

$$\delta\dot{\beta} = \mathrm{D_m}\delta\eta + I \times \delta\beta \tag{19}$$

$$\delta\dot{\alpha} = -n\delta\alpha \tag{20}$$

$$\delta\dot{\gamma} = -o\delta\gamma \tag{21}$$

where

$$\mathrm{D_l} = \begin{pmatrix} -l_1 & 0 & 0 \\ 0 & -l_2 & 0 \\ 0 & 0 & -l_3 \end{pmatrix} \quad \text{and} \quad \mathrm{D_m} = \begin{pmatrix} m_1 & 0 & 0 \\ 0 & m_2 & 0 \\ 0 & 0 & m_3 \end{pmatrix}.$$

When $I = 0$ (i.e. the system is at rest) the system completely decouples into:

- the longitudinal subsystem

$$\begin{pmatrix} \delta\dot{\eta}_2 \\ \delta\dot{\beta}_1 \end{pmatrix} = \begin{pmatrix} -l_1 & -\frac{1}{2} \\ 0 & m_1 \end{pmatrix} \begin{pmatrix} \delta\eta_2 \\ \delta\beta_1 \end{pmatrix}$$

- the lateral subsystem

$$\begin{pmatrix} \delta\dot{\eta}_1 \\ \delta\dot{\beta}_2 \end{pmatrix} = \begin{pmatrix} -l_2 & -\frac{1}{2} \\ 0 & m_2 \end{pmatrix} \begin{pmatrix} \delta\eta_1 \\ \delta\beta_2 \end{pmatrix}$$

- the heading subsystem

$$\begin{pmatrix} \delta\dot{\eta}_3 \\ \delta\dot{\beta}_3 \end{pmatrix} = \begin{pmatrix} -l_3 & -\frac{1}{2} \\ 0 & m_3 \end{pmatrix} \begin{pmatrix} \delta\eta_3 \\ \delta\beta_3 \end{pmatrix}$$

- the scaling subsystem

$$\delta\dot{\alpha} = -n\delta\alpha$$
$$\delta\dot{\gamma} = -o\delta\gamma.$$

When $I \neq 0$ the longitudinal, lateral and heading subsystems are slightly coupled by the biases errors $\delta\beta$.

We now prove $(\delta\eta, \delta\beta, \delta\alpha, \delta\gamma) \to (0,0,0,0)$ whatever $(l_1, l_2, l_3, m_1, m_2, m_3, n, o)$ > 0. The scaling subsystem obviously converges. For the other variables we consider the Lyapunov function

$$V = \frac{l_1}{2}\delta\eta_1^2 + \frac{l_2}{2}\delta\eta_2^2 + \frac{l_3}{2}\delta\eta_3^2 + \frac{1}{4}\|\delta\beta\|^2.$$

Differentiating V and using $\langle \delta\beta, I \times \delta\beta \rangle = 0$, we get

$$\dot{V} = -\left(l_1 m_1 \delta\eta_1^2 + l_2 m_2 \delta\eta_2^2 + l_3 m_3 \delta\eta_3^2\right) \leq 0.$$

Since V is bounded from below, this implies that $V(\delta\eta(t), \delta\beta(t))$ converges as $t \to \infty$. Since

$$\lim_{t\to\infty} \int_0^t \dot{V}(\delta\eta(\tau), \delta\beta(\tau))d\tau = \lim_{t\to\infty} V(\delta\eta(t), \delta\beta(t)) - V(\delta\eta(0), \delta\beta(0)),$$

we conclude $\lim_{t\to\infty} \int_0^t \dot{V}(\delta\eta(\tau), \delta\beta(\tau))d\tau$ exists and is finite. On the other hand, $\dot{V} \leq 0$ also implies

$$0 \leq V(\delta\eta(t), \delta\beta(t)) \leq V(\delta\eta(0), \delta\beta(0)).$$

Therefore $\delta\eta(t)$ and $\delta\beta(t)$ are bounded. Equation (18) implies that $\delta\dot{\eta}(t)$ is bounded too, and finally that $\ddot{V}$ is bounded. Hence $\dot{V}$ is uniformly continuous and by Barbalat's lemma

$$\dot{V} \to 0 \;\Rightarrow\; \delta\eta \to 0.$$

Integrating (18), we get

$$\int_0^t \delta\dot{\eta}(\tau)d\tau = \delta\eta(t) - \delta\eta(0) = \int_0^t (\mathrm{D}_1\delta\eta(\tau) - \frac{1}{2}\delta\beta(\tau))d\tau.$$

Since $\delta\eta(t) \to 0$, it follows

$$\lim_{t\to\infty} \int_0^t (\mathrm{D}_1\delta\eta(\tau) - \frac{1}{2}\delta\beta(\tau))d\tau = -\delta\eta(0).$$

We assume I is bounded, which is physically sensible. Since $\delta\eta(t)$ and $\delta\beta(t)$ are bounded, $\mathrm{D}_1\delta\dot{\eta}(t) - \frac{1}{2}\delta\dot{\beta}(t)$ is bounded too. Hence $\mathrm{D}_1\delta\eta(t) - \frac{1}{2}\delta\beta(t)$ is uniformly continuous. Applying Barbalat's lemma once again yelds

$$\lim_{t\to\infty} (\mathrm{D}_1\delta\eta(t) - \frac{1}{2}\delta\beta(t)) = 0.$$

Since $\delta\eta \to 0$, we conclude $\delta\beta \to 0$, which ends the proof.

4.2 Global Design

We now look for correction terms ensuring global convergence whereas preserving the previous nice first-order properties. It is useful to define the following vectors and error output:

$$\begin{aligned} D &= C \times A \\ y_D &= y_C \times y_A \\ E_D &= D - \frac{1}{\hat{a}_s\hat{c}_s}\hat{q} * y_D * \hat{q}^{-1}. \end{aligned}$$

If we define the matrix L by

$$LE := \frac{l_a}{g^2} A \times E_A + \frac{l_c}{(B_1 g)^2} C \times E_C + \frac{l_d}{(B_1 g^2)^2} D \times E_D$$

it is easy to see that the zero-order part of L is the same as the constant L of Sect. 4.1 with $l_1 = l_a + l_c$, $l_2 = l_a + l_d$, and $l_3 = l_c + l_d$.

To find a candidate Lyapunov function we also select the matrices (M, N, O) by

$$ME := \sigma LE$$
$$NE := n\left(\frac{l_a}{g^2}E_A^T(E_A - A) + \frac{l_d}{(B_1g^2)^2}E_D^T(E_D - D)\right)$$
$$OE := o\left(\frac{l_c}{(B_1g)^2}E_C^T(E_C - C) + \frac{l_d}{(B_1g^2)^2}E_D^T(E_D - D)\right)$$

with $(l_a, l_c, l_d, \sigma, n, o) > 0$. The positive function

$$V := \frac{1}{\sigma}\|\beta\|^2 + \frac{l_a}{g^2}\|E_A\|^2 + \frac{l_c}{(B_1g)^2}\|E_C\|^2 + \frac{l_d}{(B_1g^2)^2}\|E_D\|^2$$

satisfies $\dot{V} \leq 0$. The convergence proof follows the main lines of [4], with added technicalities.

With these choices the first-order behavior is the same as in Sect. 4.1. The difference is the existence of the proportional factor σ between the $l_i's$, $m_i's$ coefficients. We now have only 4 independent gains for the lateral, longitudinal and heading subsystems instead of 6. We choose to fix first the natural frequencies of the lateral (ω_l), longitudinal (ω_L) and heading (ω_h) subsystems. There is only 1 parameter left to fix the free damping ratios.

5 Effects of Disturbances

Two main disturbances may affect the model. When $\dot{V} \neq 0$, the accelerometers measure in fact $a_s q^{-1} * A^* * q$ where $A^* := -\dot{V} + A$. Magnetic disturbances will also change B into some B^*. For simplicity we consider that A^*, B^* are constant. The measured outputs now become

$$\begin{pmatrix} y_{A^*} \\ y_{C^*} \end{pmatrix} = \begin{pmatrix} a_s q^{-1} * A^* * q \\ c_s q^{-1} * C^* * q \end{pmatrix}.$$

The error system is unchanged but E is now the 5×1 vector

$$E := \Big(\langle E_A, e_1\rangle, \langle E_A, e_2\rangle, \langle E_A, e_3\rangle, \langle E_C, e_1\rangle, \langle E_C, e_2\rangle\Big)^T$$

made up of the projections of the vectors

$$E_A := A - \frac{1}{\hat{a}_s}\hat{q} * y_{A^*} * \hat{q}^{-1} \quad \text{and} \quad E_C := C - \frac{1}{\hat{c}_s}\hat{q} * y_{C^*} * \hat{q}^{-1}.$$

Let us define the points $(\overline{\eta}, \overline{\beta}, \overline{\alpha}, \overline{\gamma})$ as following

$$\begin{aligned}\overline{\beta} &= 0\\ \overline{\eta} * A^* * \overline{\eta}^{-1} &= (0\,0\ \|A^*\|)\\ \overline{\eta} * C^* * \overline{\eta}^{-1} &= (0\ \|C^*\|\ 0)\\ \overline{\alpha} = \frac{\|A^*\|}{\|A\|} \quad &\text{and} \quad \overline{\gamma} = \frac{\|C^*\|}{\|C\|}.\end{aligned}$$

Doing the frame rotation defined by $\overline{\eta}$ we can define the new variables

$$\begin{aligned}\tilde{\eta} &= \eta * \overline{\eta}^{-1} & \tilde{\beta} &= \overline{\eta} * \beta * \overline{\eta}^{-1}\\ \tilde{\alpha} &= \alpha\overline{\alpha} & \tilde{\gamma} &= \gamma\overline{\gamma}.\end{aligned}$$

The error system with these new variables writes

$$\begin{aligned}\dot{\tilde{\eta}} &= -\frac{1}{2}\tilde{\eta} * \tilde{\beta} + (\tilde{L}\tilde{E}) * \tilde{\eta}\\ \dot{\tilde{\beta}} &= (\tilde{\eta}^{-1} * \tilde{I} * \tilde{\eta}) \times \tilde{\beta} + \tilde{\eta}^{-1} * (\tilde{M}\tilde{E}) * \tilde{\eta}\\ \dot{\tilde{\alpha}} &= -\tilde{\alpha}\tilde{N}\tilde{E}\\ \dot{\tilde{\gamma}} &= -\tilde{\gamma}\tilde{O}\tilde{E}\end{aligned}$$

where the new output error $\tilde{E}$ is made up of the projections of the vectors

$$\tilde{E}_A = A - \tilde{\alpha}\tilde{\eta} * A * \tilde{\eta}^{-1} \quad \text{and} \quad \tilde{E}_C = C - \tilde{\alpha}\tilde{\eta} * C * \tilde{\eta}^{-1}.$$

So $(\tilde{\eta}, \tilde{\beta}, \tilde{\alpha}, \tilde{\gamma})$ verify the same error system as $(\eta, \beta, \alpha, \gamma)$. In the new frame (A^*, C^*) play the same role as (A, C). All the properties of the observer are therefore preserved.

An important case is when only the magnetic field is perturbed, where we consider A and $C^* = (C_1^*\ C_2^*\ 0)$ (instead of $C = (0\ gB_1\ 0)$). Expliciting the new equilibrium point $(\overline{\eta}, \overline{\beta}, \overline{\alpha}, \overline{\gamma})$ of the error system it can be seen that

$$\begin{aligned}\overline{\phi} = \overline{\theta} = 0 \text{ and } \overline{\psi} &= \arctan\frac{C_1^*}{C_2^*}\\ \overline{\beta} = \overline{\alpha} = 0 \text{ and } \overline{\gamma} &= \frac{\|C^*\|}{\|C\|},\end{aligned}$$

where $(\overline{\phi}, \overline{\theta}, \overline{\psi})$ are the Euler angles corresponding to $\overline{\eta}$. In particular only the yaw angle $\overline{\psi}$ and $\overline{\gamma}$ are affected by the magnetic disturbance.

6 Experimental Validation

We now compare the behavior of our observer with the commercial Microbotics MIDG II system used in Vertical Gyro mode. The following results have been obtained with the observer

$$\begin{aligned}
\dot{\hat{q}} &= \frac{1}{2}\hat{q} * (\omega_m - \hat{\omega}_b) + (LE) * \hat{q} + k(1 - \|\hat{q}\|^2)\hat{q} \\
\dot{\hat{\omega}}_b &= \hat{q}^{-1} * (ME) * \hat{q} \\
\dot{\hat{a}}_s &= \hat{a}_s NE \\
\dot{\hat{c}}_s &= \hat{c}_s OE
\end{aligned}$$

and the choice of matrices defined by the parameters below. The added term $k(1 - \|\hat{q}\|^2)\hat{q}$ is a well-known numerical trick to keep $\|\hat{q}\| = 1$. Notice this term is also invariant.

We feed the observer with the raw measurements from the MIDG II gyros, acceleros and magnetic sensors. The observer is implemented in Simulink and its values are compared to the MIDG II results (computed according to the user manual by some Kalman filter). To get similar behaviors, we have chosen $(l_a, l_c, l_d, m_a, m_c, m_d, n, o) = 10^{-2} \times (6, 10, 6, 0.32, 0.53, 0.32, 25, 50)$.

6.1 Comparison with a Commercial Device (Fig. 1(a)–1(d))

We put in evidence the good properties of our observer with a long experiment divided into 3 parts:

- for $t < 240$ s the system is left at rest until the biases reach constant values. Fig. 1(a) highlights the importance of the correction term in the angle estimation: without correction the estimated roll angle diverges with a slope of $-0.18\,°/\text{s}$ (bottom plot), which is indeed the final value of the estimated bias (middle plot) (Fig. 1(a) and 1(b)).
- for $240\,\text{s} < t < 293\,s$ we move the system in all directions. The observer and the MDG II give very similar results (Fig. 1(c)).
- at $t = 385$ s the system is motionless and a magnet is put close to the sensors for 10 s. As expected only the estimated yaw angle is affected by the magnetic disturbance (Fig. 1(d)); the MIDG II exhibits a similar behavior.

6.2 Influence of the Observer Correction Terms (Fig. 1(e))

We have chosen the correction terms so as the magnetic measurements correct essentially the yaw angle and its corresponding bias, whereas the accelerometers measurements act on the other variables. We highlight this property on the following experiment (Fig. 1(e)). Once the biases have reached constant values, the system is left at rest during 35 minutes:

- for $t < 600$ s the results are very similar for the observer and the MIDG II.
- at $t = 600$ s the "magnetic correction terms" are switched off, i.e. the gains l_c,l_d,m_c, m_d and o are set to 0. The yaw angle estimated by the observer diverges because the corresponding bias is not perfectly estimated. Indeed, these variables are not observable without the magnetic measurements. The other variables are not affected.
- at $t = 1300$ s the "accelerometers correction terms" are also switched off, i.e. l_a,m_a and n are set to 0. All the estimated angles now diverge.

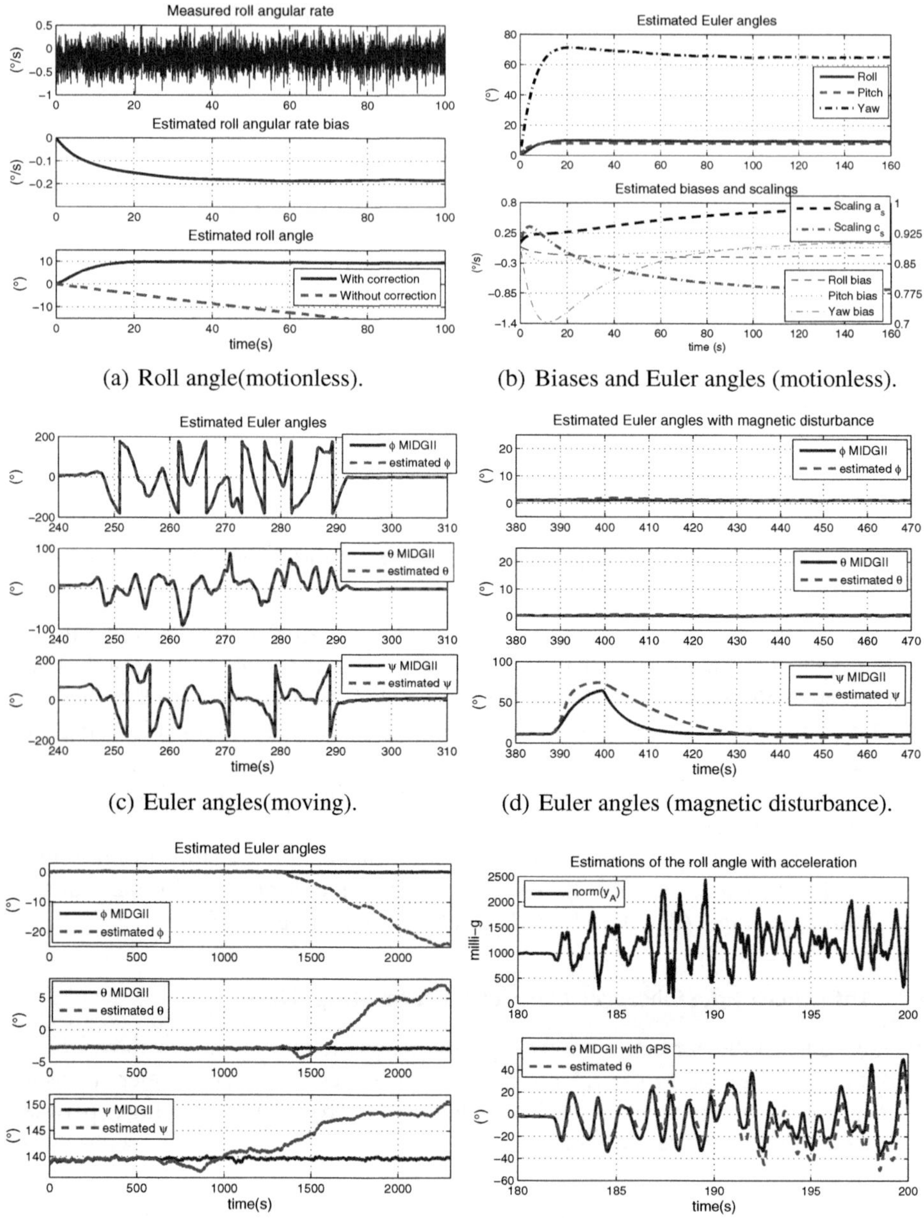

(a) Roll angle(motionless).

(b) Biases and Euler angles (motionless).

(c) Euler angles(moving).

(d) Euler angles (magnetic disturbance).

(e) Influence of correction terms.

(f) Estimated roll angle at $\dot{V} \neq 0$.

Fig. 1. Experimental validation using Matlab

6.3 Acceleration Disturbance: $\dot{V} \neq 0$ (Fig. 1(f))

The hypothesis $\dot{V} = 0$ may be wrong. In this case the observer does not converge any more to the right values. Indeed we illustrate this point on Fig. 1(f) by comparing the roll angle estimated by our observer and the roll angle estimated by the MIDG II in INS mode (in this mode the attitude and heading estimations are aided by a GPS engine, hence are close to the "true" values).

7 Implementation on an 8-Bit Microcontroller

In the previous section we validated the Simulink implementation of our observer. To demonstrate its computational simplicity we implemented it on an 8-bit microcontroller (Atmel ATmega128 running at 11.0592 MHz on development kit STK500/501). The computations are done in C with the standard floating point emulation and a simple Euler explicit approximation for the integration scheme. The microcontroller is fed with the MIDG II raw data at 50 Hz.

The experimental protocol was the following:

1. move the sensors in all directions and save the MIDG II raw measurements and estimations at 50 Hz.
2. use of xPCTarget to feed the ATmega with these data at 50Hz via a serial port and send at the same time the estimated variables. given by the microcontroller via a serial port to a computer
3. save these estimations.
4. compare offline the estimations given by the microcontroller and those given by the observer written in the Simulink embedded function.

This protocol can be illustrated by Fig. 2 (top left) where xPCTarget has been replaced by the MIDG II. We obtain the results of Fig. 2 (top right). The two estimations are very similar. We see on the bottom plot the discretization at 50 Hz due to the microcontroller.

We have also developed our own embedded device using raw sensors (first prototype: Fig. 2 (bottom)). The 8-bit microcontroller is able to interface the sensors and compute the real-time estimation at 100 Hz.

8 Appendix: Quaternions

Thanks to their four coordinates, quaternions provide a global parametrization of the orientation of a rigid body (whereas a parametrization with three Euler angles necessarily has singularities). Indeed, to any quaternion q with unit norm is associated a rotation matrix $R_q \in SO(3)$ by

$$q^{-1} * \mathbf{p} * q = R_q \cdot \mathbf{p} \quad \text{for all } \mathbf{p} \in \mathbb{R}^3.$$

A quaternion p can be seen as a scalar $p_0 \in \mathbb{R}$ together with a vector $\mathbf{p} \in \mathbb{R}^3$,

$$p = \begin{pmatrix} p_0 \\ \mathbf{p} \end{pmatrix}.$$

The (non commutative) quaternion product $*$ then reads

$$p * q \triangleq \begin{pmatrix} p_0 q_0 - \mathbf{p} \cdot \mathbf{q} \\ p_0 \mathbf{q} + q_0 \mathbf{p} + \mathbf{p} \times \mathbf{q} \end{pmatrix}.$$

The unit element is $e \triangleq \begin{pmatrix} 1 \\ \mathbf{0} \end{pmatrix}$, and $(p * q)^{-1} = q^{-1} * p^{-1}$.

Any scalar $p_0 \in \mathbb{R}$ can be seen as the quaternion $\begin{pmatrix} p_0 \\ \mathbf{0} \end{pmatrix}$, and any vector $\mathbf{p} \in \mathbb{R}^3$ can be seen as the quaternion $\begin{pmatrix} 0 \\ \mathbf{p} \end{pmatrix}$. We systematically use these identifications in the paper, which greatly simplifies the notations.

We have the useful formulas

$$p \times q \triangleq \mathbf{p} \times \mathbf{q} = \frac{1}{2}(p * q - q * p)$$

$$(\mathbf{p} \cdot \mathbf{q})\mathbf{r} = -\frac{1}{2}(p * q + q * p) * r.$$

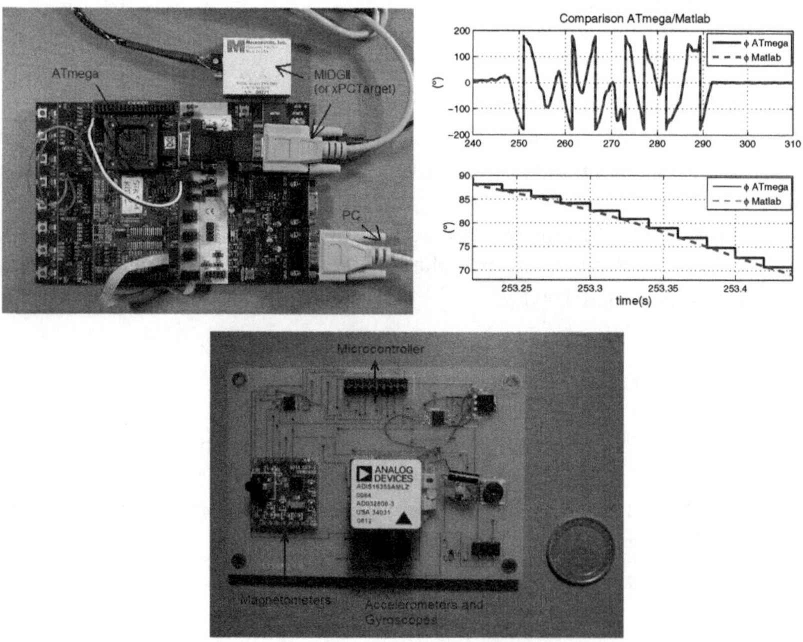

Fig. 2. Experimental setup

If q depends on time, then $\dot{q}^{-1} = -q^{-1} * \dot{q} * q^{-1}$.

Finally, consider the differential equation $\dot{q} = q * u + v * q$ where u, v are vectors $\in \mathbb{R}^3$. Let q^T be defined by $\begin{pmatrix} q_0 \\ -\mathbf{q} \end{pmatrix}$. Then $q * q^T = \|q\|^2$. Therefore,

$$\dot{\overbrace{q * q^T}} = q * (u + u^T) * q^T + \|q\|^2 (v + v^T) = 0$$

since u, v are vectors. Hence the norm of q is constant.

References

1. Bonnabel, S., Martin, Ph., Rouchon, P.: Symmetry-preserving observers. IEEE Trans. Automat. Control (2009); preliminary version arxiv.math.OC/0612193.
2. Collinson, R.P.G.: Introduction to avionics systems, 2nd edn. Kluwer Academic Publishers, Dordrecht (2003)
3. Grewal, M.S., Weill, L.R., Andrews, A.P.: Global positioning systems, inertial navigation, and integration, 2nd edn. Wiley, Chichester (2007)
4. Hamel, T., Mahony, R.: Attitude estimation on $SO(3)$ based on direct inertial measurements. In: Proc. of the 2006 IEEE International Conference on Robotics and Automation, pp. 2170–2175 (2006)
5. Kayton, M., Fried, W.R. (eds.): Avionics navigation systems, 2nd edn. Wiley, Chichester (1997)
6. Mahony, R., Hamel, T., Pfimlin, J.-M.: Non-linear complementary filters on the special orthogonal group. IEEE Trans. Automat. Control (to appear, 2008)
7. Mahony, R., Hamel, T., Pflimlin, J.-M.: Complementary filter design on the special orthogonal group $SO(3)$. In: Proc. of the 44th IEEE Conf. on Decision and Control, pp. 1477–1484 (2005)
8. Martin, Ph., Salaün, E.: Invariant observers for attitude and heading estimation from low-cost inertial and magnetic sensors. In: Proc. of the 46th IEEE Conf. on Decision and Control (2007)
9. Stevens, B.L., Lewis, F.L.: Aircraft control and simulation, 2nd edn. Wiley, Chichester (2003)
10. Thienel, J., Sanner, R.M.: A coupled nonlinear spacecraft attitude controller and observer with an unknown constant gyro bias and gyro noise. IEEE Trans. Automat. Control 48(11), 2011–2015 (2003)

Author Index

GPSR Compliance

The European Union's (EU) General Product Safety Regulation (GPSR) is a set of rules that requires consumer products to be safe and our obligations to ensure this.

If you have any concerns about our products, you can contact us on ProductSafety@springernature.com

In case Publisher is established outside the EU, the EU authorized representative is:

Springer Nature Customer Service Center GmbH
Europaplatz 3
69115 Heidelberg, Germany

Batch number: 10620143

Printed by Printforce, the Netherlands